Novel Viral Vectors for Gene Therapy
2023

Novel Viral Vectors for Gene Therapy 2023

Editors

Ottmar Herchenröder
Brigitte Pützer

Basel • Beijing • Wuhan • Barcelona • Belgrade • Novi Sad • Cluj • Manchester

Editors

Ottmar Herchenröder
Institute of Experimental
Gene Therapy and
Cancer Research
Rostock University
Rostock
Germany

Brigitte Pützer
Institute of Experimental
Gene Therapy and
Cancer Research
Rostock University
Rostock
Germany

Editorial Office
MDPI
St. Alban-Anlage 66
4052 Basel, Switzerland

This is a reprint of articles from the Special Issue published online in the open access journal *Viruses* (ISSN 1999-4915) (available at: https://www.mdpi.com/journal/viruses/special_issues/Vectors_gene_therapy).

For citation purposes, cite each article independently as indicated on the article page online and as indicated below:

Lastname, A.A.; Lastname, B.B. Article Title. *Journal Name* **Year**, *Volume Number*, Page Range.

ISBN 978-3-7258-0530-3 (Hbk)
ISBN 978-3-7258-0529-7 (PDF)
doi.org/10.3390/books978-3-7258-0529-7

Contents

Editorial

Novel Viral Vectors for Gene Therapy

Ottmar Herchenröder * and Brigitte M. Pützer *

Institute of Experimental Gene Therapy and Cancer Research, Rostock University Medical Center, 18057 Rostock, Germany
* Correspondence: herchen@med.uni-rostock.de (O.H.); brigitte.puetzer@med.uni-rostock.de (B.M.P.)

Citation: Herchenröder, O.; Pützer, B.M. Novel Viral Vectors for Gene Therapy. *Viruses* **2024**, *16*, 387. https://doi.org/10.3390/v16030387

Received: 26 February 2024
Accepted: 28 February 2024
Published: 1 March 2024

Viral vectors are gene transfer tools assembled from the backbones of naturally occurring viruses. By definition, these vehicles that transfer nucleic acids are replication-incompetent but deliver assigned payloads into eukaryotic cells by a process called transduction. For decades, numerous viral vector systems that influence cells or tissues have been used to perform basic and preclinical research. Over time, some virus-derived vectors found their way into clinical practice.

When the severe acute respiratory syndrome coronavirus type 2 (SARS-CoV-2) began to quickly spread at the turn of 2019/2020, ultimately leading to a pandemic spanning the globe and paralyzing the world, vaccines had to be developed quickly. Besides the classical methodologies for vaccine production such as inactivated viruses or preparations based on viral proteins either harvested or produced in recombinant settings, two rather novel techniques gained the upper hand. In less than twelve months, nucleic acid-based as well as adenovirus-derived vector systems were approved as vaccines to combat COVID-19. Both types of vaccines transmit a genetic blueprint into muscle tissue, enabling the organism to build the spike protein of SARS-CoV-2 and prime the immune system against the disease. Early on in 2021, both biotechnological achievements helped to control the spread of SARS-CoV-2 and to reduce the severity of the disease, thereby reducing the burden on healthcare systems. In addition, they allowed societies to return to a sense of normalcy.

On the other hand, the use of viral vectors as a vaccination tool during the COVID-19 crisis has taught us the lesson that there are still some issues to be solved to avoid unwanted serious side effects and make this next-generation gene delivery technology, proven successful in many approaches, clinically usable. The development of novel viral vectors that meet the requirements of future patients warrants their individual optimization and adaptation for different applications in gene therapy, cancer treatment, vaccine development, and cell reprogramming [1]. Since its approval as the first gene therapy product, a plethora of strategies using adenovectors emerged, including conditionally replicative oncolytic viruses [2], less-immunogenic, genetically stable high-capacity adenovirus-derived vehicles, which allow long-term episomal persistence of transgenes in non-dividing cells [3], or customized adenovectors for targeted transduction in vivo. Viral vectors are now widely used and the number of approved therapies is only expected to increase. To keep up with the increasing demand, the challenges and strategies for faster patient access need to be addressed. The editors believe that this Special Issue provides some optimistic answers.

In three articles on adenoviral vectors, we learn about the challenges in setting up robust platforms for vector production. Researchers from Witten in Germany and Bordeaux in France show that the construction and fabrication of a DNA-virus-based vector to transfer genetic information from an RNA virus are not trivial. Scientists residing in Montreal in Canada present improved methodologies to manufacture vectors that do not contaminate replication-competent entities. A team from Cardiff, UK, presents data on improving cell specificity and selectivity of adenoviral vector particles to ultimately improve oncolytic viruses.

Two papers deal with adeno-associated-virus (AAV)-derived gene ferries. The group led by Sarah Wootton from the University of Guelph, Canada, improved AAV by pseudotyping the particles with animal virus-derived capsid sequences. Victor Weiss and colleagues in Vienna, Austria, provide a new technique using a sophisticated apparatus to characterize AAVs.

A review by Wang and Shao from Tianjin, China, looks at innate immune responses to viral vectors and on how to prevent these responses. A consortium from Greece and Cyprus explains how the researchers optimized a lentiviral vector built to threat sickle cell disease.

In two back-to-back articles, both authored by Randal N. Johnston from Calgery, Canada, the researchers from Canada and Japan describe the properties of oncolytic reoviruses. These tools show anti-cancer activity in conjunction with chemotherapy. Further activity is unveiled as these oncolytic viruses can also infect and destroy pluripotent stem cells.

An illustrious panel of Central European senior scientists discusses the need for disinfectants in the context of modified viruses and vectors.

To round up this Special Issue, Kenneth Lundstrom from Switzerland presents a seminal review on the prospects of the current frontline vector systems. In his comprehensive article, the author covers most aspects of the vectors currently in use, including oncolytic viruses.

As we look forward, it is essential to continue researching viral vectors including cutting-edge technologies such as artificial intelligence for the production of targeted and more efficient gene ferries, ensuring transparency in reporting adverse events, and making data-driven decisions to balance risks and benefits. The incidents related to these vector-derived vaccines should not deter us from recognizing the immense value of viral vectors in medical applications. We believe that every author of this Special Issue would agree with this statement.

Acknowledgments: We are grateful to all contributors and especially the reviewers who dedicated valuable time and expertise to our Special Issue.

Conflicts of Interest: The authors declare no conflicts of interest.

References

1. Faustino, D.; Brinkmeier, H.; Logotheti, S.; Jonitz-Heincke, A.; Yilmaz, H.; Takan, I.; Peters, K.; Bader, R.; Lang, H.; Pavlopoulou, A.; et al. Novel integrated workflow allows production and in-depth quality assessment of multifactorial reprogrammed skeletal muscle cells from human stem cells. *Cell. Mol. Life Sci.* **2022**, *79*, 229. [CrossRef] [PubMed]
2. Nemunaitis, J.; Tong, A.W.; Nemunaitis, M.; Senzer, N.; Phadke, A.P.; Bedell, C.; Adams, N.; Zhang, Y.-A.; Maples, P.B.; Chen, S.; et al. A Phase I Study of Telomerase-specific Replication Competent Oncolytic Adenovirus (Telomelysin) for Various Solid Tumors. *Mol. Ther.* **2010**, *18*, 429. [CrossRef]
3. Wang, H.; Georgakopoulou, A.; Zhang, W.; Kim, J.; Gil, S.; Ehrhardt, A.; Lieber, A. HDAd6/35++—A new helper-dependent adenovirus vector platform for in vivo transduction of hematopoietic stem cells. *Mol. Ther. Methods Clin. Dev.* **2023**, *29*, 213. [CrossRef] [PubMed]

Article

Insights from the Construction of Adenovirus-Based Vaccine Candidates against SARS-CoV-2: Expecting the Unexpected

Denice Weklak [1], Julian Tisborn [1], Maurin Helen Mangold [1], Raphael Scheu [1], Harald Wodrich [2], Claudia Hagedorn [1], Franziska Jönsson [1,*,†] and Florian Kreppel [1,*,†]

[1] Chair of Biochemistry and Molecular Medicine, Center for Biomedical Education and Research (ZBAF), Witten/Herdecke University, Stockumer Str. 10, 58453 Witten, Germany; denice.weklak@uni-wh.de (D.W.); julian.tisborn@uni-wh.de (J.T.); maurin.mangold@t-online.de (M.H.M.); raphael.zhili@gmail.com (R.S.); claudia.hagedorn@uni-wh.de (C.H.)

[2] Microbiologie Fondamentale et Pathogénicité, MFP CNRS UMR 5234, Université de Bordeaux, 33076 Bordeaux, France; harald.wodrich@u-bordeaux.fr

[*] Correspondence: franziska.joensson@uni-wh.de (F.J.); florian.kreppel@uni-wh.de (F.K.)

[†] These authors contributed equally to this work.

Abstract: To contain the spread of the SARS-CoV-2 pandemic, rapid development of vaccines was required in 2020. Rational design, international efforts, and a lot of hard work yielded the market approval of novel SARS-CoV-2 vaccines based on diverse platforms such as mRNA or adenovirus vectors. The great success of these technologies, in fact, contributed significantly to control the pandemic. Consequently, most scientific literature available in the public domain discloses the results of clinical trials and reveals data of efficaciousness. However, a description of processes and rationales that led to specific vaccine design is only partially available, in particular for adenovirus vectors, even though it could prove helpful for future developments. Here, we disclose our insights from the endeavors to design compatible functional adenoviral vector platform expression cassettes for the SARS-CoV-2 spike protein. We observed that contextualizing genes from an ssRNA virus into a DNA virus provides significant challenges. Besides affecting physical titers, expression cassette design of adenoviral vaccine candidates can affect viral propagation and spike protein expression. Splicing of mRNAs was affected, and fusogenicity of the spike protein in ACE2-overexpressing cells was enhanced when the ER retention signal was deleted.

Keywords: adenovirus; viral vector; vaccine; SARS-CoV-2; spike protein

1. Introduction

Since emerging in Wuhan, China, in late 2019, severe acute respiratory syndrome coronavirus 2 (SARS-CoV-2) has caused over 6.9 million confirmed deaths worldwide (https://covid19.who.int/, accessed 28 August 2023) to August 2023 and severely affected the world economy [1]. Once the World Health Organization (WHO) declared SARS-CoV-2 a pandemic in March 2020, researchers strove for the quick development of novel vaccines and drugs to combat the spread of coronavirus disease 2019 (COVID-19). Because of these urgent circumstances, vaccination platforms that allow rapid engineering and simultaneously deliver a robust immune response, such as the mRNA or adenoviral vaccination platforms, asserted themselves against classical vaccination regimes such as live-attenuated vaccines. Adenoviral and mRNA vaccination platforms allow immediate reaction to new pathogens or strains by altering the sequence of the to-be-delivered nucleic acid. This diversity in vaccine regimens was one major tool controlling the spreading of the pandemic. In particular, adenoviral vaccines offer several advantages as an emergency outbreak vaccination platform, such as high target cell transduction and gene transfer efficacy, as well as triggering robust humoral and cellular immune responses and prolonged stability at room temperature for storage [2]. The design of such adenoviral vectors for the expression of a

Citation: Weklak, D.; Tisborn, J.; Mangold, M.H.; Scheu, R.; Wodrich, H.; Hagedorn, C.; Jönsson, F.; Kreppel, F. Insights from the Construction of Adenovirus-Based Vaccine Candidates against SARS-CoV-2: Expecting the Unexpected. *Viruses* **2023**, *15*, 2155. https://doi.org/10.3390/v15112155

Academic Editors: Brigitte Pützer and Ottmar Herchenröder

Received: 29 September 2023
Revised: 20 October 2023
Accepted: 23 October 2023
Published: 25 October 2023

specific transgene is not trivial, as the expression of particular transgenes may affect viral propagation and may thus require a protein-dependent design of the expression cassette. For efficient vaccine production, optimal vector design relies on the perfect balance between transgene expression and vector functionality.

Here, we want to share our knowledge gained by generating a series of human adenovirus type 5 (HAdV-C5) vector-based vaccine candidates. While a majority of adenoviral vaccines that gained conditional market approval are based on tetracycline-controlled gene expression systems [3,4], our vector design relies on constitutive protein expression in transduced target cells. This way, effects of transgene expression, such as antigen-related toxicity on human cells, can be observed, enabling early estimation of patient safety following vaccination. SARS-CoV-2 is an ssRNA virus, and the spike protein is encoded by RNA. Therefore, SARS-CoV-2 RNA is never confronted with the RNA-processing machinery in the nucleus but is immediately translated in the cytoplasm. To clone the spike gene into an adenoviral genome, a DNA sequence originating from the RNA sequence has to be generated. Consequently, this DNA (of a sequence never meant for being localized in a nucleus) needs to be transcribed and processed before leaving the nucleus. As such, transferring the sequence of a genuine RNA gene to a DNA vector bears the risk of aberrant splicing and, as a consequence, defective transgene products. Therefore, obtaining a substantial understanding regarding the contextualizing of genes from RNA viruses into a DNA background is an essential basis for future developments of adenoviral vaccines.

With this study, we want to share our knowledge gained while generating our own set of adenovirus type 5 vector-based vaccine candidates harboring a CMV (cytomegalovirus)promoter-driven *spike* expression cassette located within the HAdV-C5 E1-region. We report that physical titers, viral propagation and spike protein expression were affected based on different expression cassette designs. While mRNA splicing of transcripts was seemingly not affected, fusogenicity of ACE2 (angiotensin-converting enzyme)-overexpressing cells was enhanced, dependent on the design of adenoviral vectors. We think that these insights into our struggles with vector production will be useful for future vaccine development.

2. Materials and Methods

2.1. Cell Lines and Cell Culture

HEK293 (ATCC®, Manassas, VA, USA, CRL-1573™) and A549 (ATCC®, Manassas, VA, USA, CLL-185™) cells were cultured in MEM Eagle (PAN Biotech™, Aidenbach, Germany, P04-08500) supplemented with 10% FCS (PAN Biotech™, Aidenbach, Germany, P40-37500) and 1% penicillin–streptomycin (PAN Biotech™, Aidenbach, Germany, P06-07050). U2OS-GFP-ACE2 cells were generated following transduction and selection with a lentivirus encoding the SARS-CoV-2 receptor angiotensin-converting enzyme 2 (ACE2, addgene, Watertown, MA, USA, #145839) and maintained under selection of 5 µg/mL puromycin (Merck KGaA, Darmstadt, Germany, P7255-25MG). U2OS, as well as KM-12 (Cellosaurus, CVCL_1331), Huh-7 (Cellosaurus, CVCL_0336) and SH-SY5Y (ATCC®, Manassas, VA, USA, CRL-2266™) were maintained in DMEM (PAN Biotech™, Aidenbach, Germany, P04-03590) supplemented with 10% FCS and 1% penicillin–streptomycin (PAN Biotech™, Aidenbach, Germany, P06-07050). Cells were passaged twice per week with 0.05% trypsin–0.02% EDTA (PAN Biotech™, Aidenbach, Germany, P10-0231SP).

2.2. Vector Construction

For this study, each E1/E3-deleted vector of human adenovirus type 5 (AY339865; bp 1–440, bp 3522–28131, bp 30814–35934) was generated using the Counter-Selection BAC Modification Kit (GeneBridges GmbH, Heidelberg, Germany, K002), which is based on pRed/ET recombination. Briefly, using in vivo homologous recombination in *Escherichia coli* (*E. coli*) DH10β, a counter-selection rpsL-neo cassette with flanking homology arms was introduced in an E1-located expression cassette consisting of a CMV promoter and an SV40 intron. The rpsL-neo cassette was either replaced by the Wuhan isolate (GenScript Biotech Corporation, Piscataway Township, NJ, USA, NC_045512.2) or codon-optimized (pUC57-2019-nCOV-S, GenScript Biotech Corporation, Piscataway Township, NJ, USA) DNA sequence of the full-length spike protein or only the S_1 domain. Furthermore, the SV40 intron, the N-terminally located signal peptide (SP) of the spike protein, or the endoplasmic reticulum (ER) retention signal of the spike protein was deleted by placement of the rpsL-neo cassette and subsequently replaced by a non-coding non-intronic sequence in the case of the intron or by the IL-2 signal peptide in the case of the spike signal peptide (for reference, see Section 3.1).

2.3. Adenovirus Vector Purification

Before transfection of cells, adenoviral vectors were liberated from bacterial bacmid backbone by restriction digestion with SwaI (New England Biolabs GmbH, Ipswich, MA, USA, R0604S). In short, 2.5 µg of DNA was incubated with 3 U SwaI in supplied buffer at 25 °C overnight. HEK293 cells were seeded at a density of 1×10^5 cells/well in a 24-well plate and cultivated overnight. The next day, 500 ng DNA was premixed with a 150 mM NaCl solution to a volume of 25 µL, and 6 µL of a 7.5 mM linear polyethyleneimine (PEI) (22 kDa) solution was mixed with a 150 mM NaCl solution to a volume of 25 µL. PEI mixture was added to the DNA solution and incubated at room temperature (RT) for 10 min. Subsequently, the complete PEI-DNA mixture was added to one well of the 24-well plate. Cells were harvested once cytopathic effects became visible after roughly 12 to 14 days. Lysates were used for cycles of reinfections and vector particles were finally purified using CsCl density gradient centrifugation. Two purification steps using discontinuous CsCl gradients (1.27 g/cm^3 and 1.41 g/cm^3) were performed, and vector bands were collected after centrifugation at 4 °C and 13.200 rpm for 2 h. Preparations were desalted using PD10 columns (Cytiva, Marlborough, MA, USA, 17085101). For the determination of physical titers, 20 µL of vector preparation was mixed with 0.1% SDS and 79 µL buffer and incubated at 56 °C for 10 min. Extinction at 260 nm was measured using UV spectroscopy, and titers were calculated utilizing the following formula:

$$\mathrm{vp}/\mathrm{\mu L} = E_{260} \times 1.1 \times 10^9 \times DF \tag{1}$$

where vp = virus particles, E_{260} = extinction at 260 nm, and DF = dilution factor

2.4. Polymerase Chain Reaction (PCR)

DNA fragments for homologous recombination were generated using Q5® high-fidelity polymerase (New England Biolabs GmbH, Ipswich, MA, USA, M0491L) with primers including 50 bp homology arms to the insertion site. Thus, 10 ng DNA template were mixed with 0.04 U/µL polymerase, 0.2 mM dNTPs, 1 µM primers forward and reverse (Table S1), as well as the appropriate buffer. The following program was used: cycle 1—98 °C, 120 s; cycles 2–27—98 °C, 30 s, then 67 °C, 30 s, followed by 72 °C, 120 s; cycle 28: 72 °C, 180 s. Colony PCR was performed to verify the replacement of rpsL-neo cassette by target sequence. As such, picked colonies were cultured in 100 µL LB medium supplemented with respective antibiotics and incubated at 37 °C for 3 h. Thereafter, 3 µL of bacterial culture was added to a mixture of 1 µM forward and reverse primers (forward: 5′ GCTCGTTTAGTGAACCGTCAGA 3′, reverse: 5′ GAGGCCGAGTTTGTCAGAAAGC 3′), 0.2 mM dNTPs, 0.05 U/µL GoTaq® polymerase (Promega Corporation, Madison, WI, USA, M3001) in appropriate buffer. PCR reaction was performed with the following program: cycle 1—95 °C, 180 s; cycles 2–31—95 °C, 45 s, then 58 °C, 45 s, followed by 74 °C, 120 s; cycle 32—74 °C, 500 s. For the analysis of mRNA splicing of transcripts, a 3 µL DNA sample was mixed with 0.2 mM dNTPs, 0.05 U/µL GoTaq® polymerase (Promega Corporation, Madison, WI, USA, M3001), and 1 µM forward and reverse primers (Table S2) in appropriate buffer. PCR cycles were: cycle 1—95 °C, 180 s; cycles 2–31—95 °C, 70 s, then 59 °C, 70 s, followed by 74 °C, 300 s; cycle 32—74 °C, 600 s.

2.5. Sodium Dodecyl Polyacrylamide Gel Electrophoresis (SDS-PAGE)

For adjustment of titer from viral preparations, 5×10^9 viral particles (vp), as calculated by physical titers, were diluted in Ad buffer (150 mM NaCl, 50 mM HEPES, pH 8.0) to a final volume of 20 µL. SDS loading buffer (5 µL) was added, and samples were boiled for 5 min at 95 °C. Samples were loaded onto an 8% polyacrylamide gel, which after electrophoresis was stained using silver staining (see Section 2.6).

Spike protein expression by viral vectors was analyzed from either supernatant or cell pellet. Thus, 100 µL of supernatant after infection of HEK293 or transduction of A549 or U2OS cells was transferred to a reaction tube. Protein extraction from cells was performed according to a modified protocol of the original by K. K. Wang [5]. Cells were washed once with 1 mL DPBS (PAN Biotech™, Aidenbach, Germany, P04-36500), then treated with 300 µL RIPA-SDS buffer with protease-inhibitor cocktail (Merck KGaA, Darmstadt, Germany, 11836153001) and incubated at room temperature for 10 min. After addition of 100 µL 100% trichloroacetic acid (TCA), the supernatant was transferred to a reaction tube. The pellet after centrifugation (RT, $5000\times g$, 5 min) was washed with 1 mL 2.5% TCA and again centrifuged (RT, $5000\times g$, 5 min). Tris base (20 µL, 3 M) was added to the pellet after centrifugation, and samples were then incubated at room temperature for 30 min. Finally, samples were mixed with 20 µL ddH$_2$O and 10 µL of $5\times$ SDS sample buffer, then incubated at 95 °C for 10 min. Samples were loaded onto an 8% polyacrylamide gel, and separated proteins, after electrophoresis, were thereafter transferred to a nitrocellulose membrane (see Section 2.9).

2.6. Silver Staining of Polyacrylamide Gel

Silver staining of polyacrylamide gels was performed according to H. Blum et al. [6].

Briefly, polyacrylamide gel after electrophoresis was transferred into a fixation solution (50% (v/v) methanol, 12% (v/v) acetic acid, 13.6 mM formaldehyde) and incubated at room temperature for 30 min. The gel was then rinsed with a washing buffer (50% ethanol) for 15 min and then treated with pretreatment buffer (0.8 mM sodium thiosulfate) for 1 min. After washing thrice for 20 s with ddH$_2$O, the gel was incubated in impregnation buffer (0.2% (w/v) silver nitrate, 13.6 mM formaldehyde) for 20 min at room temperature. Gel was subsequently developed in developer solution (6% (w/v) sodium carbonate, 16 µM sodium thiosulfate, 13.6 mM formaldehyde) after washing twice with ddH$_2$O for 20 s. The reaction was stopped by treatment with stopping buffer (50% (v/v) methanol, 12% (v/v) acetic acid) after washing with ddH$_2$O twice for 2 min.

2.7. Infection of HEK293 Cells with Adenoviral Vectors

For real-time quantitative PCR (RT-qPCR) analysis, 3×10^4 HEK293 cells/well were seeded in a 96-well plate and cultured overnight. The next day, cells were infected with spike protein expressing HAdV-C5 vectors with a multiplicity of infection (MOI) of 200 according to adjusted titers and incubated for 48 h at 37 °C, 5% CO$_2$. For the analysis of mRNA splicing, 2×10^6 HEK293 were seeded in 6 cm cell culture dishes and cultured overnight. The following day, cells were infected with an MOI of 300 and incubated at 37 °C and 5% CO$_2$ for 48 h.

2.8. Transduction of A549 or U2OS Cells with Adenoviral Vectors

For the analysis of mRNA splicing, 2×10^6 A549 cells were seeded in 6 cm cell culture dishes a day prior to transduction. Cells were then transduced with an MOI of 300 and incubated at 37 °C and 5% CO$_2$ for 48 h.

For Western blot analysis of spike protein expression, 2×10^5 U2OS cells/well were seeded in a 24-well plate and cultured overnight. Cells were transduced with spike protein encoding viral vectors using an MOI of 300 according to adjusted titers and incubated for 48 h at 37 °C, 5% CO$_2$. Furthermore, to enhance understanding of spike protein expression, 2×10^5 A549 cells/well in a 24-well plate were transduced with adenoviral vectors with an MOI of 1000 according to adjusted titers and incubated for 48 h at 37 °C, 5% CO$_2$.

2.9. Fast Protein Liquid Chromatography (FPLC) for Spike Protein Purification

In order to purify glycosylated S_1 protein from media after secretion, at least 10×15 cm cell culture dishes of A549 cells were transduced with appropriate vector using an MOI of 3000. The next day, cells were washed thrice with 15 mL of warm medium and DPBS. Cells were then treated with 25 mL EX-CELL® (Merck KGaA, Darmstadt, Germany, 14571C-1000ML) medium and incubated at 37 °C, 5% CO_2 for an additional 48 h until the medium was harvested. The supernatant was mixed with cOmplete™ protease-inhibitor cocktail (Merck KGaA, Darmstadt, Germany, 11697498001), then centrifuged for 8 min at $300 \times g$. The supernatant was transferred into centrifuge tubes, and 5 µL [0.015 U/µL] of an avidin solution (IBA Lifesciences GmbH, Göttingen, Germany, 2-0204-050) was added. After incubation at room temperature for 20 min, tubes were centrifuged at $18.000 \times g$ and 4 °C for 30 min.

For protein purification, an NGC Quest 10 chromatography system (Bio-Rad Laboratories GmbH, Hercules, CA, USA, #7880001) was used. Subsequently, the supernatant was loaded onto a StrepTactin™ column (Cytiva, Marlborough, MA, USA, 29048653). Purification of S_1 protein from the column was performed according to the manufacturer's instructions. Product fractions after purification were concentrated to a total volume of 100 µL using Amicon® ultra centrifugal units (Merck KGaA, Darmstadt, Germany, UFC503096) according to the product's manual.

2.10. Western Blot

After subjecting protein samples to SDS-PAGE, proteins were transferred to a nitrocellulose membrane (Cytiva, Marlborough, MA, USA, 10600003). Membranes were blocked with 5% bovine serum albumin in TBS-T (Tris-buffered saline, 100 mM Tris and 150 mM NaCl, with 0.05% Tween-20) overnight at 4 °C. For this study, either the rabbit-α-spike antibody (Sino Biological, Beijing, China, 40592-T62; 1:1000 in 5% BSA/TBS-T) or human serum after vaccination (1:20 in 5% BSA/TBS-T) was used as primary antibody. Membranes were incubated with primary antibody for 1 h at room temperature, then washed five times for at least 5 min with TBS-T. Secondary antibodies used were either goat-α-rabbit-IRDye800CW antibody (LI-COR Biosciences, Lincoln, NE, USA, 926-32211; 1:15.000 in 5% BSA/TBS-T) or goat-α-human-IRDye680RD antibody (LI-COR Biosciences, Lincoln, NE, USA, 926-68078; 1:15.000 in 5% BSA/TBS-T). Membranes were incubated with the appropriate secondary antibody for 45 min at room temperature. After subsequent washing steps with TBS-T, membranes were documented with the LI-COR Odyssey® CLx imaging system.

2.11. Isolation of Viral DNA for Quantitative PCR Analysis

Viral DNA from supernatant 48 h after infection of HEK293 cells was used for analyzing viral particle production using qPCR. Thus, the 96-well plate was subjected to three cycles of freezing and thawing to liberate viral particles from still-adherent cells. Supernatant after centrifugation (RT, $300 \times g$, 5 min) was transferred to 1.5 mL reaction tube. Viral DNA was then isolated using a Quick-DNA Miniprep Plus Kit (Zymo Research Europe GmbH, Freiburg, Germany, D4069) following the instructions of the manual, and sample concentration was determined by UV spectroscopy. For qPCR, DNA samples were diluted to a concentration of 7 ng/µL and stored at 4 °C until measurement.

2.12. Isolation of Total RNA and cDNA Synthesis

Total RNA was isolated using TRIzol™ (Thermo Fisher Scientific, Waltham, MA, USA, 15596026) following the instructions of the provided manual. Cells seeded in 6 cm cell culture dishes were harvested in 1 mL TRIzol™. RNA samples were stored at −80 °C or immediately reverse-transcribed into cDNA using the PrimeScript™ first strand cDNA synthesis kit (Takara Bio Inc., Kusatsu, Japan, 6110A) following the manufacturer's instruction manual. Prior to cDNA synthesis, 10 µg RNA was digested with 10 U DNAse I (Thermo Fisher Scientific, Waltham, MA, USA, EN0521) in $MgCl_2$ buffer for 3 h at 37 °C. cDNA was diluted with 60 µL ddH$_2$O for the analysis of mRNA splicing by PCR and stored at $-20\,°$C until further use.

2.13. Quantitative PCR Analysis

Viral production was quantified by qPCR analysis. As such, 1.5 µL of isolated viral DNA was added to 10 µL TB Green Advantage Premix (Takara Bio Inc., Kusatsu, Japan, 639676), 0.2 µM of appropriate forward and reverse primer (E4 forward: 5′ TAGACGATCCCTACTGTACG 3′, E4 reverse: 5′ GGAAATATGACTACG TCCGG 3′, PLAT forward: 5′ AGGGCTGGAGAGAAAACCTC 3′, PLAT reverse: 5′ TTCCTTCACTG-GCTCAGCTT 3′) and adjusted with ddH$_2$O to a final volume of 20 µL. The following program was used for template amplification: cycle 1—95 °C, 1 min; cycles 2–46—95 °C, 10 s followed by 60 °C, 25 s. The E4 copy number of each sample determined by qPCR was normalized to PLAT (plasminogen activator, tissue type) concentration.

2.14. Bright-Field Microscopic Imaging

Prior to microscopic imaging, 3×10^4 ACE2-overexpressing U2OS cells were transduced with spike-expressing HAdV-C5 vectors with an MOI of 200 for 72 h. When analyzing the effect of angiotensin II (AngII) on syncytia formation, ACE2-overexpressing U2OS cells (3×10^4/well) were transduced with an MOI of 30–600 in the presence or absence of angiotensin II (2000 ng/well) for 72 h. Every 24 h, the medium was refreshed to consistently provide cells with AngII. Finally, 96-well plates were imaged using a Leica DMi8 S microscope.

2.15. Statistical Analysis

For statistical data analysis, the software RStudio was used. Normality was tested via the Shapiro–Wilk test and homogeneity of variance was investigated using Levene's test. In the case of unequal variances, a Welch ANOVA was performed, and Dunnett's T3 test was used as the post hoc test.

3. Results

3.1. Vector Design

For this study, an extensive panel of E1/E3-deleted type 5 adenoviral vectors with differences regarding their E1-located expression cassette was generated (Figure 1).

For a better overview, vectors were grouped based on functionality or properties. Most groups consist of vector variants expressing either the full-length spike protein (S$_{full}$) or only the S$_1$ domain, enabling differences in immune response. At first, the DNA sequence of the SARS-CoV-2 spike protein from the Wuhan isolate (NCBI reference sequence NC_045512.2) was inserted into an expression cassette consisting of a cytomegalovirus (CMV) promoter, simian virus (SV) 40 intron and polyadenylation signal (PolyA). In general, inserting introns upstream of a transgene coding sequence has been shown to enhance the expression of the transgene. The SV40 intron (see Table S3) has proven to enhance transgene expression level and transfection efficacy efficiently compared to other introns, such as the hCMV intron A [7]. As such, we decided to use this intron to enhance spike protein expression for our vaccine candidates. Subsequently, for the second group, the wild-type spike protein sequence was replaced with an human-expression system codon-optimized sequence harboring either the SV40 intron (S$_{full}$ and S1) or a non-coding non-intronic sequence (ΔSV40-S$_{full}$ and ΔSV40-S$_1$). Moreover, a third vector group was designed to purify codon-optimized S$_1$ protein using affinity chromatography. A StrepII tag was introduced, either N- or C-terminally (N-Strep-S$_1$ and C-Strep-S$_1$, ΔSV40-S$_1$-C-Strep—codon-optimized). Furthermore, to study the effects of different domains of the spike protein on vector production and protein secretion, a fourth group of vectors was generated, where the N-terminal signal peptide of the spike protein was either deleted (ΔSP-S$_{full}$ and ΔSP-S$_1$) or exchanged with the interleukin (IL)-2 signal peptide (IL-2-S$_{full}$ and IL-2-S$_1$). One of the goals was to use the adenovirus vectors as expression platforms in cell culture to obtain significant amounts of correctly glycosylated spike protein. Previous studies showed that adenoviral vectors can be suitable for this purpose [8]. We used the IL-2 signal peptide, as in our experience with different proteins, it turned out that for efficient secretion of the target protein, the nature of the secretion signal peptide plays an important role, and in our hands, the IL-2 signal peptide has worked best so far (unpublished results). Additionally, the endoplasmic reticulum (ER) retention signal was deleted (ΔER). Lastly, a final group of viral vectors was generated in which the spike protein coding sequence was exchanged for codon-optimized sequences encoding the β- or γ-strain variants of the spike protein (ΔSV40-S$_\beta$ and ΔSV40-S$_\gamma$).

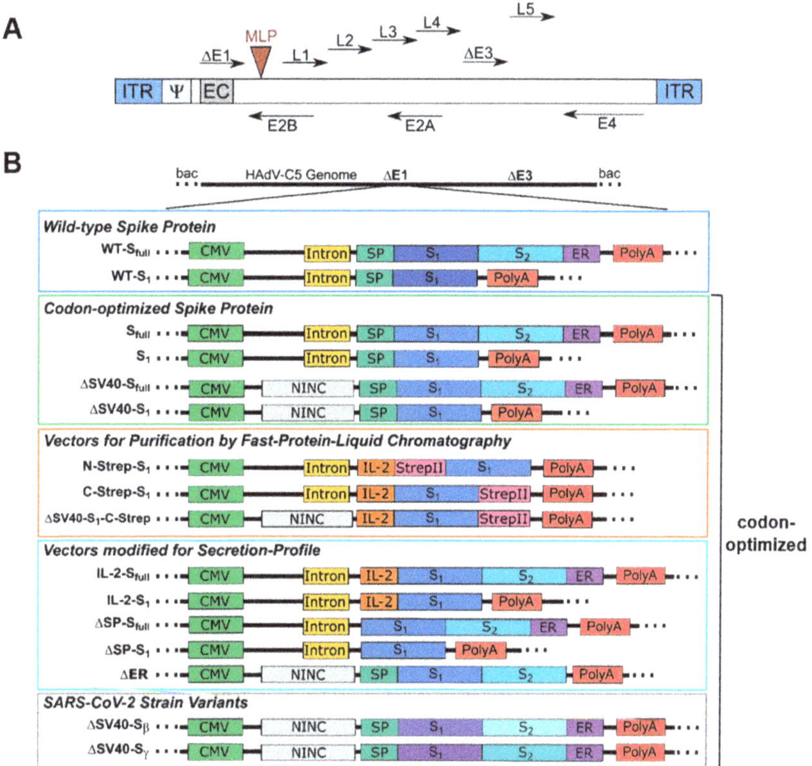

Figure 1. Design of expression cassette for all generated adenoviral vectors. (**A**) Schematic representation of human adenovirus type 5 genome used for this study. ITR: inverted terminal repeats, Ψ: packaging signal, EC: expression cassette, E1–E4: early genes 1–4, MLP: major late promoter, L1–L5: late genes 1–5. (**B**) All E1/E3-deleted type 5 adenoviral vectors were generated using homologous recombination in a bacmid context. WT: Wuhan-isolate spike protein, S$_{full}$: full-length spike, S$_1$: S$_1$ domain, N-Strep: N-terminal StrepII-Tag, C-Strep: C-terminal StrepII-Tag, ΔSV40: replacement of SV40 intron with non-intronic non-coding sequence, NINC: non-intronic non-coding sequence, IL-2: IL-2 signaling sequence, SP: N-terminal spike protein signal peptide, ΔSP: deletion of N-terminal spike protein signal peptide, ER: endoplasmic reticulum retention signal, ΔER: deletion of ER retention signal, S$_\beta$: β-variant of spike protein, S$_\gamma$: γ-variant of spike protein.

3.2. Physical Characterization of Adenoviral Vectors

To generate viral particles, E1-transcomplementing HEK293 cells were transfected with the linearized HAdV-C5 genome using PEI. While most vectors could be purified to moderate titers, some constructs were impaired in their ability to produce infectious viral particles (Figure 2). Noteworthily, viral particle production was either not possible (S_{full}, S_1, $\Delta SP\text{-}S_1$) or severely hampered (IL-2-S_1, $\Delta SP\text{-}S_{full}$, $\Delta SP\text{-}S_1$) for most SV40 intron-containing constructs. Viral particle production was also hampered for constructs with deleted ER retention signal (ΔER). Due to low particle production, purification for these vectors was prematurely stopped after the first CsCl gradient and physical titers could not be determined (Figure 2B). For the remaining viral preparations, titers were adjusted by identifying hexon band intensity after separation of viral proteins by SDS-PAGE and subsequent silver staining (Figure 2C). With the exception of C-Strep-S_1, no substantial differences in purity of viral preparations could be observed compared to AdEmpty (E1/E3-deleted vector with no transgene in expression cassette). As such, viral propagation seems to be affected by design of the $\Delta E1$-located expression cassette.

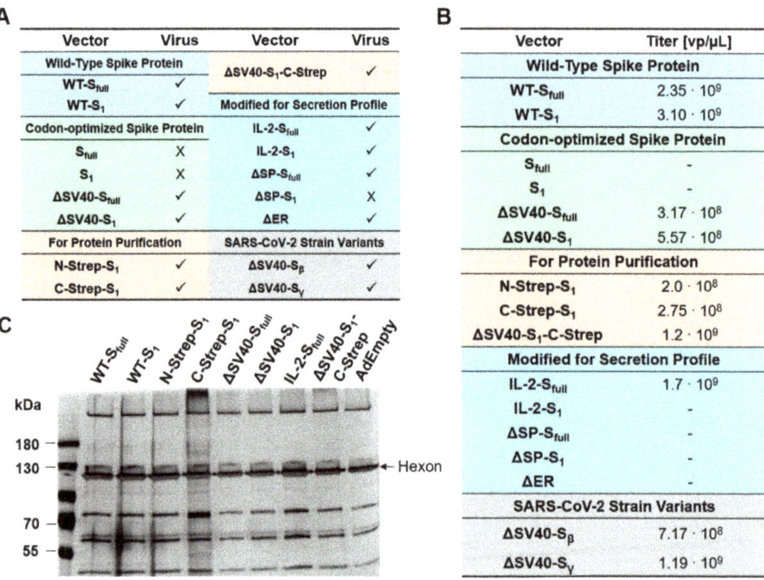

A

Vector	Virus	Vector	Virus
Wild-Type Spike Protein		ΔSV40-S₁-C-Strep	✓
WT-S_{full}	✓		
WT-S_1	✓	**Modified for Secretion Profile**	
Codon-optimized Spike Protein		IL-2-S_{full}	✓
S_{full}	X	IL-2-S_1	✓
S_1	X	$\Delta SP\text{-}S_{full}$	✓
$\Delta SV40\text{-}S_{full}$	✓	$\Delta SP\text{-}S_1$	X
$\Delta SV40\text{-}S_1$	✓	ΔER	✓
For Protein Purification		**SARS-CoV-2 Strain Variants**	
N-Strep-S_1	✓	$\Delta SV40\text{-}S_\beta$	✓
C-Strep-S_1	✓	$\Delta SV40\text{-}S_\gamma$	✓

B

Vector	Titer [vp/µL]
Wild-Type Spike Protein	
WT-S_{full}	$2.35 \cdot 10^9$
WT-S_1	$3.10 \cdot 10^9$
Codon-optimized Spike Protein	
S_{full}	-
S_1	-
$\Delta SV40\text{-}S_{full}$	$3.17 \cdot 10^8$
$\Delta SV40\text{-}S_1$	$5.57 \cdot 10^8$
For Protein Purification	
N-Strep-S_1	$2.0 \cdot 10^8$
C-Strep-S_1	$2.75 \cdot 10^8$
$\Delta SV40\text{-}S_1$-C-Strep	$1.2 \cdot 10^9$
Modified for Secretion Profile	
IL-2-S_{full}	$1.7 \cdot 10^9$
IL-2-S_1	-
$\Delta SP\text{-}S_{full}$	-
$\Delta SP\text{-}S_1$	-
ΔER	-
SARS-CoV-2 Strain Variants	
$\Delta SV40\text{-}S_\beta$	$7.17 \cdot 10^8$
$\Delta SV40\text{-}S_\gamma$	$1.19 \cdot 10^9$

Figure 2. Characterization of adenoviral vectors. (**A**) Tabular listing of spike protein encoding vector constructs generating infectious viral particles. ✓: vector growth, X: vector growth not observed. (**B**) Physical titers of adenoviral preparations as determined after calculation using excitations at 260 nm by UV spectroscopic measurement. -: physical titers could not be measured. (**C**) Polyacrylamide gel was loaded with 5×10^9 viral particles of preparations, proteins were stained using silver staining, and titers were adjusted by normalization to AdEmpty after determination of hexon band intensities.

3.3. Quantification of Viral Particle Production

Since several vectors seemed to be affected in their ability to produce infectious viral particles due to the design of their expression cassette (Figure 2), we further characterized infectious particle production by qPCR (Figure 3). As such, HEK293 cells were infected with an equal MOI of 200. Subsequently, viral DNA was isolated 48 h after infection and used for quantitative analysis. Infectivity was compared to IL-2-S_{full} because this vector showed growth behavior typical for Ad5 viral vectors (see Figure 2).

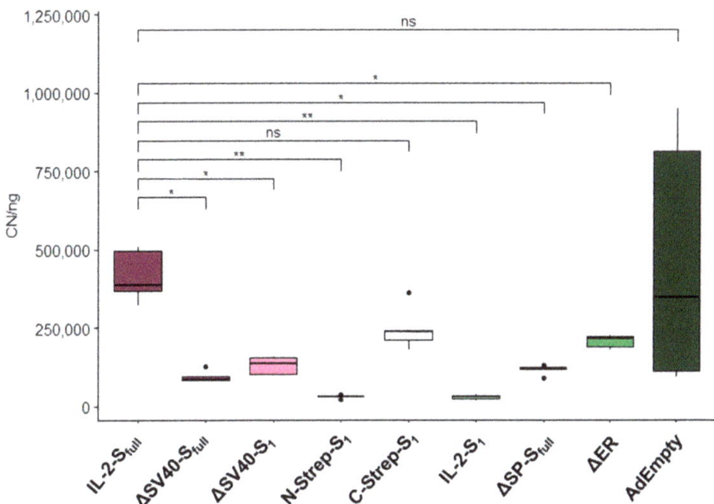

Figure 3. Characterization of infectious viral particle production. Quantification of infectious viral particle production of spike protein-encoding vectors by RT-qPCR after isolation of viral DNA from supernatant of HEK293 cells 48 h after infection. E4 copy number (CN) of each sample was normalized to PLAT gene in ng. Significance values were determined by Dunnett's T3 test ($n = 5$; ns: $p > 0.05$, *: $p < 0.05$, **: $p < 0.01$).

Compared to IL-2-S_{full} (416,800.2 ± 73,857.4), almost all vectors showed significant decreases in viral propagation based on detected isolated viral genomes (Figure 3). Although moderate titers were measured for both ΔSV40-S_{full} and ΔSV40-S_1, the number of produced viral particles was decreased in comparison to the IL-2-S_{full} vector (ΔSV40-S_{full}: 95,684.9 ± 16,736.2, $p = 0.013$, 4.35-fold decrease; ΔSV40-S_1: 130,020.3 ± 25,406.0, $p = 0.012$, 3.21-fold decrease), indicating that viral propagation of these vectors was affected. N-Strep-S_1 (30,299.7 ± 5574.5, $p = 0.006$, 13.76-fold decrease) as well as IL-2-S_1 (28,864.1 ± 6901.5, $p = 0.006$, 14.44-fold decrease) appear to be especially hampered in their ability to produce infectious particles when compared to IL-2-S_{full}. While not as pronounced as that with N-Strep-S_1 and IL-2-S_1, viral particle production for ΔSP-S_{full} (116,899.6 ± 13,954.9, $p = 0.002$, 3.57-fold decrease) and ΔER vectors (207,206.3 ± 19871.0, $p = 0.002$, 2.01-fold decrease) was reduced as well.

3.4. Protein Expression by Spike Protein-Producing Vectors

For efficient use as a vaccine, transgene expression efficiency must be ensured and verified. Spike protein expression by the panel of HAdV-C5 vectors (wild type, codon-optimized vectors, vectors for protein purification and vectors modified for secretion profile) was determined by Western blot analysis (Figure 4). Purified S_1 protein was also loaded onto the polyacrylamide gel as a positive control, whereas AdEmpty and uninfected cells served as a negative control. For a majority of constructs, such as C-Strep-S_1, ΔSV40-S_1, IL-2-S_1, ΔSP-S_{full} and ΔSV40-S_1-C-Strep, S_1 expression at the expected molecular weight could be observed (S_{full} (glycosylated) ~180–200 kDa, S_1 (glycosylated) ~120 kDa). Contrary to expectations, however, expression of the Wuhan-isolate spike protein (WT-S_{full} and WT-S_1) could not be detected in either cell lysates or supernatants of transduced U2OS cells (Figure 4A,B) and A549 cells (Figure 4C,D).

Repeated experiments with different MOIs and different primary antibodies yielded no detection of wild-type spike protein expression. Additionally, protein expression could not be observed in cell lysates or supernatants of U2OS cells transduced with N-terminally StrepII-tagged codon-optimized S_1 expressing vector (N-Strep-S_1), as well as the IL-2-S_{full} vector.

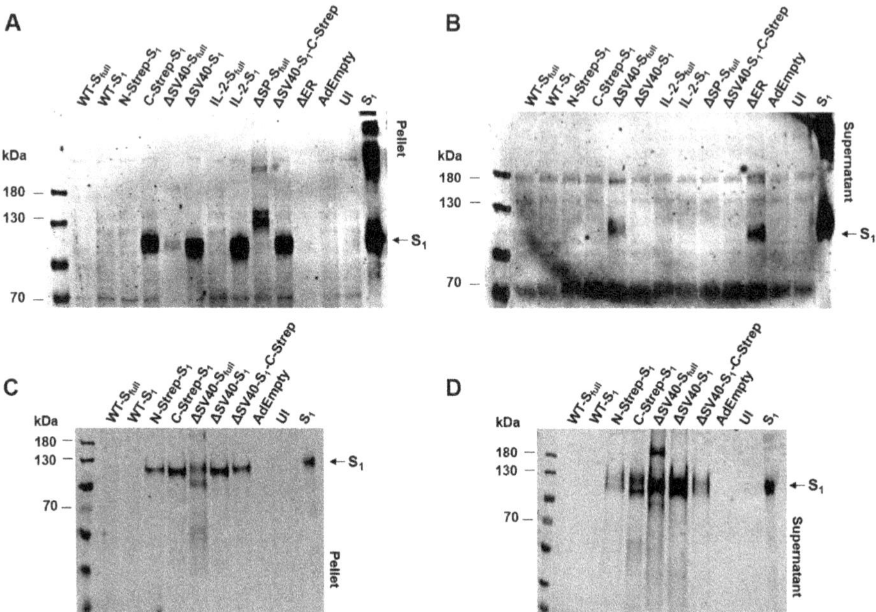

Figure 4. Spike protein expression after transduction with panel of spike protein encoding vectors. (**A**) Spike protein expression in cell lysates of U2OS cells after 48 h of transduction using rabbit polyclonal anti-S_1 antibody. UI: uninfected cells. (**B**) Spike protein expression in supernatants of U2OS cells after 48 h of transduction using rabbit polyclonal anti-S_1 antibody. Detection of spike protein expression in cell pellet (**C**) or in supernatants (**D**) of A549 cells after 48 h of transduction with a selection of vectors by incubation with human serum after boosting with mRNA vaccine.

While spike protein could not be detected in cell lysates of U2OS cells transduced with $\Delta SV40\text{-}S_{full}$ and ΔER vectors, protein expression could be discerned from supernatant. Syncytia formation was prominent in ACE2-overexpressing U2OS cells after transduction with $\Delta SV40\text{-}S_{full}$ and ΔER vectors, resulting in spike protein ending up in supernatant due to cell death. Furthermore, by incubation of membranes with human serum after boost vaccination, spike protein-specific bands could be detected in supernatants and cell lysates of transduced A549 cells for $\Delta SV40\text{-}S_{full}$ and $\Delta SV40\text{-}S_1$, as well as weakly for N-Strep-S_1, C-Strep-S_1 and $\Delta SV40\text{-}S_1$-C-Strep vectors (Figure 4C,D). Unexpectedly, the StrepII-tagged vectors showed lower protein expression compared to the $\Delta SV40\text{-}S_1$ vector, although C-Strep-S_1 and $\Delta SV40\text{-}S_1$-C-Strep vectors were successfully used for purification of codon-optimized S_1 protein by fast protein liquid chromatography (see Section 3.6).

3.5. Analysis of mRNA Splicing of Vector Transcript

Since we could not detect wild-type spike protein with our Western blot analysis and as we incorporated an RNA virus gene into a DNA background, we aimed to confirm the presence of expected mRNA transcripts expressed by different vectors using PCR. Analysis was performed by strategically positioning primers (Table S2) to amplify cDNA template after reverse transcription from isolated mRNA (Figure 5). As such, reactions were performed with forward primers being either located in the 5′ untranslated region (5′UTR) for SV40-containing vectors (#1), in the non-intronic non-coding sequence for $\Delta SV40\text{-}S_{full}$ and $\Delta SV40\text{-}S_1$ (#2) or in the S_1 (#3, #4 for WT; #6, #7 for codon-optimized S protein) or S_2 (#5 for WT; #8 for codon-optimized S protein) domain and a universal reverse primer located in the 3′UTR (#9).

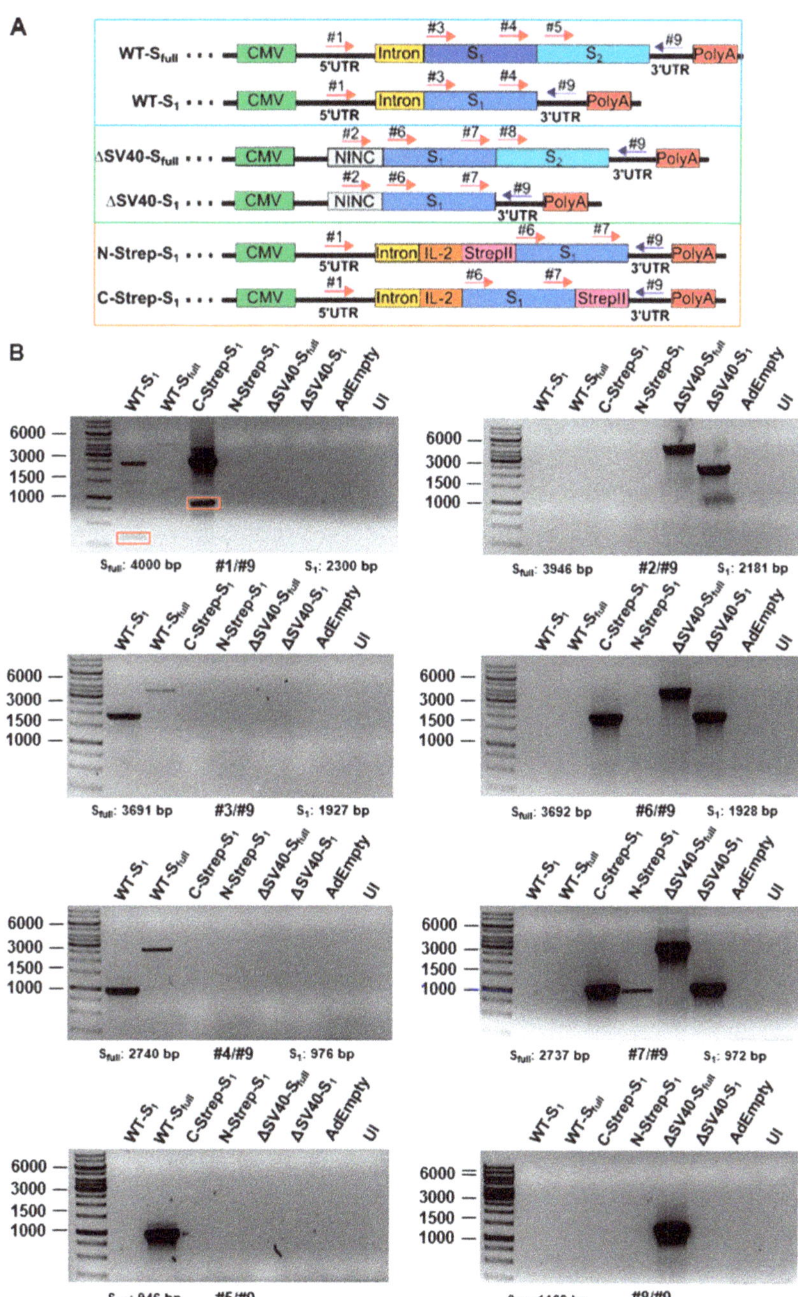

Figure 5. Analysis of mRNA splicing by PCR. (**A**) Schematic representation of primer placement for amplification of in cDNA-transcribed mRNA. (**B**) HEK293 cells were infected with a panel of viral vectors for 48 h and total RNA was isolated. RNA was reverse-transcribed into cDNA and then PCR was performed with appropriate primers (Table S2). PCR products were subsequently separated by agarose gel. Red boxes indicate bands excised for Sanger sequencing (see Figure S2).

Contrary to our expectations, the splicing of mRNA transcripts in A549 cells (Figure S1) appears to be unaffected, as PCR products in all primer combinations showed the correct length. Furthermore, no additional bands for any of the tested vectors were detected in all primer combinations. Unexpectedly, WT plasmid, which served as control, generated a PCR product with the forward primer located in the non-coding linker region while not being amplified with the 5′UTR primer (Figure S1). As for HEK293 cells, mRNA splicing seems to be affected mostly for SV40 intron-containing constructs (Figure 5). For primer combination #1/#9 (forward: 5′UTR, reverse: 3′UTR), multiple bands with lesser intensity were detected for WT-S_1, WT-S_{full} and C-Strep-S_1 vector. Also, contrary to expectations, no bands could be detected for N-Strep-S_1 (Figure 5) in most primer combinations. Sequencing of these additional bands observed for C-Strep-S_1 and WT-S_1 (Figure S2) suggests the removal of SV40 intron in addition to IL-2 signal peptide and the majority of the S_1 protein.

3.6. Purification of S_1 Using Fast Protein Liquid Chromatography

For the expression of proteins that are posttranslationally modified in mammalian cells, a mammalian expression system is preferred. Purifying S_1 from lung cells would ensure correct folding and glycosylation of the protein, which is why we aimed to accomplish the production of S_1 in A549 cells. For this purpose, a panel of vectors was generated, in which codon-optimized S_1 protein was fused to the StrepII-affinity tag (Figure 6A). Media of transduced cells with either C-Strep-S_1 or ΔSV40-S_1-C-Strep vector were collected. Prior interception of biotinylated proteins by the addition of avidin, cell debris and other impurities were removed by two centrifugation steps. Subsequently, StrepTactin™ columns (Cytiva, Marlborough, MA, USA, 29048653) were used for the purification of S_1 protein via FPLC.

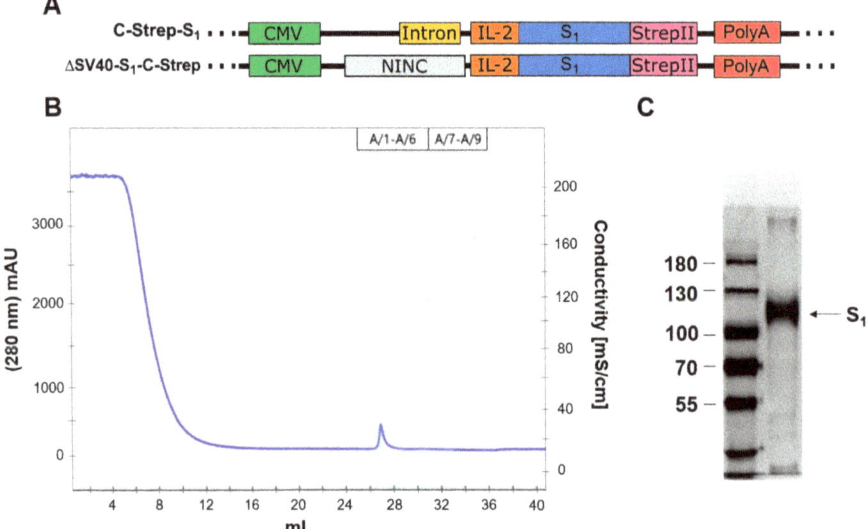

Figure 6. Purification of S_1 from A549 cell culture media. (**A**) Vector constructs used for purification of S_1 from A549. (**B**) Chromatogram of S_1 purification from A549 cell culture media after 48 h of incubation using affinity chromatography (StrepII-StrepTactin™ system). (**C**) Polyacrylamide gel was loaded with 250 ng of purified S_1 protein and stained utilizing silver staining. S_1 protein band is annotated with an arrow.

A total of 90 µg of S_1 protein could be purified from media of 10 × 15 cm A549 cell culture dishes transduced with ΔSV40-S_1-C-Strep (Figure 6B,C). Migration of the S_1 protein differs from the calculated molecular weight of around 78.5 kDa, which suggests glycosylation of protein in A549 cells (Figure 6C). The functionality of purified S_1 could be demonstrated on enzyme-linked immunosorbent assay (unpublished data).

3.7. Fusogenicity in ACE2-Overexpressing Cell Lines

In order to enter host cells, SARS-CoV-2 virus employs membrane fusion after attachment of spike protein to the foreign angiotensin-converting enzyme-2 (ACE2) receptor [9,10]. The spike protein plays a pivotal role in membrane fusion, acting as a fusogen and promoting membrane approximation by inserting its fusion protein after cleavage into S_1 and S_2 domains into the host cell membrane. To determine, whether spike protein expressed by vaccine vectors promotes cell fusion, ACE2-overexpressing U2OS cells were transduced with HAdV-C5 vectors for 72 h and subsequently imaged by bright-field microscopic imaging (Figure 7A). Compared to the control (AdEmpty) and Wuhan-isolate spike protein-expressing vector (WT-S_{full}), the vector with deleted ER retention signal (ΔER) showed enhanced syncytia formation. While this effect could also be observed for SV40-S_{full} vector, it appeared less pronounced. This effect of enhanced syncytia formation by the ΔER vector could also be witnessed for a panel of different cell lines, such as Huh-7, KM-12 and SH-SY5Y (Figure 7B).

ACE2 plays an essential role in the renin–angiotensin–aldosterone system (RAAS) by cleaving angiotensin II (AngII) into the heptapeptide angiotensin-(1–7) (Ang1–7). As such, it balances out AngII effects, which include vasoconstriction, blood volume regulation and inflammation [11]. To prove the involvement of ACE2 in the syncytia formation, ACE2-overexpressing U2OS cells were transduced with the ΔER in either the presence or absence of the natural ACE2 ligand AngII. Media were refreshed every 24 h to provide AngII continuously. Compared to the non-treated control, decreased syncytia formation of AngII-treated cells was observed (Figure 7C). This observation appears especially noticeable for U2OS cells transduced with higher MOIs (50–200) of the ΔER vector.

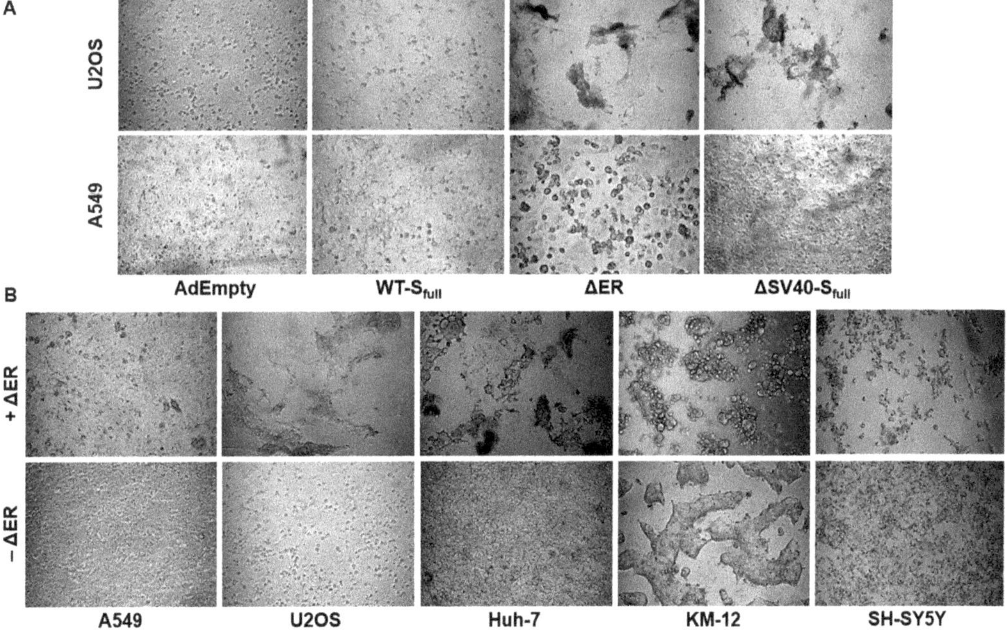

Figure 7. *Cont.*

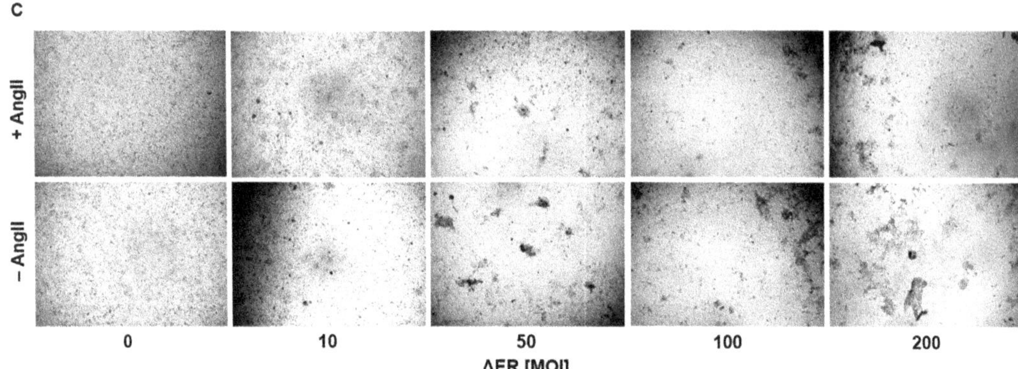

Figure 7. Microscopic Analysis of syncytia formation in ACE2-expressing cell lines. (**A**) ACE2-overexpressing U2OS cells were transduced with the aforementioned vectors with an MOI of 200. Syncytia formation was more pronounced when cells were transduced with the ΔER vector. (**B**) The ΔER vector is capable of inducing syncytia formation in a panel of different cell lines. (**C**) U2OS cells were pretreated with or without AngII, then transduced with ΔER with an MOI of 10–200. Syncytia formation is ACE2-dependent, as treatment with AngII decreases syncytia formation in ACE2-overexpressing U2OS cells. Magnification 50×.

4. Discussion

With the conditional market approval for the emergency use of several adenoviral vector vaccines (e.g., Vaxzevria, Jcovden) during the COVID-19 pandemic, worldwide awareness of viral vector therapeutics rose. Adenoviral vectors proved themselves as a reliable platform for emergency outbreak diseases, being able to be quickly modified as well as manufactured at high rates while eliciting robust immune responses. While protein subunit vaccines and inactivated viral vaccines potentially require adjuvants for efficient stimulation of immune response, immunogenicity of adenoviral vectors achieves sufficient activation of the immune system on its own [2,12,13]. Furthermore, while mRNA vaccines do not face splicing difficulties, their low stability demands storage at low temperatures, thus proving problematic for vaccination in developing countries.

On the contrary, certain adenoviral vector vaccines do not require cold chain storage [2,12,13]. Overall, the adenoviral vector platform and the LNP-RNA platform do in fact complement each other rather than compete with each other. Still, knowledge regarding efficient adenoviral vaccine design is limited. Most of the market-approved vectors employ tetracycline-regulated protein expression for antigen production, masking possible transgene-related effects on patients, as these effects cannot be observed during propagation. Additionally, to date, there has been little experience pertaining to the incorporation of genes from an RNA virus into a DNA. As such, we strove to share the lessons we learned while generating our own spike protein-encoding vaccines based on human adenovirus type 5. In our study, we focused on investigating the effects of transgene codon optimization, expression cassette building blocks, and deletion or replacement of transgene internal functionalities on viral propagation and protein expression.

As our study shows, the design of such adenoviral vector vaccines for expressing a specific transgene is no trivial matter. A delicate balance between expression cassette functionality, viral propagation and protein expression level is necessary. We demonstrated the effect of expression cassette design on viral propagation, protein expression and mRNA splicing. As such, the use of SV40 intron in expression cassettes to enhance protein expression [7] instead hampered the production of infectious viral particles considerably. Furthermore, the deletion of certain domains of the spike protein can influence infectivity and viral particle production, as observed for the ΔER vector. While deletion or replacement of the N-terminally located signal peptide of the spike protein did not seemingly interfere directly with viral propagation in this study, a precise statement cannot be provided, as vectors with deleted or replaced signal peptide also contained the SV40 intron. Still, our data suggest that careful consideration should be given to the insertion of additional sequences, such as introns or affinity tags, as minor adjustments may result in less infectious viral particles. Through qPCR analysis, these observations concerning viral propagation could be confirmed. Nearly all evaluated vectors showed a significant decrease in detected viral genomes compared to the IL-2-S_{full} vector.

Although vector growth did not appear to be affected for $\Delta SV40$-S_{full} or $\Delta SV40$-S_1 based on physical titers, viral particle production still seemed to be hampered according to qPCR results. As such, the expression of the spike protein may affect vector growth. As for protein expression analysis by Western blot, the spike protein was detected only for vectors expressing the codon-optimized version of the spike protein.Codon optimization of transgenes for vaccination approaches thus appears beneficial regarding protein expression level. Problems with mRNA splicing were expected to be the cause for undetectable spike protein expression because of SV40 intron inclusion in the $\Delta E1$-located expression cassette. While additional bands could be identified when HEK293 cells were infected with vectors for one primer combination (#1/#9), no additional bands were observed when A549 cells were transduced. However, for most SV40 intron-containing vectors (such as WT-S_1, WT-S_{full} as well as C-Strep-S_1 vector), splice variants seem to exist. Sequencing of these additional bands (Figure S2) suggests not only excision of introns from the processed mRNA but also the deletion of the IL-2 signal peptide and a major part of the S_1 protein. When predicting splice acceptor sites with the SpliceRover online tool [14], a splice site (1473–1484 bp) could be predicted at the incised location. Similar results were obtained for the band at around 500 bp for the WT-S_1 vector construct (data not shown). As such, using expression-enhancing building blocks such as introns should be considered carefully, especially when transgenes contain splice acceptor sites. Transfer of RNA proteins into a DNA background or codon optimization may introduce additional splice sites, which should be considered for vaccine development. For codon optimization, it was previously determined that potential splice sites increased compared to the Wuhan isolate in ChAdOx-1 (Vaxzevria) and Ad26.COV2.S (Jcovden) in silico [15]. The use of other introns, such as the HAdV-C5 tripartite leader intron instead of the SV40 intron, may alleviate the aberrant splicing we observed.

Furthermore, protein expression could not be detected for the secretion-modified IL-2-S_{full} vector. This could explain the growth behavior (Figure 3) of this vector in comparison to the other vectors of this group (IL-2-S_1, ΔSP-S_{full} and ΔSP-S_1), which showed considerably hampered viral production (IL-2-S_1, ΔSP-S_{full}) or no viral particle production (ΔSP-S_1). In general, vectors expressing increased levels of spike proteins, i.e., most of the vectors modified for their secretion profile (IL-2-S_1, ΔSP-S_{full}, ΔSP-S_1), appeared to be hampered in viral particle production as measured with physical titers (Figure 2) and qPCR analysis (Figure 3). As such, high expression of the spike protein also seems to affect viral propagation. This emphasizes the importance of balancing transgene expression and efficient viral vector production. Alternatively, transgene expression can be adjusted by choice of promoter [16], a system to repress transgene expression during vector production [17] or microRNA downregulation [18,19].

Unexpectedly, spike protein expression could not be detected for N-Strep-S_1 on Western blot analysis with anti-S_1 antibody (Figure 4A,B). Only a very faint signal was detected when incubating with human serum after a prime boost with an mRNA vaccine. Contrary to this observation, a signal for C-Strep-S_1 and $\Delta SV40$-S_1-C-Strep vectors could be determined by Western blot analysis (Figure 4). As such, placement of Strep-tag at either the N- or C-terminus may also influence transgene protein expression. Previous studies clarified that placement of affinity tags at the N- or C-terminus should be considered carefully with regard to protein functionality and localization [20]. As most signal peptides, including the signal peptide of the spike protein, are located at the N-terminus, placement of an affinity tag at C-terminus for spike protein expression to ensure proper localization seems appropriate.

Additionally, in this study, deletion of the ER retention signal seemed to promote syncytia formation in ACE2-overexpressing cell lines. A previous study also documented this increased formation of syncytia formation by ER retention signal-deleted vectors, when hACE2-mCherry cells were cocultured with S-Δ19-EGFP-A549 (ER retention signal-deleted vector) [21]. Compared to AdEmpty and ΔSV40-S$_{full}$, cell fusion between ACE2-expressing cells was increased when cells were transduced with the ΔER vector.

This effect could also be observed in a panel of different cell lines, such as Huh-7, KM-12 and SH-SY5Y. Treatment with AngII compared to untreated control suggests involvement of ACE2 in the fusion process, as syncytia formation is decreased in AngII-treated cells. However, due to only performing a single experiment, a specific conclusion cannot be drawn, as further experiments would have to be performed. For example, ACE2 expression could be downregulated with miRNA and be compared with a nonsense miRNA control to observe whether ΔER vector-mediated syncytia formation would decrease.

To summarize, this study proved that the design of expression cassettes in adenoviral vaccine vectors can affect viral propagation and transgene protein expression. We observed that codon optimization of transgene sequence enhances protein expression, whereas deletion and/or replacement of protein domains can promote negative effects (e.g., fusogenicity) and negatively affect vector production. Expression-enhancing elements such as introns can prove harmful to viral infectivity. Splice sites may be generated by contextualizing an RNA gene into a DNA background or codon optimization. Therefore, for future adenoviral vaccine vector design, we emphasize the consideration of four main areas: (I) codon optimization, (II) careful consideration of deletion or replacement of transgene domains, (III) careful analyses of potential novel splice sites, and (IV) careful consideration of promoter choice and inclusion of, e.g., expression-enhancing elements such as introns. Nevertheless, one may still expect the unexpected.

Supplementary Materials: The following supporting information can be downloaded at https://www.mdpi.com/article/10.3390/v15112155/s1, Table S1: Primers and oligonucleotides used for homologous recombination; Table S2: Primers used for mRNA splicing analysis; Table S3: Sequence of elements used in HAdV-C5 expression cassette; Figure S1: Analysis of mRNA splicing in A549 cells. Figure S2: Results of splicing analysis by Sanger sequencing.

Author Contributions: Conceptualization, F.J. and F.K.; investigation, D.W., M.H.M., J.T., R.S., F.J. and F.K.; resources, F.J., F.K., H.W., M.H.M., R.S. and J.T.; data curation, D.W. and F.J.; writing—original draft preparation, D.W.; writing—review and editing, C.H., F.J. and F.K.; visualization, D.W.; supervision, F.J. and F.K. All authors have read and agreed to the published version of the manuscript.

Funding: This research received no external funding.

Institutional Review Board Statement: Not applicable.

Informed Consent Statement: Not applicable.

Data Availability Statement: Not applicable.

Conflicts of Interest: The authors declare no conflict of interest.

References

1. Naseer, S.; Khalid, S.; Parveen, S.; Abbass, K.; Song, H.; Achim, M.V. COVID-19 Outbreak: Impact on Global Economy. *Front. Public Health* **2023**, *10*, 1009393. [CrossRef]
2. Coughlan, L.; Kremer, E.J.; Shayakhmetov, D.M. Adenovirus-Based Vaccines—A Platform for Pandemic Preparedness against Emerging Viral Pathogens. *Mol. Ther.* **2022**, *30*, 1822–1849. [CrossRef]
3. van Doremalen, N.; Lambe, T.; Spencer, A.; Belij-Rammerstorfer, S.; Purushotham, J.N.; Port, J.R.; Avanzato, V.A.; Bushmaker, T.; Flaxman, A.; Ulaszewska, M.; et al. ChAdOx1 NCoV-19 Vaccine Prevents SARS-CoV-2 Pneumonia in Rhesus Macaques. *Nature* **2020**, *586*, 578–582. [CrossRef]
4. Bos, R.; Rutten, L.; Van Der Lubbe, J.E.M.; Bakkers, M.J.G.; Hardenberg, G.; Wegmann, F.; Zuijdgeest, D.; De Wilde, A.H.; Koornneef, A.; Verwilligen, A.; et al. Ad26 Vector-Based COVID-19 Vaccine Encoding a Prefusion-Stabilized SARS-CoV-2 Spike Immunogen Induces Potent Humoral and Cellular Immune Responses. *NPJ Vaccines* **2020**, *5*, 91. [CrossRef] [PubMed]

5. Wang, W.; Vignani, R.; Scali, M.; Cresti, M. A Universal and Rapid Protocol for Protein Extraction from Recalcitrant Plant Tissues for Proteomic Analysis. *Electrophoresis* **2006**, *27*, 2782–2786. [CrossRef] [PubMed]
6. Blum, H.; Beier, H.; Gross, H.J. Improved Silver Staining of Plant Proteins, RNA and DNA in Polyacrylamide Gels. *Electrophoresis* **1987**, *8*, 93–99. [CrossRef]
7. Xu, D.; Wang, X.; Jia, Y.; Wang, T.; Tian, Z.; Feng, X.; Zhang, Y. SV40 Intron, a Potent Strong Intron Element That Effectively Increases Transgene Expression in Transfected Chinese Hamster Ovary Cells. *J. Cell. Mol. Med.* **2018**, *22*, 2231–2239. [CrossRef]
8. Wieland, A.; Denzel, M.; Schmidt, E.; Kochanek, S.; Kreppel, F.; Reimann, J.; Schirmbeck, R. Recombinant Complexes of Antigen with Stress Proteins Are Potent CD8 T-Cell-Stimulating Immunogens. *J. Mol. Med. Berl. Ger.* **2008**, *86*, 1067–1079. [CrossRef]
9. Li, X.; Yuan, H.; Li, X.; Wang, H. Spike Protein Mediated Membrane Fusion during SARS-CoV-2 Infection. *J. Med. Virol.* **2023**, *95*, e28212. [CrossRef]
10. Huang, Y.; Yang, C.; Xu, X.; Xu, W.; Liu, S. Structural and Functional Properties of SARS-CoV-2 Spike Protein: Potential Antivirus Drug Development for COVID-19. *Acta Pharmacol. Sin.* **2020**, *41*, 1141–1149. [CrossRef]
11. Scialo, F.; Daniele, A.; Amato, F.; Pastore, L.; Matera, M.G.; Cazzola, M.; Castaldo, G.; Bianco, A. ACE2: The Major Cell Entry Receptor for SARS-CoV-2. *Lung* **2020**, *198*, 867–877. [CrossRef] [PubMed]
12. Nagy, A.; Alhatlani, B. An Overview of Current COVID-19 Vaccine Platforms. *Comput. Struct. Biotechnol. J.* **2021**, *19*, 2508–2517. [CrossRef] [PubMed]
13. Bayani, F.; Hashkavaei, N.S.; Arjmand, S.; Rezaei, S.; Uskoković, V.; Alijanianzadeh, M.; Uversky, V.N.; Ranaei Siadat, S.O.; Mozaffari-Jovin, S.; Sefidbakht, Y. An Overview of the Vaccine Platforms to Combat COVID-19 with a Focus on the Subunit Vaccines. *Prog. Biophys. Mol. Biol.* **2023**, *178*, 32–49. [CrossRef]
14. Zuallaert, J.; Godin, F.; Kim, M.; Soete, A.; Saeys, Y.; De Neve, W. SpliceRover: Interpretable Convolutional Neural Networks for Improved Splice Site Prediction. *Bioinforma. Oxf. Engl.* **2018**, *34*, 4180–4188. [CrossRef] [PubMed]
15. Kowarz, E.; Krutzke, L.; Külp, M.; Streb, P.; Larghero, P.; Reis, J.; Bracharz, S.; Engler, T.; Kochanek, S.; Marschalek, R. Vaccine-Induced COVID-19 Mimicry Syndrome. *eLife* **2022**, *11*, e74974. [CrossRef] [PubMed]
16. Qin, J.Y.; Zhang, L.; Clift, K.L.; Hulur, I.; Xiang, A.P.; Ren, B.-Z.; Lahn, B.T. Systematic Comparison of Constitutive Promoters and the Doxycycline-Inducible Promoter. *PLoS ONE* **2010**, *5*, e10611. [CrossRef]
17. Strobel, B.; Klauser, B.; Hartig, J.S.; Lamla, T.; Gantner, F.; Kreuz, S. Riboswitch-Mediated Attenuation of Transgene Cytotoxicity Increases Adeno-Associated Virus Vector Yields in HEK-293 Cells. *Mol. Ther. J. Am. Soc. Gene Ther.* **2015**, *23*, 1582–1591. [CrossRef]
18. Reid, C.A.; Boye, S.L.; Hauswirth, W.W.; Lipinski, D.M. MiRNA-Mediated Post-Transcriptional Silencing of Transgenes Leads to Increased Adeno-Associated Viral Vector Yield and Targeting Specificity. *Gene Ther.* **2017**, *24*, 462–469. [CrossRef]
19. Kron, M.W.; Espenlaub, S.; Engler, T.; Schirmbeck, R.; Kochanek, S.; Kreppel, F. MiRNA-Mediated Silencing in Hepatocytes Can Increase Adaptive Immune Responses to Adenovirus Vector-Delivered Transgenic Antigens. *Mol. Ther. J. Am. Soc. Gene Ther.* **2011**, *19*, 1547–1557. [CrossRef]
20. Palmer, E.; Freeman, T. Investigation into the Use of C- and N-Terminal GFP Fusion Proteins for Subcellular Localization Studies Using Reverse Transfection Microarrays. *Comp. Funct. Genom.* **2004**, *5*, 342–353. [CrossRef]
21. Wang, X.; Chen, C.-H.; Badeti, S.; Cho, J.H.; Naghizadeh, A.; Wang, Z.; Liu, D. Deletion of ER-Retention Motif on SARS-CoV-2 Spike Protein Reduces Cell Hybrid during Cell–Cell Fusion. *Cell Biosci.* **2021**, *11*, 114. [CrossRef] [PubMed]

Article

A Method to Generate and Rescue Recombinant Adenovirus Devoid of Replication-Competent Particles in Animal-Origin-Free Culture Medium

Seyyed Mehdy Elahi [1,*], Jennifer Jiang [1], Nazila Nazemi-Moghaddam [1] and Rénald Gilbert [1,2]

[1] Department of Production Platforms & Analytics, National Research Council Canada, Building Montreal, 6100 Avenue Royalmount, Montreal, QC H4P 2R2, Canada; jennifer.jiang2@mail.mcgill.ca (J.J.); nazila.nazemi-moghaddam@cnrc-nrc.gc.ca (N.N.-M.); renald.gilbert@cnrc-nrc.gc.ca (R.G.)

[2] Department of Bioengineering, McGill University, Montreal, QC H3A 0E9, Canada

* Correspondence: mehdy.elahi@cnrc-nrc.gc.ca; Tel.: +1-514-496-2747; Fax: +1-514-496-5143

Abstract: Adenoviruses are promising vectors for vaccine production and gene therapy. Despite all the efforts in removing animal-derived components such as fetal bovine serum (FBS) during the production of adenovirus vector (AdV), FBS is still frequently employed in the early stages of production. Conventionally, first-generation AdVs (E1 deleted) are generated in different variants of adherent HEK293 cells, and plaque purification (if needed) is performed in adherent cell lines in the presence of FBS. In this study, we generated an AdV stock in SF-BMAdR (A549 cells adapted to suspension culture in serum-free medium). We also developed a limiting dilution method using the same cell line to replace the plaque purification assay. By combining these two technologies, we were able to completely remove the need for FBS from the process of generating and producing AdVs. In addition, we demonstrated that the purified AdV stock is free of any replication-competent adenovirus (RCA). Furthermore, we demonstrated that our limiting dilution method could effectively rescue an AdV from a stock that is highly contaminated with RCA.

Keywords: adenovirus vector; replication-competent adenovirus; limiting dilution; animal-origin-free culture media

Citation: Elahi, S.M.; Jiang, J.; Nazemi-Moghaddam, N.; Gilbert, R. A Method to Generate and Rescue Recombinant Adenovirus Devoid of Replication-Competent Particles in Animal-Origin-Free Culture Medium. *Viruses* **2023**, *15*, 2152. https://doi.org/10.3390/v15112152

Academic Editors: Ottmar Herchenröder and Brigitte Pützer

Received: 30 August 2023
Revised: 12 October 2023
Accepted: 23 October 2023
Published: 25 October 2023

1. Introduction

The replicative defective adenovirus vector (RD-AdV) with a deletion in its E1 region is one of the most useful adenovirus vectors for gene therapy and vaccination (as reviewed: [1–3]). Furthermore, since the beginning of the COVID-19 pandemic, AdVs have been the most common type of viral vector used for the production of COVID-19 vaccines [4]. To increase the safety of vaccines and gene therapies, the virus is typically rendered replication-incompetent by deleting its E1 gene. In a typical E1-deleted AdV, the gene of interest (GOI) is carried by the vector in place of the essential E1 region (E1A and E1B genes). Amplification of this type of AdV must take place in complementing cell lines, such as HEK293A cells (an adherent version of HEK293), that provide the E1 functions in *trans*. HEK293A cells were established by the stable transfection of human embryonic kidney cells with a segment of the adenovirus genome corresponding to the left end of the vector [5,6]. AdVs that possess flanking homology sequences with the adenovirus fragment are integrated into HEK293A cells, and this allows the acquisition of the E1 region through homologous recombination. This homologous recombination results in the emergence of replication-competent adenovirus (RCA) and the loss of GOI [7–9]. However, the presence of RCA in biological products, such as those used for gene therapy and vaccine development, is not desirable because it increases the product toxicity and reduces its potency. The presence of RCA is thus highly regulated. The maximum limit of RCA in biological products used for human application is 1 in 3×10^{10} viral particles [10].

Numerous efforts have been made to reduce or eliminate RCA. One of the approaches is to generate complementing cell lines that prevent the formation of RCA during vector growth. A few complementing cell lines have been generated with reduced homology between the integrated adenovirus sequence and the vector [11–18]. Three of these cell lines were also adapted to grow in suspension culture and in serum-free medium for easier scale-up during viral vector production and for regulatory compliance [11,12,19]. Previously, we have established such a complementing cell line, referred to as SF-BMAdR cells, by stable transfection of the minimal E1A and E1B genes needed for complementation into human lung carcinoma (A549), to prevent RCA formation by homologous recombination during recombinant AdV production. Despite the progress made in establishing new complementing cell lines, many scientists still use different variants of HEK293 (adherent or suspension) cell lines for the production of their recombinant adenoviruses since the cells are widely available with well-established experimental protocols.

Another issue in AdV production is the use of fetal bovine serum (FBS) in cell culture. The presence of FBS in the culture media is challenging and undesirable for clinical applications due to (I) its complex and variable composition, (II) the risk of contamination with animal viruses, and (III) ethical issues concerning animal cruelty [20]. The generation of AdV can be performed through homologous recombination in mammalian cells or by in vitro molecular cloning followed by transfection of the digested DNA into mammalian cells [1]. In both methods, FBS is normally added to the cell culture. Also, FBS is added during the plaque purification assay for isolating recombinant adenovirus from the parental virus (if recombination is performed in mammalian cell line) or for cloning the AdV.

To attempt to solve the current challenges surrounding the safety and purity requirements of AdV for clinical applications, in this study, we have developed a method to generate, isolate, and amplify clones of recombinant adenovirus devoid of RCA in the absence of FBS. In addition, we have demonstrated that the limiting dilution method developed in this study can be used to purify a recombinant adenovirus from a viral stock highly contaminated with RCA.

2. Materials and Methods

2.1. Cells Lines, Plasmids, and Viruses

Three stable cell lines were used in the current study: anchorage-dependent HEK293A cells [5] (HEK293A is another name for HEK293. The letter "A" in HEK293A cells refers to the adherent origin of cell line and differentiates it from HEK293S or HEK293SF that are adapted to suspension culture, in the presence or in the absence of FBS), A549 cells (a human lung carcinoma) (obtained from ATCC, CCL-185), and SF-BMAdR cells (produced in-house [12]). The HEK293A and A549 cell lines were grown in Dulbecco's modified Eagle's medium (Gibco, Life Technologies, Ottawa, ON, Canada), supplemented with 5% of fetal bovine serum (Hyclone, Logan, TU, USA) and 200 mM of glutamine (Gibco, Life Technologies, Ottawa, ON, Canada). The SF-BMAdR were grown in PRO293s CDM (PRO293s, Lonza, Walkersville, MD, USA) or Hycell TransFX H medium (Hycone, Logan, UT, USA) and were supplemented with 6 mM of glutamine, respectively. Adherent cells were grown in treated 96-well plates or 10 and 15 cm plates (Sarstedt, Numbrecht, North Rhine-Westphalia, Germany). Depending on the volume of cell culture, the suspension cultures were performed in 96-, 24-, and 6-well plates (Sarstedt, Numbrecht, North Rhine-Westphalia, Germany) or in 125 mL to 2 L shake flasks (Corning, Oneonta, NY, USA). Suspension culture agitation (for volume of ≥ 3 mL) was conducted at 100–120 rpm using orbital shakers (Bellco, or GL-300 Analytics).

The construction of pAdEasy-1-NS3-5a by the AdEasy plasmid system (Agilent Technologies, Santa Clara, CA, USA) which expresses hepatitis C virus NS3 to NS5a antigens was described previously [21]. This plasmid was used for the generation and large-scale production of Ad5-CMV5-NS3-5a in SF-BMAdR cell line without FBS. AdErB2, a first-generation AdV expressing the ErB2 gene [22], was used as a model for E1-deleted

AdV. AdPTG3602 [23], a wild-type adenovirus without any deletion in the genome, was used as a surrogate for RCA.

2.2. Developing a Limiting Dilution Method to Replace Plaque Purification

As the acceptable limit of RCA in biological material, in general, is extremely low, it is not possible to show the clonality and effectiveness of our limiting dilution method for isolating the AdV from RCA without artificially increasing the initial amounts of RCA in the test sample. Thus, an artificial AdV viral stock containing high amounts of wild-type adenovirus was generated to demonstrate the efficacy of our limiting dilution. The AdV:Her2 (as the surrogate for first-generation AdV), with wild-type adenovirus (as the surrogate for RCA) had a 1:1 ratio (Figure 1). Three 96-well plates were seeded with 30,000 SF-BMAdR cells in 50 μL/well. A total of 6 $\log_2$ dilutions of mixed viruses starting with 5 infectious units (ifu) to 0.16 ifu/50 μL was prepared. Immediately, 48 wells were infected with 50 μL of each $\log_2$ dilution of the mixed viruses. After 1 week of incubation, the plates underwent three rounds of freeze–thaw cycles. In the next step, the HEK293A cells were used to differentiate the infected wells from noninfected wells. Three plates of 96-well plates were seeded with 15,000 HEK293A cells/well and, the day after, plates were infected with 50 μL of the supernatant of infected SF-BMAdR cells. Each well on the infected plate of HEK293A corresponds to the same position/clone on the infected SF-BMAdR plates. Five days post-infection (p.i.), cells were fixed with methanol, and the presence of adenovirus fiber in infected cells was confirmed by immunostaining via a protocol described previously [24]. Briefly, 100% methanol was used as a fixing agent. PBS + 2% FBS was used as a washing and blocking reagent, as well as a diluent for the monoclonal antibodies used during immunostaining, and conjugated. The plates were washed three times between each step of the immunostaining. The fixed cells were blocked for 60 min, and incubated for 2 h with an in-house monoclonal antibody against adenovirus fiber (2G11-2) at a concentration of 1 μg/mL. Next, the samples were washed and incubated for 1 h at room temperature with a secondary antibody, which is a fluorescein isothiocyanate (FITC) conjugated anti-mouse antibody (Fluorescein-AffiniPure F(ab')2 Fragment Goat Anti-Mouse IgG, Fc gamma Fragment Specific, (Jackson ImmunoResearch #115-096-071, West Grove, PA, USA)) used at a dilution of 1/600.

To differentiate recombinant adenovirus from RCA, the positive colonies from infected SF-BMAdR 96-well plates were picked and amplified once in SF-BMAdR cells via the same protocol as described above to increase virus titer. In the next step, 96-well plates were seeded with 15,000 cells/well of HEK293A and A549 cells in alternating columns. The day after, 50 μL of $\log_{10}$ dilution (from 10^{-1} to 10^{-8}) of each amplified AdV clone was added to eight wells of one column for each cell line, with each dilution added to one well per cell line. The presence of a cytopathic effect (CPE) was observed and scored at 10 days p.i. in both cell lines using a bright-field microscope.

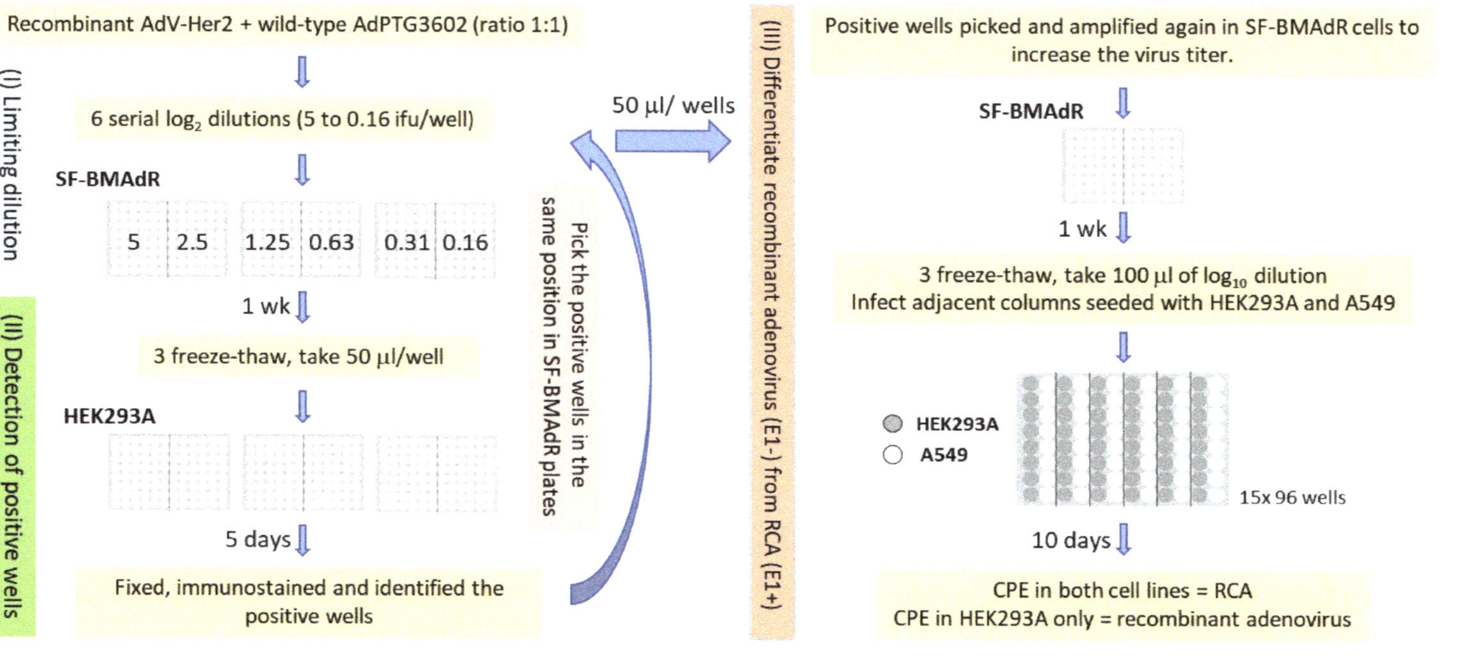

Figure 1. Workflow for rescuing recombinant adenoviruses from stock highly contaminated with RCA via limiting dilution and characterization of AdV clones. This protocol consists of three steps. (**I**) Limiting dilution step in which 48 wells were infected with 6 serial log₂ dilutions (5 to 0.16 ifu) per well of the AdV/RCA mixed stock. (**II**) Detection step in which positive wells were identified by immunostaining using HEK293A cells. (**III**) Differentiation step in which, after one additional round of clone amplification in SF-BMAdR cells, each clone infects HEK293A and A549 cells in parallel to differentiate whether they are rAdV or RCA.

2.3. Generation of rAdV without FBS

A recombinant adenovirus encoding the HCV-NS3 to NS5a antigens (Ad5-CMV5-NS3-5a) was made previously by homologous recombination in bacteria using the AdEasy plasmid system (Agilent Technologies, Santa Clara, CA, USA). The virus was made by transfecting the HEK293A with a PacI-digested shuttle vector [21]. In this study, we repeated the experiment by replacing the HEK293A with SF-BMAdR cells. A total of 3 mL of SF-BMAdR cells in Pro293s medium was seeded in 6-well plates at a concentration of 10^6 cells/mL and transfected with 3 or 5 μg of the PacI-digested shuttle vector using Polyethylenimine (PEIpro, Polyplus, Illkirch, France) with the ratio of PEIpro to DNA being 1.5 to 1. At 7 days post-transfection, the transfected cells were frozen and thawed three times. Supernatant of infected cells in both transfections was subject to two rounds of limiting dilution as described above, with only minor modifications. In the first round of limiting dilution, the starting dilution for the first log2 dilution was 2×10^{-1} and 32 wells were infected per dilution; in the second round of limiting dilution, the first dilution was 2×10^{-2} and 24 wells were infected for each dilution. After the first round of limiting dilution, one clone from each transfection was selected for the second round of limiting dilution.

2.4. Scale-Up and RCA Detection

After the second round of limiting dilution of Ad5-CMV5-NS3-5a, 12 subclones were selected to be amplified in SF-BMAdR cells for one round, in preparation for RCA detection analysis, which is performed with parallel amplification of viral subclones in HEK293A and A549. To confirm the absence of RCA at a larger scale, a subclone was amplified three times in the SF-BMAdR cell line, up to 1 L production. Briefly, the SF-BMAdR cells were cultured in PRO293s CDM or Hycell supplemented with 6 mM of glutamine. During the cell culture scale-up, cell concentration was kept between 2.5×10^5 and 1.5×10^6 cells/mL. On the day of infection, cells were diluted to 5×10^5 cells/mL with fresh medium. The three successive infections were carried out in 6-well plates (3 mL culture), 125 mL shaker flasks (20 mL culture), and, finally, in two 2 L shaker flasks (2×500 mL culture). For the first two infections, cells were harvested and freeze–thawed three times three days p.i. For the last infection, the freeze/thaw cycle was performed over cell pellets resuspended in 20 mL culture. The virus was purified by CsCl two-step-gradient (1.2 and 1.4 g/cm^3) centrifugation ($59,000 \times g$, 4 °C, 1 h 30) followed by continuous CsCl gradient centrifugation ($59,000 \times g$, 4 °C, 20 h) [24]. To quantify the viral particles (VP)/mL, the virus was diluted in a lysis solution (0.1% sodium dodecyl sulfate (SDS), 10 mM Tris-Cl pH 7.4, 1 mM EDTA) for 10 min at 56 °C. The OD260 was measured using the Nanodrop spectrophotometer (Thermo Fisher Scientific, Waltham, MA, USA) and VP/mL was calculated by multiplying the OD by the dilution factor and the extinction coefficient of 1.1×10^{12} [25].

The purified stock was tested for RCA presence. The A549 cells were seeded in 11 mm $\times$ 150 mm dishes at 6×10^6 cells/dish. The day after, they were infected with 2×10^{10} VP/dish of Ad5-CMV5-NS3-5a. To observe if there were any interference effects during the RCA detection test, three dishes (control group) were also infected with 3, 5, and 10 infectious units of AdPTG3602 (a wild-type adenovirus). At 6 dpi, 5 mL fresh medium was added to each dish and the CPE was followed, up to 10 dpi. At this time, the medium and cells were harvested. After centrifugation, cells were resuspended in 3 mL of fresh medium and freeze/thawed three times. Cell lysis was centrifuged and all of the supernatant was used for infecting 11 newly seeded 150 mm dishes of A549 cells. A total of 5 mL of fresh medium was added to each dish at 6 dpi and the CPE was followed, up to 10 dpi. Cells and medium were harvested and concentrated to 1 mL and freeze/thawed three times at 10 dpi. The presence or absence of adenovirus in each sample was confirmed in the third passage by immunostaining infected A549 cells. Briefly, 96-well plates of A549 cells at a concentration of 30,000 cells/well were infected with $\log_{10}$ dilution of infected cell lysis. At 48 hpi, the cells were fixed and immunostained by a monoclonal antibody against adenovirus fiber as described previously [24].

3. Results

3.1. Cloning AdV by Limiting Dilution

Plaque purification of AdV is one of two steps in adenovirus generation that is commonly performed in an adherent cell line in the presence of FBS. Plaque purification is essential for separating wild-type adenoviruses from recombinants when adenoviruses are generated by homologue recombination in mammalian cell lines. However, if the AdV is produced by the AdEasy system (recombination happens in *E. coli* instead), plaque purification is nonmandatory, as all the produced AdVs after transfection are recombinant [26]. However, it is better to perform at least one round of plaque purification to ensure the purity and clonality of the virus and initiate vector production scale-up with a virus clone that has a confirmed GOI expression.

In this study, we developed a limiting dilution method in suspension culture without FBS to replace the plaque purification assay in order to completely remove FBS from the virus production process. The efficacy of this method for AdV clone isolation was shown by purifying an E1-deleted AdV from a stock highly contaminated with wild-type adenovirus. However, with this method, we cannot observe visible CPE in infected SF-BMAdR cells since they are in suspension. An additional amplification step in adherent cells (HEK293A) is, therefore, essential to identify the wells infected with adenoviruses.

The number of infected wells per each $\frac{1}{2}$ of the 96-well plates post immunostaining is indicated in Table 1. There is a direct correlation between the number of positive wells and the number of infectious viral particles that were used for infecting each well. At this stage, we were not able to distinguish between RCA (E1-positive) and first-generation (E1-negative) AdV, as both viruses can replicate in HEK293A cells. A comparative infection in A549 and HEK293A was needed to differentiate RCA-infected wells from E1-deleted AdV-infected wells. However, our preliminary results showed that the titer of E1-deleted AdV in infected SF-BMAdR cells with one infectious particle was low, and we could not directly use these seed stocks. Thus, an additional round of amplification in SF-BMAdR was necessary to increase the virus' titer before comparing their CPE in A549 and HEK293A. Table 2 shows a few examples of how the result is interpreted. For example, we determined that both samples #1 and #2 contain RCA as the virus replicated in both cell lines. We determined that samples #3 and #4 are E1-deleted AdV, as they replicated only in HEK293A but not A549.

Table 1. The number of wells which were identified as either E1-deleted AdV or RCA at each concentration of the mixed seed stock. The positive wells are either scored as RCA or E1-deleted AdV or excluded from analysis. The last column presents the sum of all the analyzed wells.

Infectious Dose (ifu)/Well (SF-BMAdR)	5	2.5	1.25	0.63	0.31	0.16	Total of All Wells
Total positive wells in FFA (HEK293A)	38	27	14	7	5	1	92
Amplifed wells in SF-BMAdR cells	35	27	14	7	5	1	89
RCA (amplified in HEK293A & A549 cells)	16	10	7	2	0	0	35
AdV (amplified only in HEK293A)	13	11	6	3	5	1	39
Exclueded (Not amplified or had low titer)	6	6	1	2	0	0	15

On some occasions, when the virus titer was very high (sample #4), a cytotoxic effect could be detected in the first dilution of A549. This is caused by infecting the cells with a very high multiplicity of infection (MOI). Samples like #5 and #6 are excluded from our analysis because they did not show CPE in either cell line with dilution higher than 10^{-2} and, therefore, their comparison in both cell lines was not meaningful. Of over 74 amplified adenoviruses, 39 were E1-deleted AdV and 35 were RCA with a ratio of 1.1:1, which is roughly the same as the ratio of the mixed stock for limiting dilution that we started with (1:1) (Table 1).

Table 2. A comparative analysis of AdV (E1-negative) and RCA (E1-positive) in HEK293A and A549 cell lines. The E1-deleted AdV replicates only in HEK29A but RCA replicates in both cell lines.

	Cell Lines											
	HEK-293A	A549	HEK-293A	A549	HEK-293A	A549	HEK-293A	A549	HEK-293A	A549	HEK-293A	A549
	#1		#2		#3		#4		#5		#6	
	+	+	+	+	+	−	+	+	+	−	−	−
	+	+	+	+	+	−	+	−	−	−	−	−
	+	+	+	+	+	−	+	−	−	−	−	−
Log10 dilution 10^{-1} to 10^{-8}	+	+	−	−	+	−	+	−	−	−	−	−
	+	+	−	−	+	−	+	−	−	−	−	−
	+	+	−	−	−	−	+	−	−	−	−	−
	+	−	−	−	−	−	−	−	−	−	−	−
	−	−	−	−	−	−	−	−	−	−	−	−
	Interpretation											
	RCA		RCA		AdV		AdV		Excluded		Excluded	

3.2. Generation of rAdV in the Absence of FBS and RCA

To have an entire process of AdV generation and production devoid of FBS and RCA, we employed an alternative cell line, the SF-BMAdR, to replace HEK293A cells at the stage of transfection. Ad5-CMV5-NS3-5a was generated in the SF-BMAdR cell line in the absence of FBS. Transfection of six-well plates with 3 μg of DNA has generated two times more AdV than transfection with 5 μg of DNA. The supernatant of infected cells in both transfections was subject to two rounds of limiting dilution. All 12 of the selected clones that underwent the first round of limiting dilution were E1-deleted AdV. One clone from each transfection was selected for the second round of limiting dilution. Again, 12 subclones were selected in total and, after one cycle of amplification in SF-BMAdR cells, followed by parallel amplification in HEK293A and A549, all subclones were characterized as E1-deleted AdV.

To confirm the absence of RCA, one subclone of Ad5-CMV5-NS3-5a was amplified three times in SF-BMAdR. The final 1L stock was purified. RCA detection was performed in A549 cells. At the second passage in A549 cells, the CPE was observed in all the three samples of the control group that were co-infected with 3, 5, and 10 infectious units of wild-type adenovirus but not in the eight dishes infected with only recombinant adenovirus (test group), as expected (Figure 2A). No interference was detected when cells were infected with an MOI of 1666 VP/cell of AdV (2×10^{10} VP/dish) and 3 to 10 ifu/dish of wild-type adenovirus. The immunostaining of infected cells confirmed that the CPE observed in the control group was caused by the replication of adenovirus. No CPE or immunostaining-positive cells were detected in the test group (infected cells with only E1-deleted AdV) (Figure 2B).

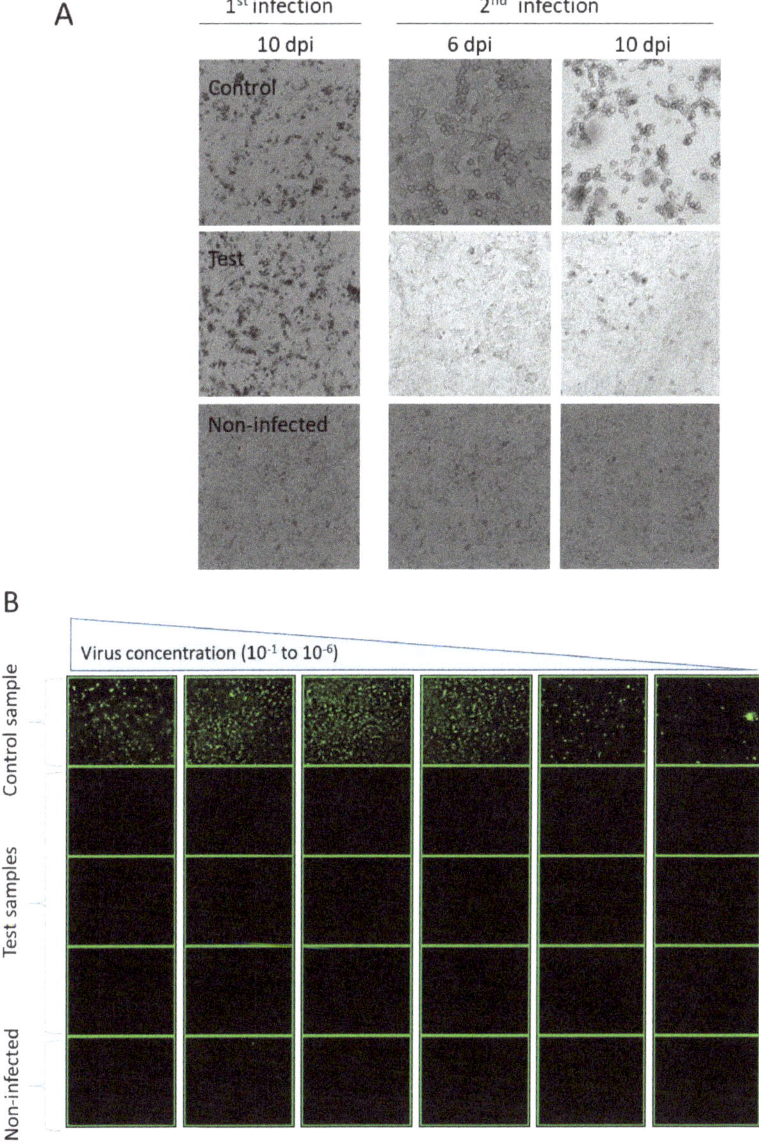

Figure 2. RCA detection in A549 cells. (**A**) Presence of RCA was evaluated in purified virus stock by two successive passages in A549 cells followed by the immunostaining over the third passage. In the second passage, the dishes in the control group showed CPE since 6 dpi (dish infected with 3 IU of wild-type shown) but no CPE was detected in the eight test dishes until the end of the experiment. (**B**) For immunostaining, the A549 cells were infected with different dilutions of cell lysis harvested from the second amplification. Only one photo from each virus dilution is shown. The first row shows the control group in the first passage co-infected with 3 IU of wild-type adenovirus. The second to fourth rows show three out of eight test dishes infected with only E1-deleted AdV at the first passage. The last row shows noninfected cells.

4. Discussion

The limiting dilution method that we developed in this paper was able to isolate E1-deleted AdV from a viral stock highly contaminated with undesirable RCA. Fortunately, in real-world applications, the level of contamination is usually much lower. This means our limiting dilution method has the potential to be even more effective.

Regardless of the method used for AdV isolation, whether it is plaque purification or limiting dilution, we could not exclude the possibility of having more than one AdV infecting the same cell or neighbouring cells (in plaque assay) or well (in limiting dilution). This could happen especially when the wells are infected with more than one infectious particle per well. However, the possibility of this happening is much lower at a higher dilution. In our limiting dilution method, if a well is infected simultaneously with an AdV and an RCA, it is identified as RCA and it will be excluded from the next step. In the case of infecting the same well with two AdVs, if the clonality of the AdV stock is imperative, another round of limiting dilution is recommended. Infecting more wells with a higher dilution in the first round of limiting dilution can be an alternative to performing a second round of limiting dilution.

The unavailability of cell lines that prevent the formation of RCA or the inability to change the AdV production process in adherent or suspension variants of the HEK293 cell line can potentially lead to RCA formation at the later stages of production. The limiting dilution method described in this paper can also be used as an alternative to plaque purification to rescue a well-characterized E1-deleted AdV stock contaminated with RCA. Specifically, the method can purify AdV stocks when the level of RCA in the stock exceeds the acceptable level for clinical trials. In addition, the lack of FBS or animal-derived components used in this process is an additional asset. To reduce the number of 96-well plates needing to be analyzed during limiting dilution, we suggest titrating the virus seed in order to infect the wells with less than 1 infectious unit per well. We used the immunostaining method for the detection of AdV-infected wells during limiting dilution to decrease the incubation period. However, immunostaining is not indispensable and, by extending the incubation period by 5–7 days, CPE will become apparent in the infected wells.

To completely remove FBS from the entire virus generation and production process, we studied the feasibility of generating AdV by transfection of SF-BMAdR cells. The virus titer produced in SF-BMAdR cells was lower than those produced in HEK293A cells. This is due to the lower transfection efficiency of large plasmids in A549-derived cells (such as SF-BMAdR cells) compared to HEK293A cells [27]. However, this is not a significant issue since we only need tens or hundreds of infectious virus particles to start the limiting dilution. If the production of the AdV stock originates from a pool of transfected cells without prior cloning, one additional passage is needed to have a high titer seed stock for scale-up.

The possible presence of RCA was evaluated in the purified virus stock by two successive passages in A549 cells followed by immunostaining over the third passage. In total, 9.6×10^7 cells were infected with 1.6×10^{11} VP of AdV resulting in an MOI of 1666 VP/cell. No interference for replication of RCA was reported in the presence of up to 2000 particles per cell of E1-deleted AdV [28]. We did not detect any CPE after the first and second passage in any of the test dishes. However, all three dishes in the control group showed CPE at the second passage (but not at the first passage). Consequently, at least two passages are needed to confirm the presence of RCA. We used a conventional cell-culture-based RCA assay for detecting RCA, and the assay needs 3 weeks to complete. However, the infectivity PCR-based RCA assays [29] are faster methods that can reach the same level of sensitivity in a shorter period of time. Immunostaining was performed in our study to confirm that CPE was indeed caused by adenovirus and to further increase the sensibility of the assay, in the case that we missed a weak CPE (difficult to observe by eye) at the second passage. The test group dishes were infected in total with 1.6×10^{11} VP. As we were able to detect CPE caused by as low as 3 infectious units of wild-type adenovirus without interference in all the control group samples, our RCA detection method reached levels of sensitivity of 1 IU of RCA in 5.33×10^{10} VP, which is slightly more sensitive than the FDA recommendation of RCA in biological materials (1 IU of RCA in 3×10^{10} VP).

5. Conclusions

The method developed in this study allows us to generate clones of recombinant adenovirus in the absence of FBS and devoid of RCA. It thus improves the safety for clinical applications of AdV. Also, this limiting dilution method allows us to rescue well-characterized E1-deleted AdV stocks contaminated with RCA and decreases the amount of time required for the generation and characterization of new viral stocks.

Author Contributions: Conceptualization, S.M.E.; Formal analysis, S.M.E.; Funding acquisition, R.G.; Investigation, S.M.E., J.J. and N.N.-M.; Supervision, S.M.E.; Validation, S.M.E.; Writing—original draft, S.M.E., J.J. and R.G.; Writing—review and editing, S.M.E., J.J. and R.G. All authors have read and agreed to the published version of the manuscript.

Funding: This work was supported by the National Research Council Canada (NRC).

Institutional Review Board Statement: Not applicable.

Informed Consent Statement: Not applicable.

Data Availability Statement: The data presented in this study are available in the current article. The raw data that support the findings of this study are available on request from the corresponding author.

Conflicts of Interest: The authors declare no conflict of interest.

References

1. Danthinne, X.; Imperiale, M.J. Production of First Generation Adenovirus Vectors: A Review. *Gene Ther.* **2000**, *7*, 1707–1714. [CrossRef] [PubMed]
2. Kratzer, R.F.; Kreppel, F. Production, Purification, and Titration of First-Generation Adenovirus Vectors. *Methods Mol. Biol.* **2017**, *1654*, 377–388. [CrossRef] [PubMed]
3. Tsilingiris, D.; Vallianou, N.G.; Karampela, I.; Muscogiuri, G.; Dalamaga, M. Use of Adenovirus Type-5 Vector Vaccines in COVID-19: Potential Implications for Metabolic Health? *Minerva Endocrinol.* **2022**, *47*, 264–269. [CrossRef]
4. World Health Organization. COVID-19 Vaccine Tracker and Landscape. Available online: https://www.who.int/publications/m/item/draft-landscape-of-covid-19-candidate-vaccines (accessed on 7 June 2023).
5. Graham, F.L.; Smiley, J.; Russell, W.C.; Nairn, R. Characteristics of a Human Cell Line Transformed by DNA from Human Adenovirus Type 5. *J. Gen. Virol.* **1977**, *36*, 59–74. [CrossRef] [PubMed]
6. Louis, N.; Evelegh, C.; Graham, F.L. Cloning and Sequencing of the Cellular-Viral Junctions from the Human Adenovirus Type 5 Transformed 293 Cell Line. *Virology* **1997**, *233*, 423–429. [CrossRef] [PubMed]
7. Lochmüller, H.; Jani, A.; Huard, J.; Prescott, S.; Simoneau, M.; Massie, B.; Karpati, G.; Acsadi, G. Emergence of Early Region 1-Containing Replication-Competent Adenovirus in Stocks of Replication-Defective Adenovirus Recombinants (Delta E1 + Delta E3) during Multiple Passages in 293 Cells. *Hum. Gene Ther.* **1994**, *5*, 1485–1491. [CrossRef]
8. Hehir, K.M.; Armentano, D.; Cardoza, L.M.; Choquette, T.L.; Berthelette, P.B.; White, G.A.; Couture, L.A.; Everton, M.B.; Keegan, J.; Martin, J.M.; et al. Molecular Characterization of Replication-Competent Variants of Adenovirus Vectors and Genome Modifications to Prevent Their Occurrence. *J. Virol.* **1996**, *70*, 8459–8467. [CrossRef]
9. Zhu, J.; Grace, M.; Casale, J.; Chang, A.T.; Musco, M.L.; Bordens, R.; Greenberg, R.; Schaefer, E.; Indelicato, S.R. Characterization of Replication-Competent Adenovirus Isolates from Large-Scale Production of a Recombinant Adenoviral Vector. *Hum. Gene Ther.* **1999**, *10*, 113–121. [CrossRef]
10. Food and Drug Administration. *Chemistry, Manufacturing, and Control (CMC) Information for Human Gene Therapy Investigational New Drug Applications (INDs) Guidance for Industry*; Food and Drug Administration: Silver Spring, MD, USA, 2020.
11. Farson, D.; Tao, L.; Ko, D.; Li, Q.; Brignetti, D.; Segawa, K.; Mittelstaedt, D.; Harding, T.; Yu, D.-C.; Li, Y. Development of Novel E1-Complementary Cells for Adenoviral Production Free of Replication-Competent Adenovirus. *Mol. Ther.* **2006**, *14*, 305–311. [CrossRef]
12. Gilbert, R.; Guilbault, C.; Gagnon, D.; Bernier, A.; Bourget, L.; Elahi, S.M.; Kamen, A.; Massie, B. Establishment and Validation of New Complementing Cells for Production of E1-Deleted Adenovirus Vectors in Serum-Free Suspension Culture. *J. Virol. Methods* **2014**, *208*, 177–188. [CrossRef]
13. Kim, J.S.; Lee, S.H.; Cho, Y.S.; Park, K.; Kim, Y.H.; Lee, J.H. Development of a Packaging Cell Line for Propagation of Replication-Deficient Adenovirus Vector. *Exp. Mol. Med.* **2001**, *33*, 145–149. [CrossRef] [PubMed]
14. Kovesdi, I.; Hedley, S.J. Adenoviral Producer Cells. *Viruses* **2010**, *2*, 1681–1703. [CrossRef] [PubMed]

15. Howe, J.A.; Pelka, P.; Antelman, D.; Wilson, C.; Cornell, D.; Hancock, W.; Ramachandra, M.; Avanzini, J.; Horn, M.; Wills, K.; et al. Matching Complementing Functions of Transformed Cells with Stable Expression of Selected Viral Genes for Production of E1-Deleted Adenovirus Vectors. *Virology* **2006**, *345*, 220–230. [CrossRef] [PubMed]

16. Xu, Q.; Arevalo, M.T.; Pichichero, M.E.; Zeng, M. A New Complementing Cell Line for Replication-Incompetent E1-Deleted Adenovirus Propagation. *Cytotechnology* **2006**, *51*, 133–140. [CrossRef]

17. Fallaux, F.J.; Bout, A.; van der Velde, I.; van den Wollenberg, D.J.; Hehir, K.M.; Keegan, J.; Auger, C.; Cramer, S.J.; van Ormondt, H.; van der Eb, A.J.; et al. New Helper Cells and Matched Early Region 1-Deleted Adenovirus Vectors Prevent Generation of Replication-Competent Adenoviruses. *Hum. Gene Ther.* **1998**, *9*, 1909–1917. [CrossRef]

18. Gao, G.P.; Engdahl, R.K.; Wilson, J.M. A Cell Line for High-Yield Production of E1-Deleted Adenovirus Vectors without the Emergence of Replication-Competent Virus. *Hum. Gene Ther.* **2000**, *11*, 213–219. [CrossRef]

19. Xie, L.; Pilbrough, W.; Metallo, C.; Zhong, T.; Pikus, L.; Leung, J.; Aunins, J.G.; Zhou, W. Serum-Free Suspension Cultivation of PER.C6(R) Cells and Recombinant Adenovirus Production under Different PH Conditions. *Biotechnol. Bioeng.* **2002**, *80*, 569–579. [CrossRef]

20. Gstraunthaler, G.; Lindl, T.; van der Valk, J. A Plea to Reduce or Replace Fetal Bovine Serum in Cell Culture Media. *Cytotechnology* **2013**, *65*, 791–793. [CrossRef]

21. Young, K.G.; Haq, K.; MacLean, S.; Dudani, R.; Elahi, S.M.; Gilbert, R.; Weeratna, R.D.; Krishnan, L. Development of a Recombinant Murine Tumour Model Using Hepatoma Cells Expressing Hepatitis C Virus Nonstructural Antigens. *J. Viral. Hepat.* **2018**, *25*, 649–660. [CrossRef]

22. Haq, K.; Jia, Y.; Elahi, S.M.; MacLean, S.; Akache, B.; Gurnani, K.; Chattopadhyay, A.; Nazemi-Moghaddam, N.; Gilbert, R.; McCluskie, M.J.; et al. Evaluation of Recombinant Adenovirus Vectors and Adjuvanted Protein as a Heterologous Prime-Boost Strategy Using HER2 as a Model Antigen. *Vaccine* **2019**, *37*, 7029–7040. [CrossRef]

23. Oualikene, W.; Lamoureux, L.; Weber, J.M.; Massie, B. Protease-Deleted Adenovirus Vectors and Complementing Cell Lines: Potential Applications of Single-Round Replication Mutants for Vaccination and Gene Therapy. *Hum. Gene Ther.* **2000**, *11*, 1341–1353. [CrossRef] [PubMed]

24. Elahi, S.M.; Nazemi-Moghaddam, N.; Gadoury, C.; Lippens, J.; Radinovic, S.; Venne, M.-H.; Marcil, A.; Gilbert, R. A Rapid Focus-Forming Assay for Quantification of Infectious Adenoviral Vectors. *J. Virol. Methods* **2021**, *297*, 114267. [CrossRef] [PubMed]

25. Mittereder, N.; March, K.L.; Trapnell, B.C. Evaluation of the Concentration and Bioactivity of Adenovirus Vectors for Gene Therapy. *J. Virol.* **1996**, *70*, 7498–7509. [CrossRef] [PubMed]

26. Reddy, P.S.; Ganesh, S.; Hawkins, L.; Idamakanti, N. Generation of Recombinant Adenovirus Using the Escherichia Coli BJ5183 Recombination System. *Methods Mol. Med.* **2007**, *130*, 61–68. [CrossRef]

27. Søndergaard, J.N.; Geng, K.; Sommerauer, C.; Atanasoai, I.; Yin, X.; Kutter, C. Successful Delivery of Large-Size CRISPR/Cas9 Vectors in Hard-to-Transfect Human Cells Using Small Plasmids. *Commun. Biol.* **2020**, *3*, 319. [CrossRef] [PubMed]

28. Murakami, P.; Havenga, M.; Fawaz, F.; Vogels, R.; Marzio, G.; Pungor, E.; Files, J.; Do, L.; Goudsmit, J.; McCaman, M. Common Structure of Rare Replication-Deficient E1-Positive Particles in Adenoviral Vector Batches. *J. Virol.* **2004**, *78*, 6200–6208. [CrossRef]

29. Schalk, J.A.C.; de Vries, C.G.J.C.A.; Orzechowski, T.J.H.; Rots, M.G. A Rapid and Sensitive Assay for Detection of Replication-Competent Adenoviruses by a Combination of Microcarrier Cell Culture and Quantitative PCR. *J. Virol. Methods* **2007**, *145*, 89–95. [CrossRef]

Article

Engineering Adenoviral Vectors with Improved GBM Selectivity

Emily A. Bates [1], Charlotte Lovatt [1], Alice R. Plein [1], James A. Davies [1], Florian A. Siebzehnrubl [2] and Alan L. Parker [1,3,*]

[1] Division of Cancer and Genetics, School of Medicine, Cardiff University, Heath Park, Cardiff CF14 4XN, UK; batese@cardiff.ac.uk (E.A.B.); lovattce@cardiff.ac.uk (C.L.); arplein@gmail.com (A.R.P.); daviesja9@cardiff.ac.uk (J.A.D.)

[2] European Cancer Stem Cell Research Institute, School of Biosciences, Cardiff University, Maindy Road, Cardiff CF24 4HQ, UK; fas@cardiff.ac.uk

[3] Systems Immunity University Research Institute, School of Medicine, Cardiff University, Heath Park, Cardiff CF14 4XN, UK

* Correspondence: parkeral@cardiff.ac.uk

Abstract: Glioblastoma (GBM) is the most common and aggressive adult brain cancer with an average survival rate of around 15 months in patients receiving standard treatment. Oncolytic adenovirus expressing therapeutic transgenes represent a promising alternative treatment for GBM. Of the many human adenoviral serotypes described to date, adenovirus 5 (HAdV-C5) has been the most utilised clinically and experimentally. However, the use of Ad5 as an anti-cancer agent may be hampered by naturally high seroprevalence rates to HAdV-C5 coupled with the infection of healthy cells via native receptors. To explore whether alternative natural adenoviral tropisms are better suited to GBM therapeutics, we pseudotyped an HAdV-C5-based platform using the fibre knob protein from alternative serotypes. We demonstrate that the adenoviral entry receptor coxsackie, adenovirus receptor (CAR) and CD46 are highly expressed by both GBM and healthy brain tissue, whereas Desmoglein 2 (DSG2) is expressed at a low level in GBM. We demonstrate that adenoviral pseudotypes, engaging CAR, CD46 and DSG2, effectively transduce GBM cells. However, the presence of these receptors on non-transformed cells presents the possibility of off-target effects and therapeutic transgene expression in healthy cells. To enhance the specificity of transgene expression to GBM, we assessed the potential for tumour-specific promoters hTERT and survivin to drive reporter gene expression selectively in GBM cell lines. We demonstrate tight GBM-specific transgene expression using these constructs, indicating that the combination of pseudotyping and tumour-specific promoter approaches may enable the development of efficacious therapies better suited to GBM.

Keywords: brain tumour; oncolytic virus; receptor; therapy

Citation: Bates, E.A.; Lovatt, C.; Plein, A.R.; Davies, J.A.; Siebzehnrubl, F.A.; Parker, A.L. Engineering Adenoviral Vectors with Improved GBM Selectivity. *Viruses* **2023**, *15*, 1086. https://doi.org/10.3390/v15051086

Academic Editors: Brigitte Pützer and Ottmar Herchenröder

Received: 30 March 2023
Revised: 21 April 2023
Accepted: 26 April 2023
Published: 28 April 2023

1. Introduction

Glioblastoma (GBM), a tumour thought to arise from neuroglial stem or progenitor cells, is the most aggressive primary adult brain cancer. It has an average incidence of 3 cases per 100,000 people per year worldwide [1], making it the most common type of malignant adult brain neoplasm. The current standard treatment procedure for GBM involves the maximum surgical resection of the tumour (where possible), followed by radiation therapy and concomitant chemotherapy using the oral alkylating agent temozolomide (TMZ). Nevertheless, due to its heterogeneity, invasiveness, and rapid growth, the prognosis for GBM patients is very poor, where the mean overall survival is only approximately 15 months [2].

Oncolytic viruses, which are viruses engineered to selectively infect and lyse tumour cells, constitute a promising alternative therapeutic agent for GBM (reviewed in [3]). Their ability to self-amplify within tumour cells not only expands the therapy at the point of need (i.e., within the tumour microenvironment, TME), but also induces immunogenic cell death (ICD) through the lytic nature of cell death. The power of oncolytic viruses can be further augmented via the engineering of potent therapeutic transgenes into the viral genome that can further enhance immune activation and immune-cell-mediated tumour cell killing. Collectively, these traits have the potential to instigate an anti-glioma immune response against both the primary tumour and metastatic growth within the brain.

Whilst the array of viruses available for oncolytic applications are wide, adenoviruses (HAdV) have proven to be the most popular as oncolytic viruses against several types of cancer, as gauged by the volume of clinical trials conducted in the area to date [4]. They have a clinically proven safety profile, are relatively amenable to genetic manipulation and are able to accommodate relatively large transgene inserts [5]. However, to date, the promising results demonstrated in vitro and in murine models of cancer have largely failed to translate into significant efficacy in clinical trials. Whilst these trials have demonstrated feasibility and safety, improvements in cancer patient outcomes have generally been modest [6,7]. Limited efficacy may be due in part to the over reliance of many clinical studies on adenovirus 5 (HAdV-C5).

There have been over 100 human adenoviral serotypes described to date with differing entry receptors as well as prevalence rates within the human population [8]. HAdV-C5 has a particularly high prevalence in the human population, especially in Africa and Asia. Neutralising antibodies against HAdV-C5, generated in response to a natural pathogenic infection, may hamper the efficacy of HAdV-C5-based therapies when deployed clinically due to immune inactivation [9,10]. Alternative serotypes including HAdV-D10 have been explored to overcome the effects of anti-HAdV-C5 neutralisation [11]. In addition, HAdV-C5 engages coxsackie and adenovirus receptor (CAR) [12] as a primary means of cell entry, a receptor that is expressed both on erythrocytes and the tight junctions of epithelial cells (reviewed in [13]), yet is commonly downregulated by some types of cancer [14,15]. Furthermore, interactions between the major HAdV-C5 capsid protein, hexon and blood clotting factors, particularly factor (F) X, result in the rapid and efficient cellular uptake of HAdV-C5 virions via widely expressed heparan sulphate proteoglycans (HSPGs) [16–18]. This results in the depletion of the therapeutic effects due to widespread off-target infection and sequestration with inefficient tumour cell infection by HAdV-C5 [19]. HAdV-C5-based therapies for recurrent GBM are often delivered via intratumoral administration which limits off-target effects; however, these therapies have so far only been able to demonstrate modest success. Whilst demonstrating the safety and oncolysis via the intratumoral delivery of the virus, the overall median patient survival was not improved due to the standard of care (13 months) as the tumour invariably returned [20]. Local delivery is likely a necessity in GBM for overcoming the physical obstacle of the blood–brain barrier and to prevent local recurrence after surgical resection. We therefore investigated whether the development of Ad-based therapeutics using alternative adenovirus entry receptors may represent a more potent virotherapy for GBM. We have focused on HAdV-C5 vectors pseudotyped with fibre knob proteins derived from serotypes HAdV-D26, HAdV-B35 and HAdV-B3, which utilise the cell entry receptors sialic acid (SA) [21] or CD46 [22] and Desmoglein 2 (DSG2) [23], respectively. We demonstrate that GBM cell lines and tissue express all three adenoviral entry receptors tested, namely CAR, CD46 and DSG2. In addition, sialic acid is known to be expressed at high levels in the brain [24]. GBM does not express $\alpha v \beta 6$ integrin, which is upregulated in many other types of aggressive cancers (e.g., ovarian, lung, skin and cervical cancer [25]), and is a promising target for novel adenoviral oncolytic therapies [11,26–28]. We also investigated the efficacy of various HAdV-C5 pseudotypes to gauge which adenoviral serotype may be better suited for oncolytic therapy in GBM. Our experiments revealed that both GBM and GBM stem cells express recognised adenoviral receptors and can be transduced by HAdV-C5 and HAdV-C5 with pseudotyped fibre knob proteins binding CD46, sialic acid and DSG2. Since these entry receptors are not unique to transformed GBM cells, and are expressed on normal cells, we additionally investigated whether improving selectivity, through the incorporation of tumour-specific promoters, could be employed to regulate the expression of potentially toxic or immunostimulatory transgenes in transformed GBM cells. Our findings indicate that a combination of alternative receptor usage, either through the use of pseudotyped or whole serotyped vectors, in combination with tumour-specific promoters such as survivin, may offer a powerful and highly selective means to deliver cytotoxic or immunostimulatory payloads selectively to GBM cells.

2. Materials and Methods

2.1. Cell Lines and Adenoviral Vectors

Human-patient-derived GBM stem cell lines L0 and L1 [29] were cultured as described previously [30,31]. Briefly, cells were cultured in DMEM-F12 supplemented with N2 (Thermo Fisher, Waltham, MA, USA, cat. no. 17502048) and 20 ng/mL EGF (R&D systems, Minneapolis, MN, USA cat. no. 236-EG-200), 20 ng/mL FGF (R&D systems Minneapolis, MN, USA, cat. No. 233-FB-010) and 100 µg/mL heparin (Sigma, Gillingham, UK cat. no. H4784). Cells were seeded at a density of 50,000 cells/mL, cultured as neurospheres and passaged every 7 days using Accutase (Sigma, Gillingham, UK cat. no. A6964) to disassociate the cells. E51 and E55 were obtained from the Glioma Cellular Genetics Resource (GCGR) from Professor Steve Pollard (University of Edinburgh). E51 and E55 were derived from GBM patients and cultured as previously described [32]. Cells were grown in DMEM-F12 (cat. no. D8437). Complete DMEM-F12 with 1.25% glucose (Sigma, Gillingham, UK G8644), 2% l MEM-NEAA (Gibco, Paisley, UK 11140-035), penicillin and streptomycin, 0.16% and 7.5% BSA solution (Gibco, Paisley, UK 15260-037), 0.2% beta mercaptoethanol (Gibco, Paisley, UK 31350-010), 1% B27 Supplement (Gibco, Paisley, UK 17504-044) and 0.5% N2 supplement (Gibco, Paisley, UK 17502-048) was used. Complete medium was supplemented with mouse EGF (10 ng/mL, peprotech, Cranbury, NJ, USA 315-09), human FGF (10 ng/mL, peprotech, Cranbury, NJ, USA 100-18b) and laminin (2 µg/mL, Sigma, Gillingham, UK, L2020). HFCAR cells were generated in house and cultured in DMEM supplemented with 10% FBS, 2% penicillin and streptomycin and 1% L-Glutamine.

Adenoviral vectors were produced as described previously using recombineering techniques [33]. Replication-deficient HAdV-C5 expressing a luciferase reporter gene provided the backbone for the replacement of fibre knob (K) protein to generate HAdV-C5/D26K, HAdV-C5/B35K and HAdV-C5/D49K [21]. HAdV-C5/B3K expressing the GFP reporter gene was provided by Professor Andre Lieber (University of Washington). HAdV-C5 vectors with luciferase expression under the control of the tumour-specific promoters hTERT and Survivin and Survivin/hTERT fusion promoter (Survivin-luciferase-BHG poly A-hTERT) were generated in house using recombineering techniques. Promoter sequences have been included in Supplementary Table S1. Viral vectors were expanded in 293-TRex cells and purified using two step CsCl gradient ultracentrifugation.

2.2. Flow Cytometry

Cells were incubated in FACS buffer (PBS) (Thermo Fisher, Waltham, MA, USA cat. no. 10010023) with 5% foetal bovine serum (Sigma, Gillingham, UK cat. no. F9665) and labelled with mouse primary antibodies: anti-$\alpha v \beta 6$ 10D5 (Millipore, Burlington, MA, USA cat no. mab20772), anti-CAR RmcB (Millipore, Burlington, MA, USA cat. no. 05-644), anti-CD46 258-MEM (Novus Biologicals, Littleton, CO, USA cat. no. LS-B5950-50) or anti-DSG2 CSTEM28 (Thermo Fisher, cat. no. 14-9159-82); all antibodies were diluted 1:500. Cells were subsequently stained with a secondary anti-mouse 647-conjugated antibody (Thermo Fisher, Waltham, MA, USA) and fixed with 4% paraformaldehyde. Appropriate fluorescence gating parameters were established using unstained and matched isotype control IgG-stained cells. In all of the experiments, doublets were eliminated using pulse geometry gates (FSC-H versus FSC-A). Single-cell suspensions were analysed using a BD

AccuriTM C6 flow cytometer; FlowJo software (FlowJo LLC., Franklin Lakes, NJ, USA) was used for subsequent analyses.

2.3. Viral Infection and Luciferase Assays

Viral transduction assays were performed as described previously [34]. Briefly, cells were seeded at a density of 20,000 cells/well in a 96-microwell plate (Nunc, Thermo Fisher, Waltham, MA, USA) and infected with concentrations of 1000, 5000 or 10,000 virus particles (vps)/cell the following day. For each experimental condition, triplicate wells were treated, and a no-virus control was included. After 72 h, cells were lysed and the luciferase activity was quantified using a BioTek microplate reader and the luciferase assay system (Promega, Madison, WI, USA cat. no. E1500) according to the manufacturer's instructions. In addition, total protein concentration per well was determined using a microplate absorbance plate reader (BioTek, Winooski, VT, USA) and the Pierce Micro BCA Protein assay (Thermo Fisher, Waltham, MA, USA cat. no. 10249133) to enable correction to relative luminescence units (RLUs) per mg of total protein. Viral infections with GFP-expressing vectors were carried out in the same manner; however, GFP levels were measured after 72 h via fixing cells in 4% paraformaldehyde and measuring fluorescence using the BD AccuriTM C6 flow cytometer, acquiring the data in the FL1 channel. Data were analysed using FlowJo software (FlowJo LLC.).

2.4. mRNA Expression Analysis

The GBM multiforme RNA-seq data from the Cancer Genome Atlas (TCGA) (n = 153) and normal brain Genotype Tissue expression (GTEx) (n = 1141) RNA-seq datasets were downloaded from the UCSC Xena RNA-seq Compendium (https://xena.ucsc.edu/), accessed on 3 March 2023, where the samples had been normalised and re-analysed using the same RNA-seq pipeline to eliminate batch effects [35]. The expression of CAR, CD46, DSG2, Survivin and hTERT was compared between glioblastoma tumour and normal brain samples.

2.5. Statistical Analysis

Statistical analyses were performed using Graphpad Prism 8 software (version 8.4.3). The normality of datasets was tested using the D'Agostino Pearson test and non-normally distributed data were compared using a Mann–Whitney U test. When comparing the means of $\geq$2 non-normally distributed groups involving one independent variable, the Kruskal–Wallis test with Dunn's multiple comparisons was used. Statistically significant differences were marked as * $p \leq 0.05$; ** $p \leq 0.01$; *** $p \leq 0.001$ and **** $p \leq 0.0001$.

3. Results

3.1. Expression of the Adenoviral Receptors CAR, CD46 and DSG2 in GBM

The species C HAdV-C5 is well described as engaging CAR and a primary cell entry receptor, which anatomically is localised in tight junctions. Species B adenovirus serotypes HAdV-B35 and HAdV-B3 use the ubiquitously expressed membrane protein CD46 and the cell–cell adhesion protein DSG2, respectively. Using mRNA expression data taken from the TCGA dataset for GBM and the GTex dataset for normal healthy brain tissue, the presence of CAR, CD46 and DSG2 was compared (Figure 1A). CAR and CD46 mRNA were abundant in both healthy brains and GBM. CAR mRNA levels in GBM were significantly higher than in healthy brains. High levels of CD46 mRNA were observed in both healthy brains and GBM, although there were significantly higher mRNA levels in normal brains. DSG2 mRNA levels were substantially lower overall compared to CAR and CD46; however, there was greater expression of DSG2 mRNA in GBM compared to normal brain tissue. To conclude, although there is differential mRNA expression of adenoviral receptors between healthy brain tissue and GBM, the presence of native adenoviral receptors on normal brain tissue represents an obstacle when considering the development of oncolytic adenoviruses using these receptors as potential GBM treatments.

The expression of the adenoviral receptor on three GBM-derived cell lines was analysed using flow cytometry (Figure 1B). L1 was a suspension stem cell line derived from a high-grade GBM patient. E51 and E55 were adherent glioma stem cell lines derived from high-grade GBM patients. Figure 1B demonstrates receptor staining for CAR, CD46 and DSG2. L1 and E51 cells express high levels of all three adenoviral receptors. E55 cells were positive for CD46 but demonstrated low levels of both CAR (42.6%) and DSG2 (0.77%).

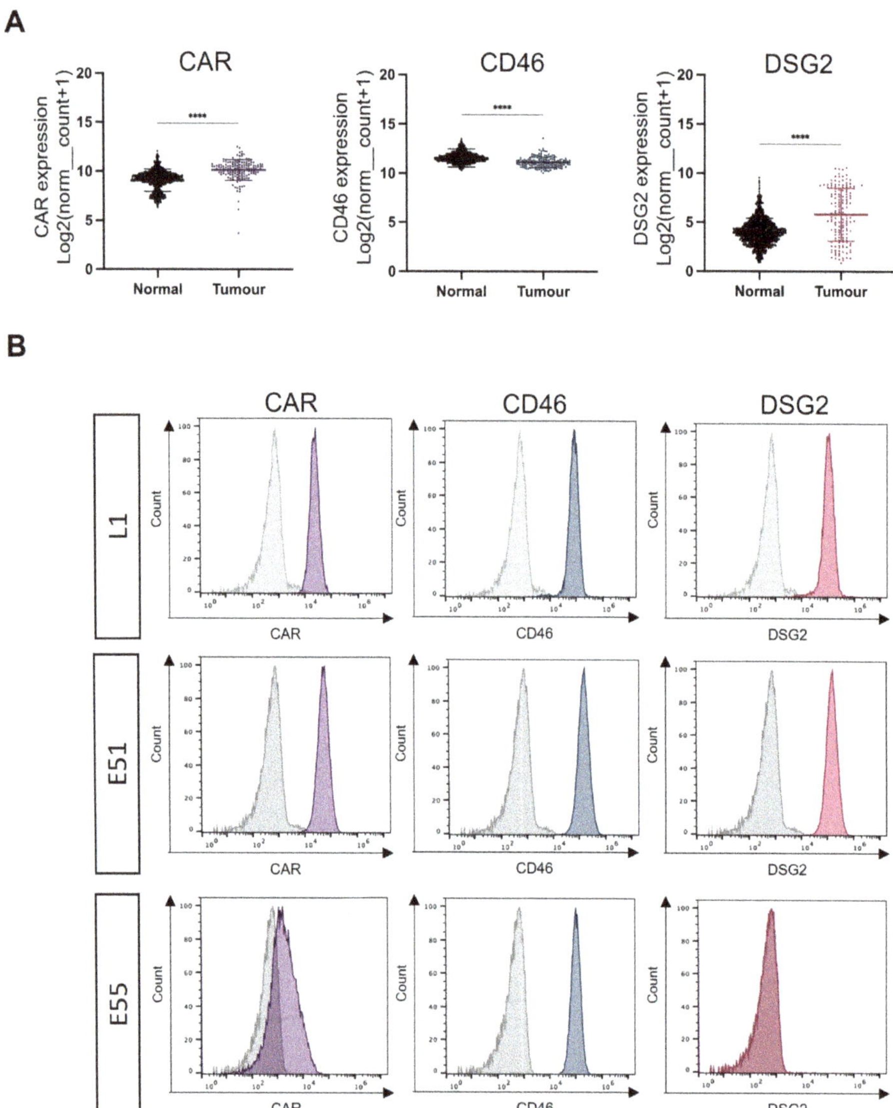

Figure 1. Expression of recognised adenoviral receptors on GBM, L1 cells and glioma stem cells (E51 and E55). (**A**) mRNA expression analysis of adenoviral receptors CAR, CD46 and DSG2 in GBM vs. normal brain. The data were collated from TCGA (tumour) and GTEx (normal) and represent $n = 153$ and $n = 1141$, respectively. Statistical significance was determined using the Mann–Whitney U test: **** $p \leq 0.0001$. (**B**) Flow cytometry was used to determine receptor staining for CAR, CD46 and DSG2 on the surface of L1 (GBM) and glioma stem cells (E51 and E55). Histograms are representative examples from an individual experiment repeated in triplicate.

3.2. Transduction of Pseudotyped Adenoviral Vectors in GBM

GBM-derived cell lines L1, E51 and E55 were transduced using an HAdV-C5 vector and HAdV-C5 pseudotypes HAdV-C5/D26K, HAdV-C5/B35K and HAdV-C5/D49K, expressing the transgene under the control of a CMV IE promoter. L1 cells were transduced efficiently by all four viral vectors (Figure 2A). Interestingly, a dependent response was not observed in this cell line, suggesting that the threshold of infection was at the lowest dose of 1000 vps/cell. E51 cells were transduced by all viral pseudotypes, where we observed a dose-dependent response with increasing transduction from 1000 vps/cell to 10,000 vps/cell (Figure 2B). Both L1 and E51 express high levels of CAR and CD46; therefore, these transduction data are in line with the expected results. E55 cells express lower levels of CAR than L1 and E51; however, it remained permissive to transduction by HAdV-C5. All pseudotype vectors were able to transduce E55 cells (Figure 2C). Overall, after a comparison of all four viral vectors tested, HAdV-C5/D49K, which uses an unknown cellular receptor [36,37], transduced all three cell lines the most efficiently. This was also observed at a lower dose of 500 vps/cell used in L1 cells (Figure 2D). L1 cells were transduced with 500 vps/cell to determine whether this was under the threshold to observe a dose-dependent response. Interestingly, at this dose, it was apparent that HAdV-C5, HAdV-C5/D26K and HAdV-C5/B35K were not transduced as efficiently as HAdV-C5/D49K.

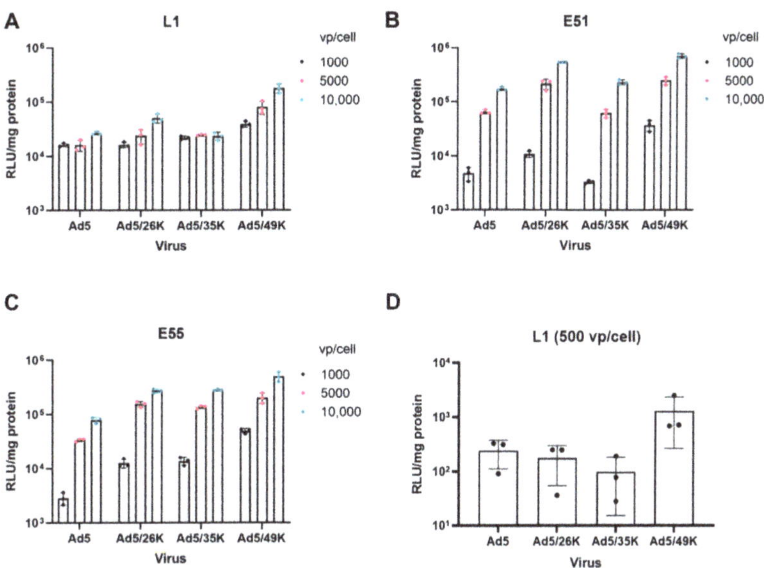

Figure 2. Transduction of L1, E51 and E55 cells via adenoviral pseudotype vectors. (**A**) L1 cells, (**B**) E51 cells and (**C**) E55 cells were transduced with 1000, 5000 and 10,000 vps/cell of adenoviral pseudotype vectors HAdV-C5 (Ad5), HAdV-C5/D26K (Ad5/26K), HAdV-C5/B35K (Ad5/35K) and HAdV-C5/D49K (Ad5/49K). (**D**) L1 cells were transduced with 500 vps/cell to investigate the threshold for infection. Luciferase activity was measured 72 h post infection. Data represent the mean of triplicates with individual data points shown. Error bars represent standard deviation, and data have been presented on a log scale.

HAdV-B3 is recognised as engaging DSG2 for cell entry and therefore an HAdV-C5/B3K pseudotype vector expressing GFP as a reporter gene was used to investigate the differential usage of DSG2 via adenoviral vectors in GBM. The transduction of HAdV-C5/B3K was quantified via the expression of the GFP reporter gene in L1 (DSG2 high) and E55 (DSG2 low) cells. L1 cells demonstrated a dose-dependent transduction of the HAdV-C5/B3K virus; however, HAdV-C5 transduction remained consistent (approximately 60%) for all three doses (Figure 3A). HAdV-C5/B3K did not transduce L1 as efficiently as HAdV-C5, despite the presence of DSG2 on this cell line (Figure 1B). E55 cells were negative for DSG2 (Figure 1B), and consequently HAdV-C5/B3K did not transduce E55 cells efficiently even at the highest dose of 10,000 vps/cell (Figure 3B). Representative microscopic images were taken (Figure 3C). These data suggest that although there is a degree of receptor specificity, it cannot be relied upon for tumour-specific uptake into GBM cells. Post entry selectivity must also be considered to improve the safety of adenoviral vectors for the treatment of GBM.

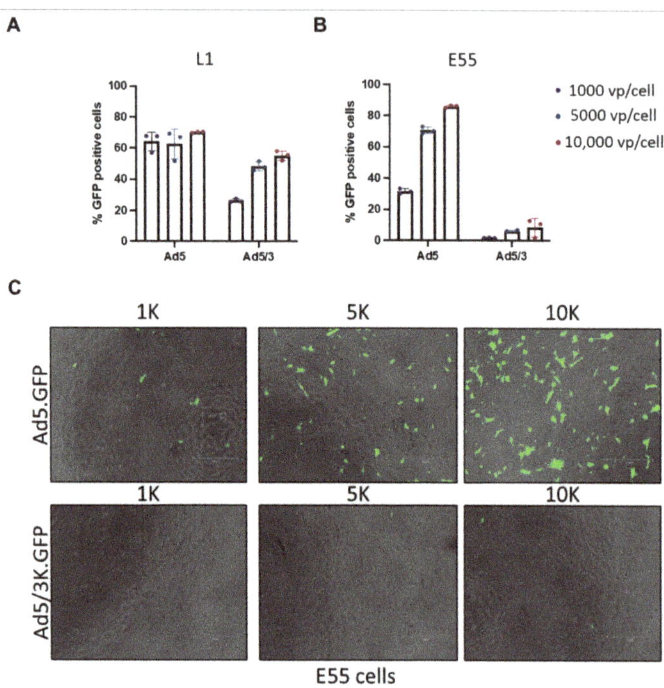

Figure 3. L1 and E55 cells were transduced with HAdV-C5 (Ad5) and HAdV-C5/B3K (Ad5/3) vectors expressing GFP. GFP levels were measured via flow cytometry and data were analysed using FlowJo. (**A**) L1 cells and (**B**) E55 were transduced with HAdV-C5.GFP and HAdV-C5/B3K.GFP at concentrations of 1000, 5000 and 10,000 vps/cell. Data are presented as mean of triplicate values of an independent experiment and individual data points are shown. (**C**) Representative images of E55 cells taken at 10X magnification bars using the EVOS cell imaging system (ThermoFisher). Images were taken both in TRANS and GFP channels and merged. Scale bar represents 300 μm.

3.3. Tumour-Specific Promoter hTERT and Survivin Offer Benefits in GBM Cells over Normal Cells

It has been reported that 75% of GBMs contain mutations in the human telomerase reverse transcriptase gene (hTERT) [38]. Similarly, survivin, an inhibitor of apoptosis, is expressed in nearly 80% of GBMs [39]. Adenoviral vectors that contain tumour-specific promoters exploiting the presence of hTERT and survivin may offer potential for the treatment of GBM. The TCGA dataset was used to evaluate the expression levels of these mRNA expressions for survivin and hTERT in GBM compared to normal brains (GTex). Figure 4A demonstrates that the mRNA expression of survivin is significantly higher in GBM compared to normal brain tissue. hTERT expression levels are also significantly higher in tumoral compared to healthy tissue (Figure 4B), supporting the reported literature. We therefore evaluated whether survivin or hTERT promoters demonstrated elevated selectivity of expression of a reporter gene in GBM-derived cells compared to human-derived fibroblasts expressing CAR (HF-CAR cells). Two GBM patient-derived cell lines L0 and L1, as well as HFCARs, were transduced using HAdV-C5 vectors containing luciferase under the control of either survivin, hTERT promoter or a fusion promoter that required both survivin and hTERT [40]. Cells were infected with 1000 vps/cell and luciferase activity was evaluated 72 h post infection (Figure 4C). HFCARs demonstrated levels of transduction that were barely above the background for all three adenoviral vectors. Conversely, when L0 and L1 were transduced efficiently by all three viral constructs containing tumour-specific promoters, HAdV-C5.survivin was the most efficient for both GBM cell lines. Interestingly, HAdV-C5.hTERT and HAdV-C5.hTERT/survivin demonstrated similar levels of luciferase activity, indicating that the combination of both hTERT and survivin did not improve the efficacy and may have hampered the effects when compared to survivin alone.

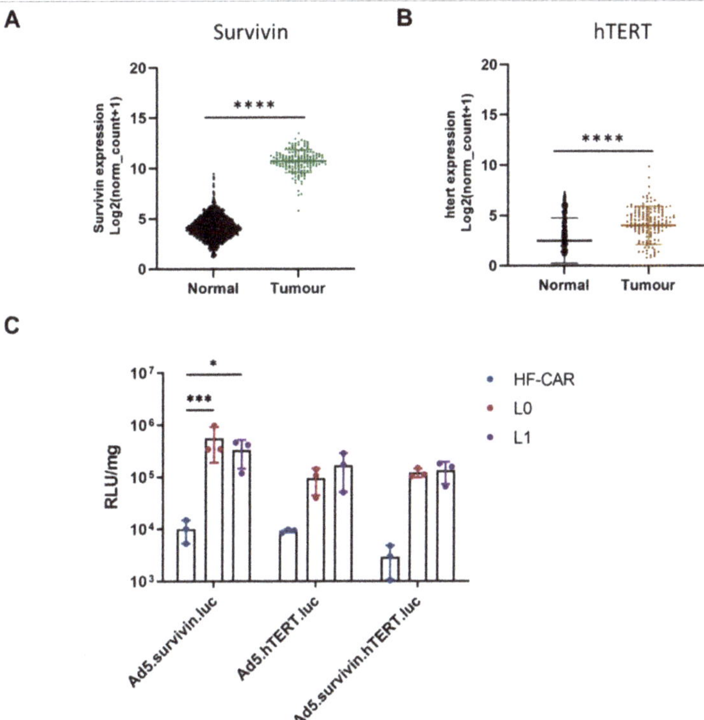

Figure 4. Evaluation of tumour-specific promoters hTERT and survivin. mRNA expression of (**A**) survivin and (**B**) hTERT were compared for GBM vs. normal brain. The data were collated from TCGA (tumour) and GTEx (normal) tissue and represent $n = 153$ and $n = 1141$, respectively. Statistical significance was determined using the Mann–Whitney U test: **** $p \leq 0.0001$. (**C**) Adenoviral vectors based on HAdV-C5 (Ad5)-expressing luciferase under control of three different tumour-specific promoters: survivin, hTERT and a combination of survivin and hTERT. The virus was added to human-derived fibroblasts (HFCARs) and GBM-derived L0 and L1 cells at a concentration of 1000 vps/cell and luciferase activity was measured 72 h after infection. The data were plotted on a log scale and represent the mean of triplicate values. Statistical significance was determined using the Kruskal–Wallis test with Dunn's multiple comparisons. Only significant values are shown: * $p \leq 0.05$ and *** $p \leq 0.001$.

4. Discussion

GBM is a common and aggressive brain cancer with poor prognosis and significant unmet clinical needs. Oncolytic viruses have previously been evaluated for the treatment of GBM and a recombinant herpesvirus, Teserpaturev, has been licensed for the treatment of GBM in Japan [41]. Adenoviral vectors have demonstrated safety in clinical trials; however, the presence of native adenoviral receptors has yet to be described in GBM [20].

We have evaluated the expression of three well-described adenoviral receptors, CAR, CD46 and DSG2, in GBM compared to healthy brain tissue. The analysis of mRNA expression indicated that all three receptors were expressed in GBM and healthy brains; however, CAR and DSG2 were significantly upregulated in GBM. Adenoviral vectors such as HAdV-C5 and HAdV-B3 which use CAR and DSG2, respectively, could be utilised to target GBM via local delivery into the tumour. It is important to consider whether the lack of specificity may impact the healthy brain tissue surrounding the tumour.

We confirmed the expression of CAR, CD46 and DSG2 on the surface of patient-derived GBM stem cell lines L1, E51 and E55. Both L1 and E51 were positive for CAR and DSG2, whereas E55 expressed lower levels of CAR and was negative for DSG2. All three cell lines express high levels of CD46. These data are consistent with the mRNA expression levels reported in GBM. Native adenoviral receptors must be considered when engineering oncolytic adenoviral therapies for GBM.

We assessed the transduction of GBM using HAdV-C5-based pseudotype vectors. HAdV-C5 binds CAR via the engagement of the fibre knob protein. The fibre knob protein of HAdV-C5 was replaced with the fibre knob proteins of HAdV-D26 (known to use sialic acid), HAdV-B35 (binds CD46) and HAdV-D49 (receptor usage unknown [36,37,42]) to generate HAdV-C5-pseudotyped vectors. These vectors were then used to determine the transduction of these serotypes based on the fibre knob engagement. L1 cells were transduced consistently by all vectors at every concentration of the virus. When the concentration of the virus was reduced to 500 vps/cell, it was possible to determine that HAdV-C5/D49K demonstrated the highest transduction. This was also observed in E51 and E55 cells. Interestingly, HAdV-C5-pseudotyped vectors demonstrated improved transduction compared to HAdV-C5 at all concentrations of the virus. This may result from the availability of each receptor on the surface of the cells or additional factors such as engagement of the co-receptors $\alpha V\beta 3$ and $\alpha v\beta 5$ integrin that mediate internalisation. E55 cells express low levels of CAR; however, they demonstrate the similar transduction of HAdV-C5 when compared with E51, which suggests that there may be enough CAR present on the surface of these cells to mediate transduction. Differences between HAdV-C5 transduction in E51 and E55 may be discerned if a lower viral concentration is used. L1 and E55 cells were transduced with HAdV-C5 and HAdV-C5/B3K which engage DSG2 as a cell entry receptor [23]. HAdV-C5/B3K has previously been reported as demonstrating superior infectivity in glioma cells [43]. As previously demonstrated, HAdV-C5 was able to transduce both cell lines efficiently. HAdV-C5/B3K demonstrated a dose-dependent transduction in L1 cells; however, the transduction efficiency was lower than HAdV-C5. Additional work is required to determine whether this is due to viral receptor usage or the processing of the virus post entry. E55 cells were not efficiently transduced by HAdV-C5/B3K, consistent with the lack of DSG2 on this cell line. Considering the transduction data, HAdV-C5/D26K, HAdV-C5/B35K and HAdV-C5/D49K in particular appear to be superior candidates for the pseudotype oncolytic virotherapy vector for GBM compared to HAdV-C5 or HAdV-C5/B3K.

Finally, given the lack of selectivity of CAR, CD46 and DSG2 expression in GBM, we considered the need for additional methods of intrinsic cellular selectivity when designing oncolytic virotherapies targeting GBM. The use of a tumour-specific promoter, Glial fibrillary acidic protein (GFAP), has previously been described in a HAdV-C5 background and demonstrated selectivity towards glial tumours [44]. We evaluated the post entry selectivity of transgene expression mediated by tumour-specific promoters utilising the tumour markers hTERT and survivin. mRNA analysis confirmed that both hTERT and survivin were significantly upregulated in GBM compared to normal brain tissue; however, the difference was more marked in the case of survivin. We designed three adenoviral vectors, based on HAdV-C5, expressing luciferase under an hTERT promoter, survivin promoter and a fusion hTERT and survivin promoter [40]. These vectors were used to transduce GBM L0 and L1 cells as well as non-transformed human fibroblasts, HF-CARs. As expected, the adenoviral vectors were not able express luciferase in the HF-CAR cells due to the absence of hTERT and survivin. Both L0 and L1 cells were efficiently transduced by all three vectors; however, HAdV-C5.survivin was able to transduce these cells more efficiently than the HAdV-C5.hTERT and HAdV-C5.hTERT/survivin fusion, suggesting that survivin may be the most promising tumour-specific promoter for future therapeutic applications in GBM both in terms of the quantity and the selectivity of transgene expression in GBM cell lines. The levels of survivin and hTERT were not evaluated in these cells, and the levels could vary between patients; however, we have demonstrated that these vectors can selectively express transgenes in GBM cells. Future work will include developing therapeutics under the control of the survivin promoter for the treatment of GBM, potentially coupled as part of an HAdV-C5/D49K-pseudotyped vector to maximise the uptake and activity in GBM cells.

5. Conclusions

To conclude, we have evaluated the expression of well-described adenoviral vectors in GBM. We identified that CAR, CD46 and DSG2 expression in GBM and patient-derived GBM cells and GBM stem cells can be transduced by HAdV-C5-based adenoviral serotypes, in particular HAdV-C5/D49K. Given the lack of natural GBM selectivity of adenoviral receptors, we highlighted the need for additional selectivity and demonstrated that tumour-specific promoters can enhance the specificity of GBM cells, where the survivin promoter appears to be particularly well suited. These data inform the design of future oncolytic adenoviral therapies expressing therapeutic transgenes for the treatment of GBM.

Supplementary Materials: The following supporting information can be downloaded at https://www.mdpi.com/article/10.3390/v15051086/s1: Table S1: Promoter sequences inserted into Ad5 vector expressing luciferase.

Author Contributions: Conceptualisation, E.A.B., F.A.S. and A.L.P.; methodology, E.A.B., C.L., A.R.P. and J.A.D.; formal analysis, E.A.B., C.L. and A.R.P.; investigation, E.A.B., C.L. and A.R.P.; resources, F.A.S.; writing—original draft preparation, E.A.B. and A.L.P.; writing—review and editing, A.L.P., C.L., A.R.P., J.A.D., F.A.S. and A.L.P.; supervision, A.L.P.; funding acquisition, A.L.P. All authors have read and agreed to the published version of the manuscript.

Funding: E.A.B. and A.R.P. were funded by the Experimental Cancer Medicine Centre award to Cardiff University (reference C7838/A25173). E.A.B. was additionally funded by Cancer Research Wales (reference 516619). C.L. was funded by a grant from the Department of Health and Social Care (DHSC) and supported by the National Institute for Health Research (NIHR; NIHR135073). J.A.D. was funded by Cancer Research UK/Stand Up to Cancer Biotherapeutics Programme grant award (reference C52915/A29104). F.A.S. and A.L.P. are funded by the Higher Education Funding Council for Wales (HEFCW).

Institutional Review Board Statement: Not applicable.

Informed Consent Statement: Not applicable.

Data Availability Statement: The data presented in this study are available upon request from the corresponding author. Publicly available datasets were also analysed in this study. These data can be found here https://xena.ucsc.edu/ accessed on 3 March 2023.

Acknowledgments: We thank Steve Pollard and Gillian Morrison of Edinburgh University and the Glioma Cellular Genetics Resource (GCGR) for the E51 and E55 cell lines.

Conflicts of Interest: ALP is the Founder and Chief Scientific Officer of Trocept Therapeutics. All of the other authors declare no conflict of interest.

References

1. Ostrom, Q.T.; Gittleman, H.; Liao, P.; Rouse, C.; Chen, Y.; Dowling, J.; Wolinsky, Y.; Kruchko, C.; Barnholtz-Sloan, J. CBTRUS statistical report: Primary brain and central nervous system tumors diagnosed in the United States in 2007–2011. *Neuro-Oncology* **2014**, *16* (Suppl. 4), iv1–iv63. [CrossRef] [PubMed]
2. Di Carlo, D.T.; Cagnazzo, F.; Benedetto, N.; Morganti, R.; Perrini, P. Multiple high-grade gliomas: Epidemiology, management, and outcome: A systematic review and meta-analysis. *Neurosurg. Rev.* **2019**, *42*, 263–275. [CrossRef] [PubMed]

3.   Lim, M.; Xia, Y.; Bettegowda, C.; Weller, M. Current state of immunotherapy for glioblastoma. *Nat. Rev. Clin. Oncol.* **2018**, *15*, 422–442. [CrossRef] [PubMed]
4.   Macedo, N.; Miller, D.M.; Haq, R.; Kaufman, H.L. Clinical landscape of oncolytic virus research in 2020. *J. Immunother. Cancer* **2020**, *8*, e001486. [CrossRef] [PubMed]
5.   Bett, A.J.; Prevec, L.; Graham, F.L. Packaging capacity and stability of human adenovirus type 5 vectors. *J. Virol.* **1993**, *67*, 5911–5921. [CrossRef]
6.   Sterman, D.H.; Treat, J.; Litzky, L.A.; Amin, K.M.; Coonrod, L.; Molnar-Kimber, K.; Recio, A.; Knox, L.; Wilson, J.M.; Albelda, S.M.; et al. Adenovirus-mediated herpes simplex virus thymidine kinase/ganciclovir gene therapy in patients with localized malignancy: Results of a phase I clinical trial in malignant mesothelioma. *Hum. Gene Ther.* **1998**, *9*, 1083–1092. [CrossRef]
7.   Ram, Z.; Culver, K.W.; Oshiro, E.M.; Viola, J.J.; DeVroom, H.L.; Otto, E.; Long, Z.; Chiang, Y.; McGarrity, G.J.; Muul, L.M.; et al. Therapy of malignant brain tumors by intratumoral implantation of retroviral vector-producing cells. *Nat. Med.* **1997**, *3*, 1354–1361. [CrossRef]
8.   HAdV Working Group. Available online: http://hadvwg.gmu.edu/ (accessed on 29 March 2023).
9.   Parker, A.L.; Waddington, S.N.; Buckley, S.M.K.; Custers, J.; Havenga, M.J.E.; van Rooijen, N.; Goudsmit, J.; McVey, J.H.; Nicklin, S.A.; Baker, A.H. Effect of neutralizing sera on factor x-mediated adenovirus serotype 5 gene transfer. *J. Virol.* **2009**, *83*, 479–483. [CrossRef]
10.   Mast, T.C.; Kierstead, L.; Gupta, S.B.; Nikas, A.A.; Kallas, E.G.; Novitsky, V.; Mbewe, B.; Pitisuttithum, P.; Schechter, M.; Vardas, E.; et al. International epidemiology of human pre-existing adenovirus (Ad) type-5, type-6, type-26 and type-36 neutralizing antibodies: Correlates of high Ad5 titers and implications for potential HIV vaccine trials. *Vaccine* **2010**, *28*, 950–957. [CrossRef]
11.   Bates, E.A.; Davies, J.A.; Váňová, J.; Nestić, D.; Meniel, V.S.; Koushyar, S.; Cunliffe, T.G.; Mundy, R.M.; Moses, E.; Uusi-Kerttula, H.K.; et al. Development of a low-seroprevalence, $\alpha v \beta 6$ integrin-selective virotherapy based on human adenovirus type 10. *Mol. Ther. Oncolytics* **2022**, *25*, 43–56. [CrossRef]
12.   Bergelson, J.M.; Cunningham, J.A.; Droguett, G.; Kurt-Jones, E.A.; Krithivas, A.; Hong, J.S.; Horwitz, M.S.; Crowell, R.L.; Finberg, R.W. Isolation of a common receptor for Coxsackie B viruses and adenoviruses 2 and 5. *Science* **1997**, *275*, 1320–1323. [CrossRef]
13.   Coyne, C.B.; Bergelson, J.M. CAR: A virus receptor within the tight junction. *Adv. Drug Deliv. Rev.* **2005**, *57*, 869–882. [CrossRef]
14.   Anders, M.; Vieth, M.; Röcken, C.; Ebert, M.; Pross, M.; Gretschel, S.; Schlag, P.M.; Wiedenmann, B.; Kemmner, W.; Höcker, M. Loss of the coxsackie and adenovirus receptor contributes to gastric cancer progression. *Br. J. Cancer* **2009**, *100*, 352–359. [CrossRef]
15.   Matsumoto, K.; Shariat, S.F.; Ayala, G.E.; Rauen, K.A.; Lerner, S.P. Loss of coxsackie and adenovirus receptor expression is associated with features of aggressive bladder cancer. *Urology* **2005**, *66*, 441–446. [CrossRef]
16.   Shayakhmetov, D.M.; Gaggar, A.; Ni, S.; Li, Z.-Y.; Lieber, A. Adenovirus binding to blood factors results in liver cell infection and hepatotoxicity. *J. Virol.* **2005**, *79*, 7478–7491. [CrossRef]
17.   Waddington, S.N.; McVey, J.H.; Bhella, D.; Parker, A.L.; Barker, K.; Atoda, H.; Pink, R.; Buckley, S.M.K.; Greig, J.A.; Denby, L.; et al. Adenovirus serotype 5 hexon mediates liver gene transfer. *Cell* **2008**, *132*, 397–409. [CrossRef]
18.   Parker, A.L.; Waddington, S.N.; Nicol, C.G.; Shayakhmetov, D.M.; Buckley, S.M.; Denby, L.; Kemball-Cook, G.; Ni, S.; Lieber, A.; McVey, J.H.; et al. Multiple vitamin K-dependent coagulation zymogens promote adenovirus-mediated gene delivery to hepatocytes. *Blood* **2006**, *108*, 2554–2561. [CrossRef]
19.   Carlisle, R.C.; Di, Y.; Cerny, A.M.; Sonnen, A.F.-P.; Sim, R.B.; Green, N.K.; Subr, V.; Ulbrich, K.; Gilbert, R.J.C.; Fisher, K.D.; et al. Human erythrocytes bind and inactivate type 5 adenovirus by presenting Coxsackie virus-adenovirus receptor and complement receptor 1. *Blood* **2009**, *113*, 1909–1918. [CrossRef]
20.   Lang, F.F.; Conrad, C.; Gomez-Manzano, C.; Yung, W.K.A.; Sawaya, R.; Weinberg, J.S.; Prabhu, S.S.; Rao, G.; Fuller, G.N.; Aldape, K.D.; et al. Phase I Study of DNX-2401 (Delta-24-RGD) Oncolytic Adenovirus: Replication and Immunotherapeutic Effects in Recurrent Malignant Glioma. *J. Clin. Oncol. Off. J. Am. Soc. Clin. Oncol.* **2018**, *36*, 1419–1427. [CrossRef]
21.   Baker, A.T.; Mundy, R.M.; Davies, J.A.; Rizkallah, P.J.; Parker, A.L. Human adenovirus type 26 uses sialic acid-bearing glycans as a primary cell entry receptor. *Sci. Adv.* **2019**, *5*, eaax3567. [CrossRef]
22.   Gaggar, A.; Shayakhmetov, D.M.; Lieber, A. CD46 is a cellular receptor for group B adenoviruses. *Nat. Med.* **2003**, *9*, 1408–1412. [CrossRef] [PubMed]
23.   Wang, H.; Li, Z.-Y.; Liu, Y.; Persson, J.; Beyer, I.; Möller, T.; Koyuncu, D.; Drescher, M.R.; Strauss, R.; Zhang, X.-B.; et al. Desmoglein 2 is a receptor for adenovirus serotypes 3, 7, 11 and 14. *Nat. Med.* **2011**, *17*, 96–104. [CrossRef] [PubMed]
24.   Rawal, P.; Zhao, L. Sialometabolism in Brain Health and Alzheimer's Disease. *Front. Neurosci.* **2021**, *15*, 648617. Available online: https://www.frontiersin.org/articles/10.3389/fnins.2021.648617 (accessed on 27 March 2023). [CrossRef] [PubMed]
25.   Van Aarsen, L.A.K.; Leone, D.R.; Ho, S.; Dolinski, B.M.; McCoon, P.E.; LePage, D.J.; Kelly, R.; Heaney, G.; Rayhorn, P.; Reid, C.; et al. Antibody-mediated blockade of integrin alpha v beta 6 inhibits tumor progression in vivo by a transforming growth factor-beta-regulated mechanism. *Cancer Res.* **2008**, *68*, 561–570. [CrossRef] [PubMed]
26.   Uusi-Kerttula, H.; Davies, J.A.; Thompson, J.M.; Wongthida, P.; Evgin, L.; Shim, K.G.; Bradshaw, A.; Baker, A.T.; Rizkallah, P.J.; Jones, R.; et al. Ad5NULL-A20: A Tropism-Modified, $\alpha v \beta 6$ Integrin-Selective Oncolytic Adenovirus for Epithelial Ovarian Cancer Therapies. *Clin. Cancer Res. Off. J. Am. Assoc. Cancer Res.* **2018**, *24*, 4215–4224. [CrossRef]

27. Davies, J.A.; Marlow, G.; Uusi-Kerttula, H.K.; Seaton, G.; Piggott, L.; Badder, L.M.; Clarkson, R.W.E.; Chester, J.D.; Parker, A.L. Efficient Intravenous Tumor Targeting Using the $\alpha v \beta 6$ Integrin-Selective Precision Virotherapy Ad5NULL-A20. *Viruses* **2021**, *13*, 864. [CrossRef]

28. Uusi-Kerttula, H.; Davies, J.; Coughlan, L.; Hulin-Curtis, S.; Jones, R.; Hanna, L.; Chester, J.D.; Parker, A.L. Pseudotyped $\alpha v \beta 6$ integrin-targeted adenovirus vectors for ovarian cancer therapies. *Oncotarget* **2016**, *7*, 27926–27937. [CrossRef]

29. Piccirillo, S.G.M.; Reynolds, B.A.; Zanetti, N.; Lamorte, G.; Binda, E.; Broggi, G.; Brem, H.; Olivi, A.; Dimeco, F.; Vescovi, A.L. Bone morphogenetic proteins inhibit the tumorigenic potential of human brain tumour-initiating cells. *Nature* **2006**, *444*, 761–765. [CrossRef]

30. Hoang-Minh, L.B.; Siebzehnrubl, F.A.; Yang, C.; Suzuki-Hatano, S.; Dajac, K.; Loche, T.; Andrews, N.; Schmoll Massari, M.; Patel, J.; Amin, K.; et al. Infiltrative and drug-resistant slow-cycling cells support metabolic heterogeneity in glioblastoma. *EMBO J.* **2018**, *37*, e98772. [CrossRef]

31. Siebzehnrubl, F.A.; Silver, D.J.; Tugertimur, B.; Deleyrolle, L.P.; Siebzehnrubl, D.; Sarkisian, M.R.; Devers, K.G.; Yachnis, A.T.; Kupper, M.D.; Neal, D.; et al. The ZEB1 pathway links glioblastoma initiation, invasion and chemoresistance. *EMBO Mol. Med.* **2013**, *5*, 1196–1212. [CrossRef]

32. Pollard, S.M.; Yoshikawa, K.; Clarke, I.D.; Danovi, D.; Stricker, S.; Russell, R.; Bayani, J.; Head, R.; Lee, M.; Bernstein, M.; et al. Glioma stem cell lines expanded in adherent culture have tumor-specific phenotypes and are suitable for chemical and genetic screens. *Cell Stem Cell* **2009**, *4*, 568–580. [CrossRef]

33. The AdZ Adenovirus Cloning System | AdZ.cf.ac.uk. Available online: https://adz.cf.ac.uk/ (accessed on 27 March 2023).

34. Uusi-Kerttula, H.; Legut, M.; Davies, J.; Jones, R.; Hudson, E.; Hanna, L.; Stanton, R.J.; Chester, J.D.; Parker, A.L. Incorporation of Peptides Targeting EGFR and FGFR1 into the Adenoviral Fiber Knob Domain and Their Evaluation as Targeted Cancer Therapies. *Hum. Gene Ther.* **2015**, *26*, 320–329. [CrossRef]

35. Goldman, M.J.; Craft, B.; Hastie, M.; Repečka, K.; McDade, F.; Kamath, A.; Banerjee, A.; Luo, Y.; Rogers, D.; Brooks, A.N.; et al. Visualizing and interpreting cancer genomics data via the Xena platform. *Nat. Biotechnol.* **2020**, *38*, 675–678. [CrossRef]

36. Baker, A.T.; Davies, J.A.; Bates, E.A.; Moses, E.; Mundy, R.M.; Marlow, G.; Cole, D.K.; Bliss, C.M.; Rizkallah, P.J.; Parker, A.L. The Fiber Knob Protein of Human Adenovirus Type 49 Mediates Highly Efficient and Promiscuous Infection of Cancer Cell Lines Using a Novel Cell Entry Mechanism. *J. Virol.* **2021**, *95*, e01849-20. [CrossRef]

37. Dakin, R.S.; Parker, A.L.; Delles, C.; Nicklin, S.A.; Baker, A.H. Efficient transduction of primary vascular cells by the rare adenovirus serotype 49 vector. *Hum. Gene Ther.* **2015**, *26*, 312–319. [CrossRef]

38. Nguyen, H.N.; Lie, A.; Li, T.; Chowdhury, R.; Liu, F.; Ozer, B.; Wei, B.; Green, R.M.; Ellingson, B.M.; Wang, H.-J.; et al. Human TERT promoter mutation enables survival advantage from MGMT promoter methylation in IDH1 wild-type primary glioblastoma treated by standard chemoradiotherapy. *Neuro-Oncology* **2017**, *19*, 394–404. [CrossRef]

39. Das, A.; Tan, W.-L.; Teo, J.; Smith, D.R. Expression of survivin in primary glioblastomas. *J. Cancer Res. Clin. Oncol.* **2002**, *128*, 302–306. [CrossRef]

40. Alekseenko, I.V.; Pleshkan, V.V.; Kopantzev, E.P.; Stukacheva, E.A.; Chernov, I.P.; Vinogradova, T.V.; Sverdlov, E.D. Activity of the Upstream Component of Tandem TERT/Survivin Promoters Depends on Features of the Downstream Component. *PLoS ONE* **2012**, *7*, e46474. [CrossRef]

41. Fukuhara, H.; Ino, Y.; Todo, T. Oncolytic virus therapy: A new era of cancer treatment at dawn. *Cancer Sci.* **2016**, *107*, 1373–1379. [CrossRef]

42. Bates, E.A.; Counsell, J.R.; Alizert, S.; Baker, A.T.; Suff, N.; Boyle, A.; Bradshaw, A.C.; Waddington, S.N.; Nicklin, S.A.; Baker, A.H.; et al. In Vitro and In Vivo Evaluation of Human Adenovirus Type 49 as a Vector for Therapeutic Applications. *Viruses* **2021**, *13*, 1483. [CrossRef]

43. Stepanenko, A.A.; Sosnovtseva, A.O.; Valikhov, M.P.; Chernysheva, A.A.; Cherepanov, S.A.; Yusubalieva, G.M.; Ruzsics, Z.; Lipatova, A.V.; Chekhonin, V.P. Superior infectivity of the fiber chimeric oncolytic adenoviruses Ad5/35 and Ad5/3 over Ad5-delta-24-RGD in primary glioma cultures. *Mol. Ther. Oncolytics* **2022**, *24*, 230–248. [CrossRef] [PubMed]

44. Ter Horst, M.; Brouwer, E.; Verwijnen, S.; Rodijk, M.; de Jong, M.; Hoeben, R.; de Leeuw, B.; Smitt, P.S. Targeting malignant gliomas with a glial fibrillary acidic protein (GFAP)-selective oncolytic adenovirus. *J. Gene Med.* **2007**, *9*, 1071–1079. [CrossRef] [PubMed]

Article

AAV Vectors Pseudotyped with Capsids from Porcine and Bovine Species Mediate In Vitro and In Vivo Gene Delivery

Darrick L. Yu [1], Laura P. van Lieshout [1], Brenna A. Y. Stevens [1], Kelsie J. (Jagt) Near [1], Jenny K. Stodola [1], Kevin J. Stinson [1], Durda Slavic [2] and Sarah K. Wootton [1,*]

[1] Department of Pathobiology, University of Guelph, Guelph, ON N1G 2W1, Canada
[2] Animal Health Laboratory, Laboratory Services Division, University of Guelph, Guelph, ON N1G 2W1, Canada
* Correspondence: kwootton@uoguelph.ca

Abstract: Adeno-associated virus (AAV) vectors are among the most widely used delivery vehicles for in vivo gene therapy as they mediate robust and sustained transgene expression with limited toxicity. However, a significant impediment to the broad clinical success of AAV-based therapies is the widespread presence of pre-existing humoral immunity to AAVs in the human population. This immunity arises from the circulation of non-pathogenic endemic human AAV serotypes. One possible solution is to use non-human AAV capsids to pseudotype transgene-containing AAV vector genomes of interest. Due to the low probability of human exposure to animal AAVs, pre-existing immunity to animal-derived AAV capsids should be low. Here, we characterize two novel AAV capsid sequences: one derived from porcine colon tissue and the other from a caprine adenovirus stock. Both AAV capsids proved to be effective transducers of HeLa and HEK293T cells in vitro. In vivo, both capsids were able to transduce the murine nose, lung, and liver after either intranasal or intraperitoneal administration. In addition, we demonstrate that the porcine AAV capsid likely arose from multiple recombination events involving human- and animal-derived AAV sequences. We hypothesize that recurrent recombination events with similar and distantly related AAV sequences represent an effective mechanism for enhancing the fitness of wildtype AAV populations.

Keywords: adeno-associated virus (AAV); novel serotype; porcine capsid; bovine capsid; luciferase imaging; transduction

Citation: Yu, D.L.; van Lieshout, L.P.; Stevens, B.A.Y.; Near, K.J.; Stodola, J.K.; Stinson, K.J.; Slavic, D.; Wootton, S.K. AAV Vectors Pseudotyped with Capsids from Porcine and Bovine Species Mediate In Vitro and In Vivo Gene Delivery. *Viruses* **2024**, *16*, 57. https://doi.org/10.3390/v16010057

Academic Editor: Joe S. Mymryk

Received: 5 November 2023
Revised: 19 December 2023
Accepted: 27 December 2023
Published: 29 December 2023

1. Introduction

AAV vectors are widely regarded as ideal candidates for in vivo gene therapy because of their biology; they readily transduce a wide range of dividing and non-dividing cells, they are replication-defective and their genomes reside episomally in transduced cells, they are minimally immunogenic, they elicit sustained transgene expression, and they have a strong safety profile [1]. However, because of the high prevalence of adenovirus infections in the human population, there is a correspondingly high level of seropositivity against AAVs of different serotypes [2]. This high rate of pre-existing immunity can limit the usefulness of many human AAV serotypes, as this can potentially compromise transgene expression by blocking transduction [3,4] and, in many cases, is an exclusion criterion for clinical trials, particularly those in which the vector is to be delivered systemically [5,6]. To circumvent this, recombinant AAV (rAAV) vectors may be altered by pseudotyping. In this process, a vector can be packaged into the capsid of a heterologous AAV. For example, a transgene flanked by AAV2 ITRs can be packaged into a capsid from bovine AAV [7,8]. The result is an rAAV particle with a relatively low to nonexistent seroprevalence in a target population, with the potential for vastly improved transduction efficiency. Since the capsid determines tissue tropism, it is possible to alter the range of tissues targeted simply by changing the vector's capsid.

Significant advancements have been made in the field of AAV capsid engineering using rational design, directed evolution, combinatorial libraries, and in silico reconstruction of ancestral capsid variants [9]. These approaches have led to the development of capsids that evade recognition by the anti-AAV antibodies typically found in the human population and have enhanced transduction of target tissues of human origin. Another approach to evading pre-existing immunity is to isolate AAVs from non-human species [10]. For instance, AAVrh10 and AAVrh74 are recombinant vectors originating from rhesus macaques. A survey of healthy adults revealed that 59% of individuals had detectable antibodies against AAVrh10, and 21% had neutralizing antibodies [11]. Analysis of antibodies against AAVrh74 in patients with Duchenne muscular dystrophy and limb girdle muscular dystrophy revealed that 83% were seronegative [12]. However, the isolation of novel AAVs from more distant species could yield AAVs that are not neutralized by human sera. Animal AAVs have been detected in canine, bovine, ovine, caprine, porcine, and avian species, although when considering the prevalence of adenovirus in animal species, there remains a wealth of as yet undiscovered AAV isolates [13–17]. As such, the purpose of this research was to identify and isolate novel AAVs from animal tissues for use in the development of new gene therapy vectors with low seropositivity in the human population.

2. Materials and Methods

Sources of samples. Fresh lung, liver, ileum, and distal colon samples from a range of mammals were collected through a partnership with the University of Guelph Animal Health Laboratory (AHL) post-mortem (PM) and bacteriology departments. Animals were necropsied on site and sampled at the time of necropsy in the PM room. Samples which were necropsied off site were sampled upon arrival to the bacteriology lab; however, only exotic species not regularly seen in the PM room were taken from this source. Genomic DNA from the colon of 164 healthy pigs, part of a study on sub-clinical infections and biomarkers in Ontario swine, was also provided by the AHL bacteriology department.

DNA extraction—QIAGEN kit. Except for the porcine colon samples, DNA extractions from all samples were performed using the QIAGEN DNeasy Blood and Tissue extraction kit, following the animal tissue spin-column protocol described by the manufacturer. Briefly, tissues were lysed by adding 25 mg of tissue to 180 μL of buffer ATL, as per the QIAGEN kit protocol, with an overnight incubation at 56 °C after addition of 20 μL proteinase K. After extraction, DNA was quantified using a NanoDrop 2000 (Thermo Scientific, Mississauga, ON, Canada) and found to be, on average, 100 ng/μL.

DNA extraction—MagNA Pure robotic extraction. Porcine colon samples were extracted using the MagNA Pure extraction robot (Roche, Mississauga, ON, Canada). Tissues were lysed by adding 100 mg of tissue to 900 μL of Tripure with 2 stainless steel grinding balls (Montreal Biotech, Dorval, QC, Canada) and homogenized for 3 min at 30 hz using a MM 300 mixer mill (Retsch, Newtown, PA, USA). Sample tubes were briefly centrifuged to sediment large pieces of tissue, and 200 μL of supernatant was loaded into the MagNA pure. The extraction was performed using the Roche DNA Isolation Kit I and the High-Performance Blood Cells protocol on the MagNA Pure, with a 200 μL sample volume and a 100 μL elution volume. After extraction, DNA was quantified using a NanoDrop 2000 (Thermo Scientific) and found to be, on average, 200 ng/μL.

Screening of genomic DNA for presence of AAV genomes by polymerase chain reaction (PCR). A hemi-nested PCR reaction using degenerate primers was used to screen for AAV genomes in the genomic DNA samples. This method was an adaptation of that employed by Katano et al. (2004), implementing the use of a nested forward primer in a second PCR reaction and allowing for the detection of the low concentrations of AAV DNA integrated with a host genome [18]. Table 1 shows the PCR conditions and primers used for both reactions. PCR Platinum Supermix (Invitrogen, Burlington, ON, Canada) was used as the reaction master mix. The plasmid pDGM6 [8] containing the rep region of AAV2 and the cap region of AAV6 were used as a positive control, and RNAse free water was used as a negative control. Analysis of PCR products was performed by gel electrophoresis on a 0.8%

agarose gel stained with ethidium bromide. The desired amplicon, containing the 3′ end of the rep region and the 5′ end of the cap region, was 1500 bp (Figure 1). PCR products with a band between 1400 and 1600 bp were selected as potential positive samples. For samples with high levels of non-specific binding, the target band was excised and recovered using the Direct-gel-spin DNA recovery kit (Applied Biological Materials).

Table 1. Primers and PCR conditions used for screening samples for the presence of AAV sequences.

	Round 1—Initial PCR	Round 2—Hemi-Nested PCR
Forward Primer	5′-ATGNTNATNTGGTGGGA GGA-3′	5′-ACCTTNGAACACCAGCAGC-3′
Reverse Primer	5′-CCANNNGGAATCGCAATGCCAAT-3′	
PCR Components	20 μL Platinum PCR Supermix 1 μL (2 μM) Primers (Each) 200 ng DNA	20 μL Platinum PCR Supermix 1 μL (2 μM) Primers (Each) 2 uL of PCR Product
PCR Conditions	94 °C—2 min 94 °C—30 s × 40 cycles 55 °C—30 s × 40 cycles 72 °C—30 s × 40 cycles 72 °C—2 min	94 °C—2 min 94 °C—30 s × 30 cycles 55 °C—30 s × 30 cycles 72 °C—30 s × 30 cycles 72 °C—2 min

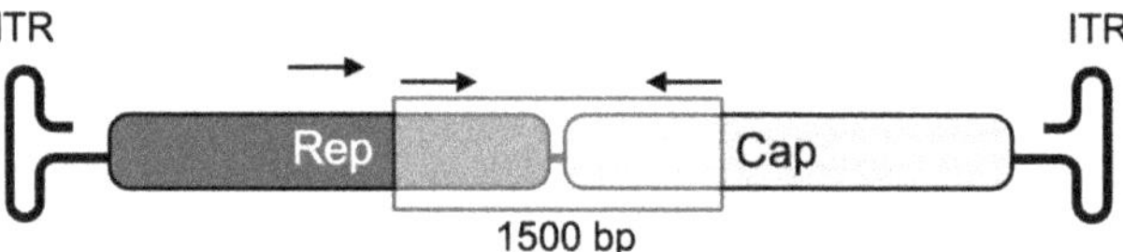

Figure 1. Location of PCR primers used for detecting AAV sequences. Primers used for screening genomic DNA samples were designed to amplify a 1500 base-pair amplicon, which consisted of the 3′ end of the rep gene and the 5′ end of the cap gene. Hemi-nested PCR using degenerate primers for a highly conserved region of the genome allowed for the detection of low quantities of AAV DNA.

pGEM-T easy vector cloning of PCR products. PCR products of the correct size were directly ligated into the pGEM-T easy vector (Promega), as per the manufacturer's instructions, using 3.5 μL of PCR product with incubation at 4 °C overnight. After incubation, 4 μL of ligation reaction mixture was transformed into Escherichia coli GT116 competent cells (InvivoGen), and plasmid DNA was extracted from cultures of isolated colonies using the Plasmid DNA Miniprep Kit from Bio Basic.

DNA sequencing. Successful ligation was confirmed by digestion of the plasmid extraction with EcoRI and analysis by gel electrophoresis. If bands were observed at 3000 bp and 1500 bp, representing the vector and insert, respectively, the plasmid extraction was sent to the University of Guelph Advanced Analysis Center for Sanger sequencing using the M13 forward and reverse primers (located ~60 bp upstream and downstream from the vector insertion site).

Sequence analysis. Forward and reverse sequences were aligned in SeqMan Pro (DNAStar) and the plasmid sequence was removed at the EcoRI cut sites flanking the insertion. The generated consensus sequences for the entire amplicon and the rep and cap regions individually were analyzed against the NCBI nucleotide database using the BLAST Megablast algorithm. Analysis of recovered AAV isolates was performed using the integrated bioinformatics software Geneious version R11 (Biomatters, Boston, MA, USA), which was used for sequence manipulation, alignment with known sequences (ClustalW algorithm), and development of phylogenetic trees (UPGMA algorithm). Geneious was also used to generate and align protein sequences in order to analyze novel AAVs at the amino acid level. The Dual Brothers recombination plugin for Geneious was used to analyze the AAV.Po.Guelph capsid sequence for recombination events [19,20].

Cloning of the full-length porcine and bovine AAV capsids. The forward and degenerate reverse primers and PCR conditions used to amplify the full-length porcine and bovine capsid genes are shown in Table 2. Note that DNA samples were stabilized by the addition of trehalose to a final concentration of 200 mM. PCR products were pGEM-Teasy cloned and sequenced using universal T7 forward and reverse primers.

Table 2. Primers and PCR conditions used for amplifying full-length capsid genes for porcine and bovine AAVs.

	Round 1—Initial PCR	Round 2—Hemi-Nested PCR
Porcine Capsid Forward Primer	5′-ATGTCGTTTGTTGATCACC-3′	5′-TTACAGGTTGCGGGTGAAGGTAGC-3′
Porcine Capsid Reverse Primer	5′-TTACAGRTTRCGRGTRAGGTAGC-3′	
PCR Components	20 μL Platinum PCR Supermix 1 μL (2 μM) Primers (Each) 200 ng DNA	20 μL Platinum PCR Supermix 1 μL (2 μM) Primers (Each) 4 μL of PCR Product
PCR Conditions	94 °C—2 min 94 °C—15 s 70 °C (−0.3 °C per cycle) 30 s 72 °C—2.5 min	34 cycles
	94 °C—15 s 55 °C—30 s 72 °C—2.5 min	14 cycles
	72 °C—5 min	

AAV vector production and purification. HEK 293 cells were transfected utilizing a PEI-based transfection protocol as previously described [21]. A four-plasmid system, including plasmids encoding rep and helper functions, pMTRep6 [22] and pLadeno5 [23], respectively, was used to generate AAV particles encoding the reporter gene human placental alkaline phosphatase (hPLAP) driven by the CAG promoter. This AAV vector genome was encoded within plasmid pACAGAP [24]. A fourth plasmid encoding AAV.Po.Guelph or AAV.Bov.Guelph capsids (pCAGαPo.GuelphCap, or pCAGαBov.GuelphCap) expressed from a truncated CAG promoter [25] provided the capability of packing the AAV vector genome into AAV.Po.Guelph or AAV.Bov.Guelph capsids, respectively. pCAGαPo.GuelphCap and pCAGαBov.GuelphCap plasmids were constructed by digesting plasmid pCAGαeGFP [25], as well as PCR products encoding AAV.Po.Guelph and AAV.Bov.Guelph, with EcoRI and BglII (NEB, Whitby, ON, Canada). An optimized Kozak sequence was inserted in front of the AAV.Po.Guelph and AAV.Bov.Guelph capsid open reading frames in order to promote translation. The porcine and bovine capsid genes were cloned into the EcoRI and BglII sites downstream of the CAGα promoter using the Capsid-EcoRI forward 5′-GCG**GAATTC**GCCACCATGTCGTTTGTTGATCACCC-3′ (where the bold text represents the EcoRI restriction site and underlined text indicates the Kozak sequence) and Capsid-BglII reverse 5′-GC**AGATCT**TTACAGATTGCGGGTAAGG-3′ (where the bold text indicates the BglII restriction site and the underlined text denotes the stop codon) primers and the pGEM-Teasy plasmid containing the full-length capsid gene as template DNA using standard molecular cloning techniques.

To determine virus distribution in the cells and supernatant, 1.2×10^6 HEK293 cells were plated in 6-well dishes and allowed to adhere overnight. Cells were transfected with 2 μg of AAV genome (pCASI-Luciferase), 2 μg of adenovirus helper plasmid (pAdΔF6), and 2 μg of AAV helper plasmid (pAAV.Po.Guelph or pAAV.Bov.Guelph) using PEI (Polysciences Inc., Warrington, PA, USA). Supernatant and cell lysate were harvested separately on days 3 and 6 post-transfection, and vector DNA was extracted and quantified by Taqman PCR as previously described [26] to determine the relative distribution of virions between the two fractions. No significant differences were observed between vector distribution in the supernatant or lysate on day 3 or 6 post-transfection for either AAV.Po.Guelph or AAV.Bov.Guelph (Figure S1).

For iodixanol gradient purified vector preparations, cells and supernatant were pooled and freeze–thawed three times at −80 °C to lyse cells and release virus particles. In between freeze–thaw cycles, the cell lysate and supernatant were vortex mixed to facilitate separation of the virus from cell debris. After completion of the freeze–thaws, lysate was spun at $3000\times g$ to pellet cellular debris, and supernatant was filtered through a Millipore Stericup 0.22 µm filter (EMD Millipore, Etobicoke, ON, Canada). Benzonase nuclease ($\geq$500 units) (Sigma, St. Louis, MO, USA) was added to digest nucleic acids and help reduce viscosity by treatment at 37 °C for 30 min. A 100 k Omega Centramate T-SERIES 0.019 m^2 cassette (Pall Canada Direct Ltd., Mississauga, ON, Canada) was used in an LV centramate holder to concentrate the supernatant from over 500 mL to approximately 25 mL, according to the manufacturer's instructions. A 100 k Amicon Ultra-15 Centrifugal Filter Unit (EMD Millipore) was used to further concentrate the sample to a volume of approximately 5 mL. The concentrated sample was loaded onto an iodixanol gradient as described previously [27] and ultracentrifuged at 41,000 rpm for 3 h at 4 °C in an SW41 rotor. The fraction containing the vector was harvested by puncturing the tube slightly below (3–5 mm) the 60–40% interface with an 18-gauge needle (bevel up) attached to a 1 mL syringe [27]. A 100 k Amicon filter unit was used to exchange the iodixanol gradient solution for HBSS.

Cell culture. HEK293 (ATCC® CRL-3216™), HEK293T (ATCC® CRL-3216™), HeLa (ATCC® CCL-2™), HT1080 (ATCC® CCL-121™), PK-15 (ATCC® CCL-33™), BHK-21 (ATCC® CCL-10™), Vero (ATCC® CCL-81™), A549 (ATCC® CRM-CCL-185™), MDBK (ATCC® CCL-22™), MDCK (ATCC® CCL-34™), NRLE, and Dulac cells were maintained in Dulbecco's modified Eagle's medium (DMEM) supplemented with 10% FBS and 2 mm L-glutamine. AML12 (ATCC® CRL-2254™) cells were supplemented with 10% fetal bovine serum, 10 µg/mL insulin, 5.5 µg/mL transferrin, 5 ng/mL selenium, and 40 ng/mL dexamethasone (insulin–transferrin–selenium (ITS-G); ThermoFisher, Mississauga, ON, Canada). Primary culture bovine tracheal epithelial cells were established as previously described [28]. Briefly, tracheas were obtained from market-weight cattle immediately following slaughter. The tracheal mucosa was removed and incubated overnight at 4 °C in phosphate-buffered saline wash solution containing 0.1% protease (Dispase, Invitrogen, San Diego, CA, USA), 0.1 mg/mL penicillin–streptomycin, 0.5 mg/mL gentamicin, and 10 µg/mL amphotericin B. The following day, cells were harvested by scraping the mucosal surface with a scalpel blade to release the epithelial cells. Cell viability was assessed with trypan blue, and 3×10^5 viable cells were suspended in 1 mL of culture medium consisting of DMEM and Ham's F-12 medium (DMEM/F12), 5% sterile fetal bovine serum, 0.1 mg/mL penicillin–streptomycin, 5 µg/mL amphotericin B, 0.5 mg/mL gentamicin, 25 µg/mL bovine pituitary extract, 10 ng/mL epidermal growth factor, and 1% insulin–transferrin–selenium (Sigma).

In Vitro transduction assay. Cells were plated in 12-well dishes at a density of 1×10^5 per well, and the next day, AAV vectors were added at an MOI of 20,000. After 72 h, cell lysates were harvested using Passive Lysis Buffer (Promega, Madison, WI, USA), and luciferase expression was quantified by luciferase assay (Promega) and normalized by protein concentration determined by Bradford assay (Fisher Scientific, Mississauga, ON, Canada).

Animal ethics statement. All experiments involving mice were approved by the University of Guelph Animal Care Committee (AUP #3827) and conducted in accordance with the Canadian Council on Animal Care (CCAC). Balb/c mice (Strain code 028) were purchased from Charles River Laboratories, and after a one-week acclimation period, AAV vectors were administered to mice at 6 weeks of age.

AAV administration in vivo. AAV vectors were produced by quadruple transfection as outlined above and purified by iodixanol gradient centrifugation and further concentrated using Amicon Ultra-15 centrifugal filter units. Six-week-old female Balb/c mice were purchased from Charles River. Intranasal (IN) delivery was performed as previously described [29]. Intraperitoneal (IP) injections were completed using a 27-gauge tuberculin syringe, and vector was diluted in PBS to 100 µL. IV administrations were performed by tail-vein injection using a 29-gauge needle with vector diluted in PBS to 100 µL.

In Vivo luciferase imaging. Six-week-old Balb/c mice were administered 1×10^{11} vg of AAV.Po.Guelph-Luciferase or AAV.Bov.Guelph-Luciferase by either IN or IP routes. Mice were imaged using the IVIS Spectrum CT instrument 20 min following injection with D-luciferin given IN or IP. Living Image version 4.7.3. software was used to quantify luciferase expression.

Genbank accession numbers. The nucleotide sequences for AAV.Bov.Guelph (AAV.Bov.G-1) and AAV.Po.Guelph (AAV.Po.G-1) capsid genes were submitted to Genbank with accession numbers OR759014 (AAV.Bov.G-1) and OR759015 (AAV.Po.G-1), respectively.

Statistics. All statistical analysis was completed by two-way ANOVA using GraphPad Prism 7 software. All error bars represent standard deviation of the mean.

3. Results

3.1. Analysis of Porcine AAV Screening Amplicon

A total of 104 samples were screened for the presence of AAVs by PCR, the majority which were porcine colon samples from animals that showed no outward signs of disease. All other samples were from diseased animals submitted for necropsy. A summary of these samples is found in Table 3. The 104 PCR screenings showed 51 positive samples and included one canine, one hippopotamus, one equine, and the remainder porcine AAVs. Analysis of a rep gene fragment derived from a screening product amplified from a porcine colon sample showed 92% homology with the previously characterized porcine isolate AAV.Po1 [30]. Furthermore, AAV.Po.Guelph contained a 15-base intergenic spacer between the rep and cap fragments, shared by the AAV.Po1 clade, whereas the AAV2 clade contains the 16-base intergenic spacer more commonly observed with human-derived AAVs.

Table 3. An overview of tissue samples collected for AAV sequence isolation.

Type of Sample	Canine	Feline	Bovine	Porcine	Ovine	Equine	Exotic	Total
Extraction Method	Qiagen	Qiagen	Qiagen	MagNA Pure	Qiagen	Qiagen	Qiagen	–
Number Screened	10	4	8	74	2	3	3	104
Positive PCR Screening	1	0	0	46	2	1	1	51

A high degree of similarity exists between the AAV.Po.Guelph isolate and previously characterized porcine isolates for the first 413 base pairs (bp) of the capsid gene (Figure 2). Alignment of DNA sequences containing this region from AAV.Po.Guelph, AAVpo2.1, AAVpo4, and AAVpo6 demonstrated a high degree of conservation [31]. Overall, across the entire capsid gene, AAVPo1 and AAV.Po.Guelph share only 65.1% nucleotide identity; however, within this 413 bp segment, sequence identity was unusually high at 90.4%. This was also true of the previously discussed AAVpo2.1, AAVpo4, and AAVpo6. These isolates had a high degree of shared sequence identity with Po1 within the 413 bp region (90.4%) but, overall, only shared 65% to 66% sequence identity with Po1. In addition, AAVpo2.1, AAVpo4, and AAVpo6 demonstrated relatively high capsid gene shared sequence identity with the AAV.Po.Guelph capsid, with 88.6%, 93.6%, and 87.7% sequence identity, respectively (Figure 2A). On the other hand, another porcine isolate—AAVpo5—demonstrates a much greater level of homology with AAVPo1, at 89% sequence identity, than it has with any of the other capsids sequenced so far. A phylogenetic analysis using goose parvovirus as the outgroup demonstrated a high degree of sequence similarity between AAV.Po.Guelph and the human AAV2 capsid (Figure 2B). Interestingly, AAVPo1 and AAVpo5 cluster more closely with AAV5 and AAVGo.1, suggesting a possible shared lineage.

origin. Penultimately, region J shares much similarity with AAV8 (tree topology 6, yellow, Figure S3), suggesting that another possible recombination event gave rise to this part of the AAV.Po.Guelph capsid gene. Lastly, region K, like region E, shares similarity with AAV2, AAV3, and AAV13, suggesting a possible recombination event that had occurred in the past with an ancestor of these sequences.

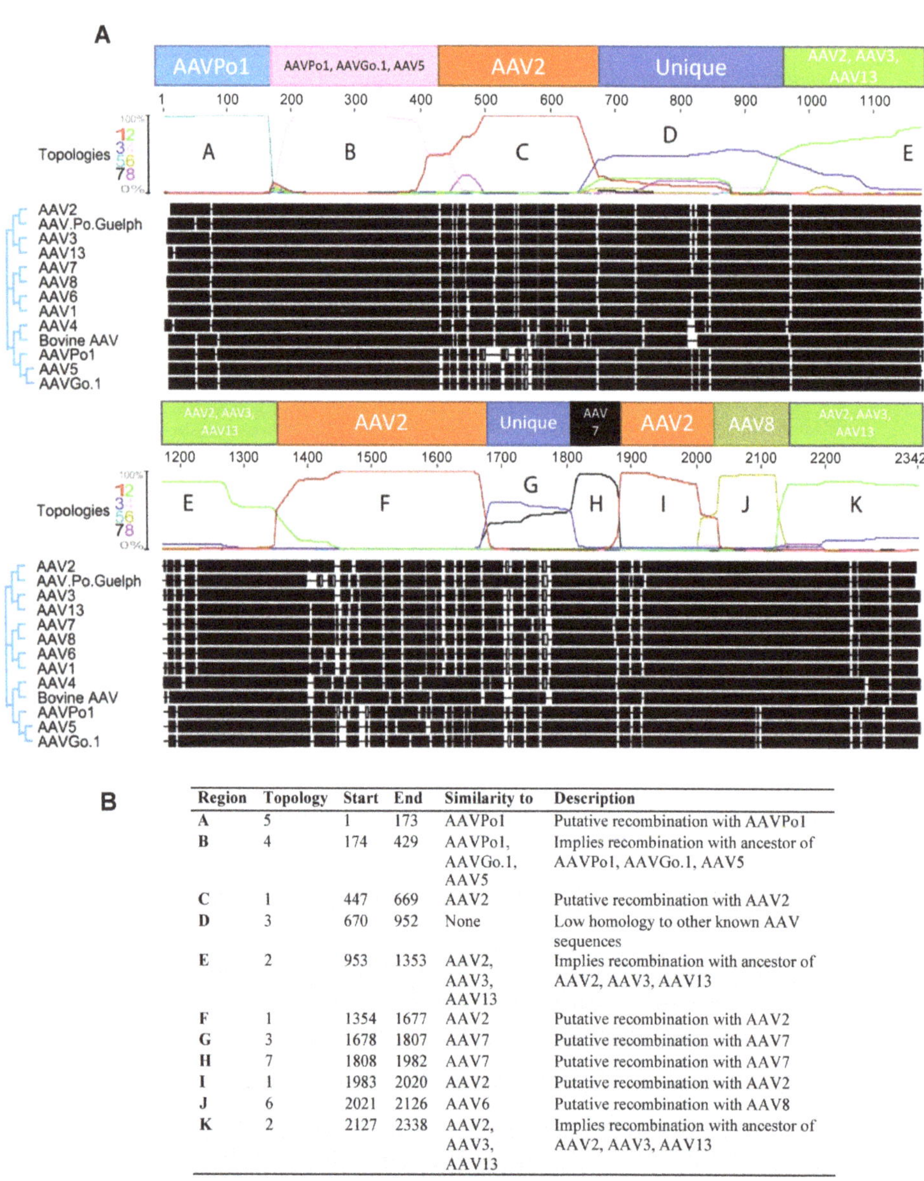

Region	Topology	Start	End	Similarity to	Description
A	5	1	173	AAVPo1	Putative recombination with AAVPo1
B	4	174	429	AAVPo1, AAVGo.1, AAV5	Implies recombination with ancestor of AAVPo1, AAVGo.1, AAV5
C	1	447	669	AAV2	Putative recombination with AAV2
D	3	670	952	None	Low homology to other known AAV sequences
E	2	953	1353	AAV2, AAV3, AAV13	Implies recombination with ancestor of AAV2, AAV3, AAV13
F	1	1354	1677	AAV2	Putative recombination with AAV2
G	3	1678	1807	AAV7	Putative recombination with AAV7
H	7	1808	1982	AAV7	Putative recombination with AAV7
I	1	1983	2020	AAV2	Putative recombination with AAV2
J	6	2021	2126	AAV6	Putative recombination with AAV8
K	2	2127	2338	AAV2, AAV3, AAV13	Implies recombination with ancestor of AAV2, AAV3, AAV13

Figure 3. Probability of various tree topologies for AAV.Po.Guelph as predicted by the Geneious Dual Brothers recombination plugin. (**A**) The rectangular overlay above the tree topology probability graph indicates the likely origin of the different sequence regions of the capsid. For specific color-coded tree topologies (1 through 8), see Figures S2 and S3. (**B**) Description and location of topology regions A through K indicating the likely lineage of specific capsid regions.

3.3. AAV.Bov.Guelph Capsid Protein Is Highly Similar to the Bovine AAV Capsid

We obtained a caprine adenovirus isolate from the Animal Health lab, and using our screening primers, we found it to be positive for AAV. The amplified PCR product was highly similar to a previously characterized bovine AAV capsid sequence (Genbank: AY388617.1), with only two amino acid changes distinguishing it from bovine AAV: I124L and R499G. Although isolated from a caprine adenovirus stock, this capsid, named AAV.Bov.Guelph, was found to be unrelated to the previously characterized caprine AAV, AAVGo.1 (GenBank: AY724675.1).

3.4. AAV.Bov.Guelph Transduces Mammalian Cell Lines More Efficiently Than AAV.Po.Guelph

The in vitro transducing properties of AAV.Bov.Guelph and AAV.Po.Guelph were evaluated in a panel of mammalian cell lines including HeLa, HEK 293T, RLE, HT1080, PK-15, HEK293, BHK-21, Vero, A549, Dulac, MDBK, AML-12, MDBK, and polarized bovine epithelial cells. The human AAV serotype 6 was included as a benchmark for in vitro transduction performance, since AAV6 and AAV1 are considered among the best serotypes for the transduction of a wide variety of cell types [34]. The panel of cell lines was highly varied and consisted of several different species and cell types, including human-derived cell lines HeLa (cervical cancer), HEK 293T (kidney), HT1080 (fibrosarcoma), HEK293 (kidney), and A549 (lung adenocarcinoma); porcine cell lines PK-15 and Dulac (both kidney); a bovine kidney cell line (MDBK); and bovine polarized tracheal epithelial cells. In addition, Vero cells (African green monkey kidney), MDCK (canine kidney), and rodent BHK-21 (Syrian golden hamster kidney) and AML-12 (mouse liver) cells were also included in the panel. Interestingly, when infected at the same multiplicity of infection (MOI), AAV.Bov.Guelph demonstrated superior transduction in the majority of cell lines tested compared to both AAV6 and AAV.Po.Guelph. AAV.Bov.Guelph showed relative luciferase expression of at least an order of magnitude higher than AAV6 in many of the cell lines tested, including HeLa, 293T, RLE, HT1080, HEK293, Vero, Dulac, and polarized bovine trachea cells (Figure 4). In other cases, AAV.Bov.Guelph exhibited greater expression than AAV6 but with less than a 10-fold increase. This was evident for PK-15, BHK-21, A549, and AML-12 cell lines, with AAV.Bov.Guelph showing a modest improvement over AAV6 in these lines. Interestingly, luciferase expression in MDBK bovine cells was lower than that in human cell lines 293T, HeLa, and HT1080 for AAV.Bov.Guelph. However, this was still at least an order of magnitude higher than AAV6 luciferase expression in this cell line. In addition, AAV.Bov.Guelph transduced polarized bovine trachea cells, while neither AAV6 nor AAV.Po.Guelph was able to transduce these cells to any degree.

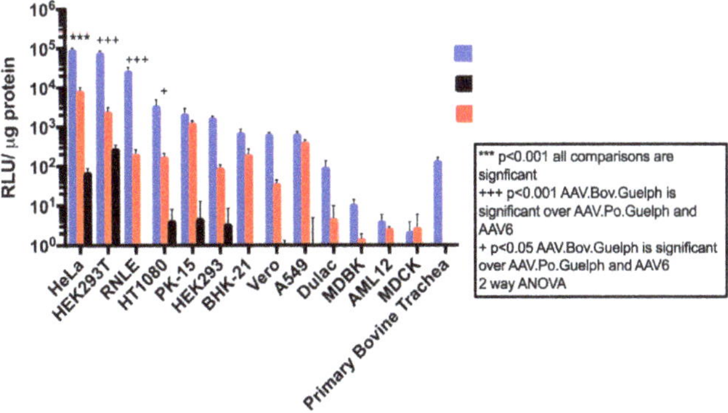

Figure 4. In vitro transduction profile of AAV.Po.Guelph, AAV.Bov.Guelph, and AAV6 in a panel of mammalian cell lines. A total of 50,000 cells in a 24-well plate were transduced with AAV at an MOI of 20,000, and 72 h post-transduction, cell lysate was harvested using passive lysis buffer and luciferase expression was quantified via luciferase assay. Bradford assays were used to normalize protein concentration. A two-way ANOVA was used to determine statistical significance.

Overall, AAV.Po.Guelph was a poor in vitro transducer, even in porcine cell lines PK-15 and Dulac cells. Notable expression for AAV.Po.Guelph was only observed for HeLa, HEK 293T, HT1080, PK-15, and HEK 293 cells. Interestingly, most of these cells were human derived, whereas no transduction was observed for any non-human cell line except for the porcine PK-15 cells.

3.5. Analysis of AAV.Po.Guelph and AAV.Bov.Guelph Expressing Luciferase in Mice following Intranasal Administration

AAV.Po.Guelph and AAV.Bov.Guelph expressing firefly luciferase were administered to mice intranasally at a dose of 1×10^{11} vg. Luciferase expression in the lungs and the nose was quantified on days 1, 3, 7, 14, 21, 28, and 56 after AAV administration (Figure 5). In the lungs, both vectors resulted in peak luciferase expression on day 7 post-administration. However, AAV.Bov.Guelph significantly outperformed AAV.Po.Guelph on days 3 to 21 post-administration, with its peak luciferase expression being almost 10-fold higher than that of AAV.Po.Guelph (Figure 5A). In the nose, a similar pattern was observed, with peak luciferase expression occurring on day 7 post-administration for both vectors. As in the lungs, AAV.Bov.Guelph resulted in significantly higher luciferase expression in the nose as compared to AAV.Po.Guelph from day 7 to 21 post-administration (Figure 5B). An in vivo luciferase imaging system was used to visualize luciferase expression in the mice and confirmed this trend (Figure 5C).

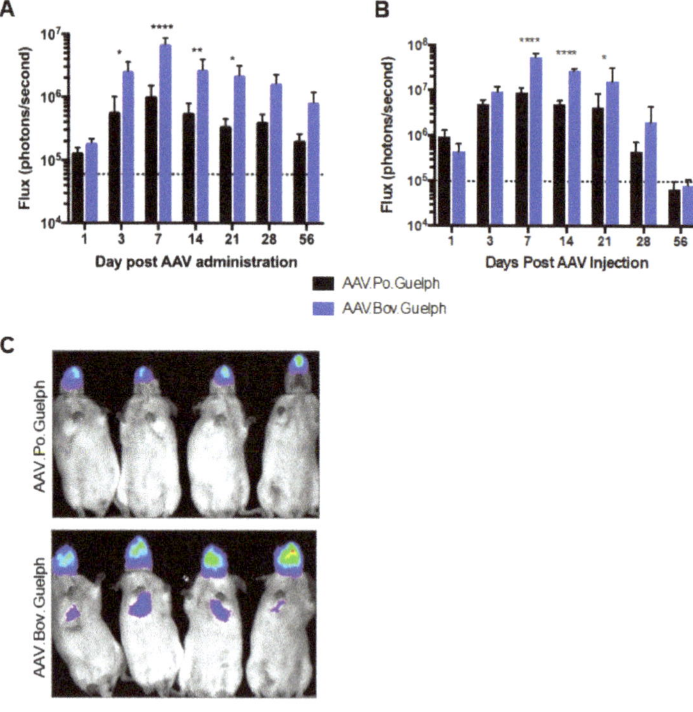

Figure 5. AAV.Po.Guelph and AAV.Bov.Guelph transduction of the mouse airway. A total of 1×10^{11} vg of AAV.Po.Guelph-Luciferase or AAV.Bov.Guelph-Luciferase was administered intranasally to six-week-old female Balb/c mice (n = 4/group). Luciferase expression in the (**A**) lungs and (**B**) nose was quantified on days 1, 3, 7, 14, 21, 28, and 56 after AAV administration. The dashed lines indicate the background threshold. **** $p < 0.0001$, ** $p < 0,01$, and * $p < 0.05$. (**C**) Images from three days after intranasal administration of AAV.

3.6. Analysis of AAV.Po.Guelph and AAV.Bov.Guelph Expressing Luciferase in Mice following Intraperitoneal Administration

To evaluate the transducing properties of AAV.Po.Guelph and AAV.Bov.Guelph expressing firefly luciferase when administered systemically, groups of mice (n = 4) were intraperitoneally injected with 1×10^{11} vg of vector, and luciferase expression in the peritoneal cavity was quantified on days 14 and 28 post-administration (Figure 6A), with the in vivo image obtained on day 28 shown (Figure 6B). No significant differences were observed between vectors at either timepoint. Confirmation of the liver as the target organ for these capsids was confirmed histologically (Figure S4).

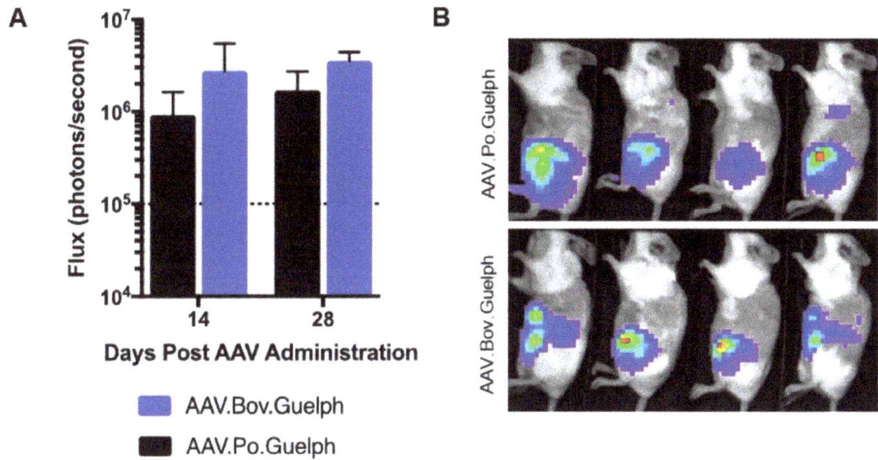

Figure 6. AAV.Po.Guelph and AAV.Bov.Guelph in vivo transduction profile following intraperitoneal administration. A total of 1×10^{11} vg of AAV.Po.Guelph-Luciferase or AAV.Bov.Guelph-Luciferase was injected IP to BALB/c mice (n = 4/ group). (**A**) The luciferase signal (left lateral) was quantified on days 14 and 28 after AAV administration. The dashed lines indicate the background threshold. (**B**) Images demonstrating distribution of luciferase expression on day 28 after IP injection.

4. Discussion

Here, we describe the isolation of two novel AAV capsids from non-human species and characterize their in vitro and in vivo transducing properties. The porcine AAV capsid, designated AAV.Po.Guelph, appears to be the product of multiple recombination events between both porcine and human AAVs, whereas AAV.Bov.Guelph was isolated as a contaminant of a caprine adenovirus stock, demonstrating the potential for bovine AAVs to infect caprine species.

As demonstrated by the numerous recombination events described herein, the AAV capsid may readily undergo recombination. It has previously been shown that defects in the AAV capsid may be complemented by a co-infecting AAV, as was the case when Bowles and colleagues were able to rescue mutant AAV2 capsids with AAV3 [35]. Recombination may be particularly advantageous in the case of the cap gene, where the exchange of genetic material with other AAV species may yield novel capsid characteristics. Such characteristics might include the ability to bind to heparin sulfate proteoglycan; the switching of epitopes from one recognized by pre-existing antibodies to one that the host might be naive to; changes in the ability to bind receptors or co-receptors, thereby resulting in novel cell or tissue tropisms; and potentially other useful properties. A previous study investigating the diversity of primate-derived AAVs has shown that even minor changes in the capsid might lead to divergent serological reactivity or transduction efficiency [36]. Furthermore, the authors concluded that homologous recombination is a powerful force shaping the evolution and sequence diversity of AAV capsid sequences. The apparently frequent recombination events between human and porcine AAV capsids suggest recurrent co-infection of porcine and human serotypes in the same cell. This likely occurs in pigs more frequently than it does in humans since pre-existing immunity to AAVpo1 did not appear to occur in pooled human sera [30], whereas porcine sera robustly prevented transduction by all human AAV serotypes tested [37]. This suggests that the pig may be acting as a sort of mixing vessel for various human and non-human AAV species and may represent a treasure trove for the isolation of novel AAV capsid sequences.

In the case of AAV.Bov.Guelph, two amino acid changes differentiated it from bovine AAV: I124L and R499G. The change from isoleucine to leucine conserves the non-polar properties of isoleucine and thus would not be expected to have a substantial effect on capsid protein folding. On the other hand, the change from the positively charged amino acid arginine to glycine, an uncharged and relatively small amino acid, may influence the local protein secondary structure. In addition, the arginine-to-glycine change occurs in a variable region known to play a role in virus neutralization, heparin binding, and cell transduction [33].

In vitro transduction assays using a panel of mammalian cells demonstrated AAV.Bov. Guelph's superior transduction properties in many commonly used cell lines including HeLa, HEK293T, HEK293, and Vero. This was expected, as Schmidt et al. found that the closely related bovine AAV was a strong transducer of HEK293T and HeLa cells [13]. In addition, AAV.Bov.Guelph demonstrated excellent transduction of respiratory tract-derived cell lines RNLE, A549, and polarized bovine tracheal epithelial cells. The diversity of transducible cell lines can be explained by their expected receptors. Schmidt and Chiorini previously identified gangliosides as bovine AAV's most probable receptor [38]. Given how closely related AAV.Bov.Guelph is to bovine AAV, it is likely that gangliosides are also AAV.Bov.Guelph's receptor. Since gangliosides are ubiquitously expressed in all mammalian cells, it is unsurprising that AAV.Bov.Guelph was able transduce the panel of mammalian cells [39].

In vivo, AAV.Bov.Guelph delivered the luciferase gene efficiently to the lungs and noses of mice after intranasal administration. Additionally, AAV.Bov.Guelph mediated luciferase expression following intraperitoneal injection. Interestingly, when Pasquale et al. used bovine AAV administered intranasally to deliver luciferase to mice, only modest luciferase activity was observed [40]. They were able to significantly increase luciferase expression by administering tannic acid prior to bovine AAV, which they attributed to tannic acid's ability to redirect AAV particles entering the transcytosis pathway to the transduction pathway [41]. Given that AAV.Bov.Guelph was able to achieve moderate luciferase expression without tannic acid, it is possible that the slight changes in capsid amino acids could impact its ability to undergo the transduction pathway as opposed to the transcytosis pathway; however, a head-to-head comparison of the two bovine capsids is needed in order to draw any conclusions. Overall, AAV.Bov.Guelph may be a useful capsid for respiratory tract directed gene therapy.

AAV.Po.Guelph possesses a four-amino-acid insertion at position 529 relative to AAVpo4, starting with two amino acids with positively charged side chains, followed by a polar amino acid, and ending with a non-polar amino acid. It is unclear what the significance of this insertion might be, but the insertion occurs in a region implicated in cell transduction, neutralization by pooled human sera, and neutralization by intravenous immunoglobulin.

In vitro, AAV.Po.Guelph performed relatively poorly, with the highest transduction observed in the human-derived cell lines HeLa and HEK293T and low to no transduction in the rest of the cell lines tested. Its closest relative capsid-wise, AAVpo4, possessed a similar transduction profile when tested on a panel of mammalian cells. Like AAV.Po.Guelph, AAVpo4 did not significantly transduce MDCK, Vero, HEK293, or A549 cells but showed strong transduction in the porcine retina cell line, VR1BL [42].

In vivo, AAV.Po.Guelph was able to transduce the lung, nose, and peritoneal cavity of mice after intranasal or intraperitoneal administration. Higher luciferase expression was observed in the nose than in the lung. Interestingly, AAVpo4 also showed poor lung transduction after intranasal administration [31]. AAV.Po.Guelph-mediated luciferase expression was observed in the peritoneal cavity and was not significantly different from AAV.Bov.Guelph-mediated expression. Bello et al. found that AAVpo4 can transduce the liver efficiently [31], and histological images from AAV.Po.Guelph-administered mice showed evidence of liver transduction. Indeed, most AAV serotypes show liver transduction when administered intravenously and intraperitoneally [43,44].

5. Conclusions

Overall, two novel capsids, AAV.Bov.Guelph and AAV.Po.Guelph, were identified. Phylogenetic analyses were performed, and transduction profiles were established both in vitro and in vivo. Future biodistribution studies should be performed to elucidate specific capsid tropism, and these capsids should be compared to commonly used rAAV capsids to evaluate their potential as novel gene therapy vectors.

Supplementary Materials: The following supporting information can be downloaded at: https://www.mdpi.com/article/10.3390/v16010057/s1, Figure S1: Distribution of AAV.Po.Guelph and AAV.Bov.Guelph vectors during production; Figure S2: Predicted phylogenic relationships between AAV.Po.Guelph capsid and other AAV capsid isolates; Figure S3: Predicted phylogenic relationships between AAV.Po.Guelph capsid and other AAV capsid isolates, continued; Figure S4: Alkaline phosphatase expression in liver tissue mediated by AAV.Po.Guelph and AAV.Bov.Guelph vectors following intravenous injection.

Author Contributions: Conceptualization, D.L.Y. and S.K.W.; methodology, D.L.Y., L.P.v.L., B.A.Y.S., K.J.N., J.K.S. and K.J.S.; formal analysis, D.L.Y., L.P.v.L. and S.K.W.; resources, D.S. and S.K.W.; data curation, D.L.Y., L.P.v.L. and S.K.W.; writing—original draft preparation, D.L.Y.; writing—review and editing, B.A.Y.S. and S.K.W.; supervision, S.K.W.; funding acquisition, S.K.W. All authors have read and agreed to the published version of the manuscript.

Funding: This research was funded by the Natural Sciences and Engineering Research Council of Canada (NSERC) Discover Grant #499834 to S.K.W. Stipend support was provided by the Ontario Veterinary College (OVC) (D.L.Y., L.P.v.L., B.A.Y.S.), and the NSERC Postgraduate Scholarship-Doctoral program (B.A.Y.S.).

Institutional Review Board Statement: This study was conducted according to the guidelines set forth by the Canadian Council on Animal Care (CCAC) and approved by the Animal Care Committee of the University of Guelph (Animal Use Protocol number 3827).

Informed Consent Statement: Not applicable.

Data Availability Statement: All data generated or analyzed during this study can be found within the published article and its Supplementary Files. Sequences for AAV.Bov.Guelph (OR759014) and AAV.Po.Guelph OR759015) have been deposited in Genbank.

Acknowledgments: We thank the technicians at the University of Guelph Animal Isolation Unit for their animal care services.

Conflicts of Interest: The authors declare no conflicts of interest. S.K.W. is an inventor on a US patent for the AAV6.2FF capsid. This patent (US20190216949) is licensed to Avamab Pharma Inc., where S.K.W. is co-founder. The funders had no role in the design of the study; in the collection, analyses, or interpretation of data; in the writing of the manuscript; or in the decision to publish the results.

References

1. Kuzmin, D.A.; Shutova, M.V.; Johnston, N.R.; Smith, O.P.; Fedorin, V.V.; Kukushkin, Y.S.; van der Loo, J.C.M.; Johnstone, E.C. The clinical landscape for AAV gene therapies. *Nat. Rev. Drug Discov.* **2021**, *20*, 173–174. [CrossRef] [PubMed]
2. Calcedo, R.; Vandenberghe, L.H.; Gao, G.; Lin, J.; Wilson, J.M. Worldwide epidemiology of neutralizing antibodies to adeno-associated viruses. *J. Infect. Dis.* **2009**, *199*, 381–390. [CrossRef] [PubMed]
3. Grimm, D.; Kay, M.A. From virus evolution to vector revolution: Use of naturally occurring serotypes of adeno-associated virus (AAV) as novel vectors for human gene therapy. *Curr. Gene Ther.* **2003**, *3*, 281–304. [CrossRef] [PubMed]

4. Jiang, H.; Couto, L.B.; Patarroyo-White, S.; Liu, T.; Nagy, D.; Vargas, J.A.; Zhou, S.; Scallan, C.D.; Sommer, J.; Vijay, S.; et al. Effects of transient immunosuppression on adenoassociated, virus-mediated, liver-directed gene transfer in rhesus macaques and implications for human gene therapy. *Blood* **2006**, *108*, 3321–3328. [CrossRef] [PubMed]

5. Shen, W.; Liu, S.; Ou, L. rAAV immunogenicity, toxicity, and durability in 255 clinical trials: A meta-analysis. *Front. Immunol.* **2022**, *13*, 1001263. [CrossRef] [PubMed]

6. Mendell, J.R.; Connolly, A.M.; Lehman, K.J.; Griffin, D.A.; Khan, S.Z.; Dharia, S.D.; Quintana-Gallardo, L.; Rodino-Klapac, L.R. Testing preexisting antibodies prior to AAV gene transfer therapy: Rationale, lessons and future considerations. *Mol. Ther. Methods Clin. Dev.* **2022**, *25*, 74–83. [CrossRef] [PubMed]

7. Gonçalves, M.A. Adeno-associated virus: From defective virus to effective vector. *Virol. J.* **2005**, *2*, 43. [CrossRef]

8. Gregorevic, P.; Blankinship, M.J.; Allen, J.M.; Crawford, R.W.; Meuse, L.; Miller, D.G.; Russell, D.W.; Chamberlain, J.S. Systemic delivery of genes to striated muscles using adeno-associated viral vectors. *Nat. Med.* **2004**, *10*, 828–834. [CrossRef]

9. Zolotukhin, S.; Vandenberghe, L.H. AAV capsid design: A Goldilocks challenge. *Trends Mol. Med.* **2022**, *28*, 183–193. [CrossRef]

10. Lisowski, L.; Tay, S.S.; Alexander, I.E. Adeno-associated virus serotypes for gene therapeutics. *Curr. Opin. Pharmacol.* **2015**, *24*, 59–67. [CrossRef]

11. Thwaite, R.; Pagès, G.; Chillón, M.; Bosch, A. AAVrh.10 immunogenicity in mice and humans. Relevance of antibody cross-reactivity in human gene therapy. *Gene Ther.* **2015**, *22*, 196–201. [CrossRef] [PubMed]

12. Goedeker, N.L.; Dharia, S.D.; Griffin, D.A.; Coy, J.; Truesdale, T.; Parikh, R.; Whitehouse, K.; Santra, S.; Asher, D.R.; Zaidman, C.M. Evaluation of rAAVrh74 gene therapy vector seroprevalence by measurement of total binding antibodies in patients with Duchenne muscular dystrophy. *Ther. Adv. Neurol. Disord.* **2023**, *16*, 17562864221149781. [CrossRef] [PubMed]

13. Schmidt, M.; Katano, H.; Bossis, I.; Chiorini, J.A. Cloning and characterization of a bovine adeno-associated virus. *J. Virol.* **2004**, *78*, 6509–6516. [CrossRef] [PubMed]

14. Bossis, I.; Chiorini, J.A. Cloning of an avian adeno-associated virus (AAAV) and generation of recombinant AAAV particles. *J. Virol.* **2003**, *77*, 6799–6810. [CrossRef] [PubMed]

15. Clarke, J.K.; McFerran, J.B.; McKillop, E.R.; Curran, W.L. Isolation of an adeno associated virus from sheep. *Arch. Virol.* **1979**, *60*, 171–176. [CrossRef] [PubMed]

16. Olson, E.J.; Haskell, S.R.; Frank, R.K.; Lehmkuhl, H.D.; Hobbs, L.A.; Warg, J.V.; Landgraf, J.G.; Wünschmann, A. Isolation of an adenovirus and an adeno-associated virus from goat kids with enteritis. *J. Vet. Diagn. Investig.* **2004**, *16*, 461–464. [CrossRef]

17. Hsi, J.; Mietzsch, M.; Chipman, P.; Afione, S.; Zeher, A.; Huang, R.; Chiorini, J.; McKenna, R. Structural and antigenic characterization of the avian adeno-associated virus capsid. *J. Virol.* **2023**, *97*, e0078023. [CrossRef]

18. Katano, H.; Afione, S.; Schmidt, M.; Chiorini, J.A. Identification of adeno-associated virus contamination in cell and virus stocks by PCR. *Biotechniques* **2004**, *36*, 676–680. [CrossRef]

19. Suchard, M.A.; Weiss, R.E.; Dorman, K.S.; Sinsheimer, J.S. Inferring Spatial Phylogenetic Variation Along Nucleotide Sequences. *J. Am. Stat. Assoc.* **2003**, *98*, 427–437. [CrossRef]

20. Minin, V.N.; Dorman, K.S.; Fang, F.; Suchard, M.A. Dual multiple change-point model leads to more accurate recombination detection. *Bioinformatics* **2005**, *21*, 3034–3042. [CrossRef]

21. Huang, X.; Hartley, A.V.; Yin, Y.; Herskowitz, J.H.; Lah, J.J.; Ressler, K.J. AAV2 production with optimized N/P ratio and PEI-mediated transfection results in low toxicity and high titer for in vitro and in vivo applications. *J. Virol. Methods* **2013**, *193*, 270–277. [CrossRef] [PubMed]

22. Allen, J.M.; Halbert, C.L.; Miller, A.D. Improved adeno-associated virus vector production with transfection of a single helper adenovirus gene, E4orf6. *Mol. Ther.* **2000**, *1*, 88–95. [CrossRef] [PubMed]

23. Halbert, C.L.; Metzger, M.J.; Lam, S.L.; Miller, A.D. Capsid-expressing DNA in AAV vectors and its elimination by use of an oversize capsid gene for vector production. *Gene Ther.* **2011**, *18*, 411–417. [CrossRef] [PubMed]

24. Halbert, C.; Lam, S.; Miller, A. High-efficiency promoter-dependent transduction by adeno-associated virus type 6 vectors in mouse lung. *Hum. Gene Ther.* **2007**, *18*, 344–354. [CrossRef] [PubMed]

25. Richardson, J.S.; Yao, M.K.; Tran, K.N.; Croyle, M.A.; Strong, J.E.; Feldmann, H.; Kobinger, G.P. Enhanced protection against Ebola virus mediated by an improved adenovirus-based vaccine. *PLoS ONE* **2009**, *4*, e5308. [CrossRef]

26. Aurnhammer, C.; Haase, M.; Muether, N.; Hausl, M.; Rauschhuber, C.; Huber, I.; Nitschko, H.; Busch, U.; Sing, A.; Ehrhardt, A.; et al. Universal real-time PCR for the detection and quantification of adeno-associated virus serotype 2-derived inverted terminal repeat sequences. *Hum. Gene Ther. Methods* **2012**, *23*, 18–28. [CrossRef]

27. Khan, I.F.; Hirata, R.K.; Russell, D.W. AAV-mediated gene targeting methods for human cells. *Nat. Protoc.* **2011**, *6*, 482–501. [CrossRef]

28. Berghuis, L.; Abdelaziz, K.T.; Bierworth, J.; Wyer, L.; Jacob, G.; Karrow, N.A.; Sharif, S.; Clark, M.E.; Caswell, J.L. Comparison of innate immune agonists for induction of tracheal antimicrobial peptide gene expression in tracheal epithelial cells of cattle. *Vet. Res.* **2014**, *45*, 105. [CrossRef]

29. van Lieshout, L.P.; Domm, J.M.; Wootton, S.K. AAV-Mediated Gene Delivery to the Lung. *Methods Mol. Biol.* **2019**, *1950*, 361–372. [CrossRef]

30. Bello, A.; Tran, K.; Chand, A.; Doria, M.; Allocca, M.; Hildinger, M.; Beniac, D.; Kranendonk, C.; Auricchio, A.; Kobinger, G.P. Isolation and evaluation of novel adeno-associated virus sequences from porcine tissues. *Gene Ther.* **2009**, *16*, 1320–1328. [CrossRef]

31. Bello, A.; Chand, A.; Aviles, J.; Soule, G.; Auricchio, A.; Kobinger, G.P. Novel adeno-associated viruses derived from pig tissues transduce most major organs in mice. *Sci. Rep.* **2014**, *4*, 6644. [CrossRef] [PubMed]

32. Puppo, A.; Bello, A.; Manfredi, A.; Cesi, G.; Marrocco, E.; Della Corte, M.; Rossi, S.; Giunti, M.; Bacci, M.L.; Simonelli, F.; et al. Recombinant vectors based on porcine adeno-associated viral serotypes transduce the murine and pig retina. *PLoS ONE* **2013**, *8*, e59025. [CrossRef]

33. Govindasamy, L.; Padron, E.; McKenna, R.; Muzyczka, N.; Kaludov, N.; Chiorini, J.A.; Agbandje-McKenna, M. Structurally mapping the diverse phenotype of adeno-associated virus serotype 4. *J. Virol.* **2006**, *80*, 11556–11570. [CrossRef] [PubMed]

34. Ellis, B.L.; Hirsch, M.L.; Barker, J.C.; Connelly, J.P.; Steininger, R.J.; Porteus, M.H. A survey of ex vivo/in vitro transduction efficiency of mammalian primary cells and cell lines with Nine natural adeno-associated virus (AAV1-9) and one engineered adeno-associated virus serotype. *Virol. J.* **2013**, *10*, 74. [CrossRef] [PubMed]

35. Bowles, D.E.; Rabinowitz, J.E.; Samulski, R.J. Marker rescue of adeno-associated virus (AAV) capsid mutants: A novel approach for chimeric AAV production. *J. Virol.* **2003**, *77*, 423–432. [CrossRef]

36. Gao, G.; Alvira, M.R.; Somanathan, S.; Lu, Y.; Vandenberghe, L.H.; Rux, J.J.; Calcedo, R.; Sanmiguel, J.; Abbas, Z.; Wilson, J.M. Adeno-associated viruses undergo substantial evolution in primates during natural infections. *Proc. Natl. Acad. Sci. USA* **2003**, *100*, 6081–6086. [CrossRef]

37. Rapti, K.; Louis-Jeune, V.; Kohlbrenner, E.; Ishikawa, K.; Ladage, D.; Zolotukhin, S.; Hajjar, R.J.; Weber, T. Neutralizing antibodies against AAV serotypes 1, 2, 6, and 9 in sera of commonly used animal models. *Mol. Ther.* **2012**, *20*, 73–83. [CrossRef]

38. Schmidt, M.; Chiorini, J.A. Gangliosides are essential for bovine adeno-associated virus entry. *J. Virol.* **2006**, *80*, 5516–5522. [CrossRef]

39. Yu, R.K.; Tsai, Y.T.; Ariga, T.; Yanagisawa, M. Structures, biosynthesis, and functions of gangliosides--an overview. *J. Oleo Sci.* **2011**, *60*, 537–544. [CrossRef]

40. Di Pasquale, G.; Ostedgaard, L.; Vermeer, D.; Swaim, W.D.; Karp, P.; Chiorini, J.A. Bovine AAV transcytosis inhibition by tannic acid results in functional expression of CFTR in vitro and altered biodistribution in vivo. *Gene Ther.* **2012**, *19*, 576–581. [CrossRef]

41. Di Pasquale, G.; Chiorini, J.A. AAV transcytosis through barrier epithelia and endothelium. *Mol. Ther.* **2006**, *13*, 506–516. [CrossRef] [PubMed]

42. Bello, A.J.A. Linking the Tropism and Transduction Efficiency of Porcine-Derived Adeno-Associated Viruses to Their Transgene-Mediated Protective Efficacy. 2014. Available online: https://mspace.lib.umanitoba.ca/server/api/core/bitstreams/8ad9bf19-411b-4c60-a8d9-08ef2d1ccdb4/content (accessed on 28 December 2023).

43. Zincarelli, C.; Soltys, S.; Rengo, G.; Rabinowitz, J.E. Analysis of AAV serotypes 1–9 mediated gene expression and tropism in mice after systemic injection. *Mol. Ther.* **2008**, *16*, 1073–1080. [CrossRef] [PubMed]

44. Ai, J.; Li, J.; Gessler, D.J.; Su, Q.; Wei, Q.; Li, H.; Gao, G. Adeno-associated virus serotype rh.10 displays strong muscle tropism following intraperitoneal delivery. *Sci. Rep.* **2017**, *7*, 40336. [CrossRef] [PubMed]

 *viruses*

Article

Adeno-Associated Virus-like Particles' Response to pH Changes as Revealed by nES-DMA

Samuele Zoratto [1], Thomas Heuser [2], Gernot Friedbacher [1], Robert Pletzenauer [3], Michael Graninger [3], Martina Marchetti-Deschmann [1] and Victor U. Weiss [1,*]

[1] Institute of Chemical Technologies and Analytics, TU Wien, A-1060 Vienna, Austria; samuele.zoratto@tuwien.ac.at (S.Z.); gernot.friedbacher@tuwien.ac.at (G.F.); martina.marchetti-deschmann@tuwien.ac.at (M.M.-D.)
[2] Electron Microscopy Facility, Vienna BioCenter Core Facilities GmbH, A-1030 Vienna, Austria; thomas.heuser@vbcf.ac.at
[3] Pharmaceutical Sciences, Baxalta Innovations GmbH (Part of Takeda), A-1221 Vienna, Austria; robert.pletzenauer@takeda.com (R.P.); michael.graninger@takeda.com (M.G.)
* Correspondence: victor.weiss@tuwien.ac.at; Tel.: +43-1-58801-151611

Abstract: Gas-phase electrophoresis on a nano-Electrospray Gas-phase Electrophoretic Mobility Molecular Analyzer (nES GEMMA) separates single-charged, native analytes according to the surface-dry particle size. A volatile electrolyte, often ammonium acetate, is a prerequisite for electrospraying. Over the years, nES GEMMA has demonstrated its unique capability to investigate (bio-)nanoparticle containing samples in respect to composition, analyte size, size distribution, and particle numbers. Virus-like particles (VLPs), being non-infectious vectors, are often employed for gene therapy applications. Focusing on adeno-associated virus 8 (AAV8) based VLPs, we investigated the response of these bionanoparticles to pH changes via nES GEMMA as ammonium acetate is known to exhibit these changes upon electrospraying. Indeed, slight yet significant differences in VLP diameters in relation to pH changes are found between empty and DNA-cargo-filled assemblies. Additionally, filled VLPs exhibit aggregation in dependence on the applied electrolyte's pH, as corroborated by atomic force microscopy. In contrast, cryogenic transmission electron microscopy did not relate to changes in the overall particle size but in the substantial particle's shape based on cargo conditions. Overall, we conclude that for VLP characterization, the pH of the applied electrolyte solution has to be closely monitored, as variations in pH might account for drastic changes in particles and VLP behavior. Likewise, extrapolation of VLP behavior from empty to filled particles has to be carried out with caution.

Keywords: AAV8; VLP; nES GEMMA; DMA; cryo TEM; gene therapy

Citation: Zoratto, S.; Heuser, T.; Friedbacher, G.; Pletzenauer, R.; Graninger, M.; Marchetti-Deschmann, M.; Weiss, V.U. Adeno-Associated Virus-like Particles' Response to pH Changes as Revealed by nES-DMA. *Viruses* **2023**, *15*, 1361. https://doi.org/10.3390/v15061361

Academic Editors: Ottmar Herchenröder, Brigitte Pützer and Kenneth Lundstrom

Received: 3 May 2023
Revised: 29 May 2023
Accepted: 8 June 2023
Published: 13 June 2023

1. Introduction

Gene therapy alters the genetic information of an organism, replacing, supplementing, or modifying its genomic profile to enable hereditary disease treatment, as lately reviewed for hemophilia [1], for example. To convey new genomic information to a cell, cargo carriers are necessary to shield nucleotides from the environment and achieve their targeted transport. One possible carrier type is virus-like particles (VLPs), macromolecular assemblies resembling their parent virus but no longer being infectious due to a lack of viral genomic information [2]. Hence, VLPs can be engineered to encapsulate other genomic information in question.

AAV are viruses of the genus *Dependoparvovirus* within the family *Parvoviridae*. They are approximately 26 nm in size, non-enveloped, non-pathogenic, and endemic in humans as well as several vertebrate species. To date, 13 serotypes have been identified in nature, each with a different tissue tropism. The capsid comprises 60 copies of three types of subunits: VP1, VP2, and VP3, arranged in a T = 1 icosahedral symmetry with a molar ratio of 1:1:10. Due to its small size, only 4.7 kb of single-strand DNA (ssDNA) can be encapsulated [3].

In this work, we took an interest in adeno-associated virus serotype 8 (AAV8) modified to generate VLPs for gene therapy applications. AAV8 VLPs were provided in two different preparations: (i) lacking any genomic cargo (i.e., an 'empty' VLP preparation), and (ii) carrying a non-viral engineered genome (i.e., a 'filled' VLP preparation).

A possible technique for (bio-)nanoparticle characterization in general and AAV8 in particular is gas-phase electrophoresis on a nano Electrospray Gas-phase Electrophoretic Mobility Molecular Analyzer (nES GEMMA), also known as nES Differential Mobility Analyzer (nES-DMA) [4]. This technique is based on the size separation of single-charged, surface-dry analytes in the gas phase. Transfer of analytes from a liquid sample—a volatile electrolyte solution is a necessary prerequisite for the technique—to the gas-phase is achieved via a nES process, followed by drying of droplets and concomitant charge equilibration of particles in a bipolar atmosphere induced by, for example, a ^{210}Po α-particle source, a soft X-ray charger, or an alternating corona discharge process [5–8]. The majority of particles lose all charges upon passage of the bipolar atmosphere and are not regarded further, whereas a certain percentage of aerosolized analytes remains single-charged and are introduced to the DMA unit of the instrumentation. There, two forces act on the particles during size separation. Firstly, analytes are transported via a high laminar sheath flow of compressed, filtered ambient air through the DMA. At the same time, a tunable electric field exerts a force on charged particles in an orthogonal direction. By variation of the applied field strength and based on electrophoretic principles, only particles of a corresponding surface dry particle diameter (electrophoretic mobility, EM diameter) have the correct trajectory to continue to the detector unit of the instrument. There, particles are counted after a nucleation process in a supersaturated atmosphere as they pass a focused laser beam. This setup guarantees true particle-number-based detection of smaller-sized sample components next to larger ones in accordance with the recommendations of the European Commission (2011/696/EU, 18 October 2011, updated 2022) for nanoparticle characterization. The same instrumentation is also known under several other names, e.g., nES-DMA, macroIMS, SMPS or liquiScan ES (e.g., [9,10]).

Since its first reference in the literature by Kaufman et al. in 1996 [4], gas-phase electrophoresis has been applied for the characterization of a multitude of (bio-)nanoparticle materials, such as viruses [11–13], VLPs [14–17], liposomes and lipoprotein particles [18–22], extracellular vesicles [23–25], and organic and inorganic nanoparticles [26–28]. Frequently, ammonium acetate is employed as an electrolyte solution for the nES process. However, ammonium acetate is no buffer per definition (i.e., ammonium acetate solution), and especially at neutral pH, it exhibits no buffering capacity. Hence, significant changes in pH occur upon drying of droplets generated in the nES process, as reported by Konermann in 2017 [29]. We now took interest in the question how AAV8-based VLPs would behave to such pH changes, focusing on a well-defined in-solution setup. Analyses were carried out via gas-phase electrophoresis, and we corroborated our findings by atomic force microscopy (AFM) and cryogenic transmission electron microscopy (cryo-TEM) as orthogonal analytical techniques. All applied analytical techniques relate a pH dependency of AAV8 particles in aqueous ammonium acetate.

2. Materials and Methods

2.1. Chemicals

Ammonium acetate (NH_4OAc, $\geq$99.99%), ammonium hydroxide (ACS reagent), and acetic acid (ACS reagent) were purchased from Sigma-Aldrich (Steinheim, Germany). The electrolyte solution was prepared by dissolving 40 mM of ammonium acetate with water of ultrahigh quality (UHQ) delivered by a Simplicity UV apparatus (18.2 MΩ $\times$ cm at 25 °C, Millipore, Billerica, MA, USA). The solution was adjusted to different pH levels ranging from 4.0 to 9.0 (with 1.0 step increase) with ammonium hydroxide or acetic acid. Lastly, the solution was filtered through a surfactant-free cellulose acetate membrane with 0.20 μm pore size syringe filters (Sartorius, Göttingen, Germany).

2.2. Sample Description

Purified AAV8 VLP samples were provided by Baxalta Innovations GmbH (Orth/Donau, Austria, part of Takeda). Two different batches were provided: (i) so-called empty AAV8 VLPs (3776 μg/mL, i.e., 7.3×10^{14} capsids/mL) with 97% of capsids not carrying any genomic information, and (ii) so-called filled AAV8 VLPs (85 μg/mL, i.e., 1.6×10^{13} capsids/mL), where 66% of all the capsids were carrying a genomic load. The percentage of capsid filling was assessed via cryogenic transmission electron microscopy (cryo-TEM) by Baxalta Innovations GmbH.

2.3. Sample Preparation

Buffer exchange against 40 mM NH_4OAc at pH 7.0 was carried out by means of 10 kDa MWCO centrifugal filters (polyethersulfone membrane from VWR, Vienna, Austria). After three repetitions of spin filtration at 9000 g each, the retentate (approx. 10 μL in volume) was reconstituted in a pH-adjusted electrolyte solution (i.e., ranging from 4.0 to 9.0) to yield 22 μg/mL and 8.5 μg/mL final concentration for empty and filled AAV8 VLPs, respectively. Before analysis, the samples were further diluted to yield 250–300 particle counts for the signal at 25 nm at pH 7.0 with nES GEMMA. The respective dilution ratio was then employed for AFM and cryo-TEM analysis.

2.4. Instrumentation

nES GEMMA analyses were carried out on a TSI Inc instrument (Shoreview, MN, USA), which consisted of a nanoelectrospray charge reduction source unit (model 3480) equipped with a ^{210}Po charge equilibration device, an electrostatic classifier control unit employing a nano differential mass analyzer (nano DMA; model 3080) and an n-butanol driven ultrafine condensation particle counter (CPC; model 3025A) for AAV8 VLP detection. The nES unit is equipped with a 24 cm long polyimide-coated fused-silica capillary with an inner diameter of 25 μm (Molex, Lincolnshire, IL, USA). The capillary is manually cut and tapered with a home-built grinding machine based on the work of Tycova et al. [30].

Nanoparticle separation and detection were achieved with the following settings: the filtered airflow on the nES generator was set to 1.6×10^{-5} m^3/s (1 L per minute, Lpm), the CO_2 gas flow to 1.6×10^{-6} m^3/s (0.1 Lpm, 99.5% from Messer, Gumpoldskirchen, Austria) and the differential capillary pressure at 27.58 kPa (4 pounds per square inch differential, PSID). Capillary conditioning was performed by pre-spraying each sample for at least 3 min before starting the measurement. Capillary rinsing was performed by infusing the electrolyte solution until the signal from the previous sample was no longer detectable. The sample was infused at a flow rate of 70 nL/min. The capillary tip voltage was set to have a stable Taylor cone (approx. 2 kV voltage and -380 nA current). The electrostatic classifier was set in automatic scanning mode (up scan time 120 s, retrace time 30 s) with a sheath gas flow rate of 2.5×10^{-4} m^3/s (15 Lpm), which yielded a range of measurable EM diameters between 1.95 nm and 64.9 nm. A total of 10 scans for each sample were used to generate a median spectrum. Mathematical and statistical calculations on the nES GEMMA spectra were made with the software OriginPro ver. 2021 (OriginLab Corporation, Northampton, MA, USA).

AFM experiments were carried out on a NanoScope VIII Multimode SPM instrument (Bruker, Bremen, Germany) using silicon cantilevers with integrated silicon tips (model NCHV from Bruker, resonance frequency: 320 kHz, L = 125 μm, k = 42 N/m). The images were acquired in tapping, constant amplitude mode at a scanning rate of 1.99 Hz over a scan area of 5 and 1 μm^2.

The homogeneity of the mica platelet was tested prior to sample analysis. Subsequently, 10–20 μL of sample (5–20 μg/mL solutions) were spotted on the platelet's surface at room temperature. The sample was allowed to adsorb for 5 min undisturbed before being gently rinsed with UHQ water and dried with nitrogen gas (outlet pressure 350 kPa).

Lastly, the AFM images have been analyzed by NanoScope Analysis 1.5 software (Bruker, Santa Barbara, CA, USA). Longitudinal and medial plane sections were used to generate profiles of the imaged particles.

Cryo-TEM samples were prepared using Quantifoil (Großlöbichau, Germany) Cu 200 mesh R2/2 holy carbon grids which were glow discharged for 60 s at −25 mA with a Bal-Tec (Balzers, Liechtenstein) SCD005 glow discharger. Grids were mounted on forceps and loaded into a Leica GP (Leica Microsystems, Vienna, Austria) grid plunger with the climate chamber set to 4 °C and 75% relative humidity. Sample aliquots of 4 µL were applied to the carbon side of the grid and front-side blotted for 2 s with Whatman filter paper #1 (Little Chalfont, Great Britain) and plunge frozen into liquid ethane at approximately −180 °C for instant vitrification. Cryo-samples were transferred to a Glacios cryo-transmission microscope (Thermo Scientific, Waltham, MA, USA) equipped with a X-FEG and a Falcon3 direct electron detector. The microscope was operated in low-dose mode and digital images were recorded using the SerialEM software [31] and the Falcon3 camera in linear mode at a defocus of −6 µm and 36,000 fold magnification corresponding to a pixel size of 4.124 Å.

2.5. Cryo-TEM Particles Diameter Determination

Individual VLP particles were picked manually from suitable cryo-TEM micrographs and their coordinates were saved as model points using the open-source IMOD processing tool [32]. The particle models were used for alignment and 2D averaging with the Particle Estimation for Electron Tomography (PEET) software, an open-source tool typically used for 3D subtomogram averaging [33,34]. Unbiased diameter determination was achieved by an in-house ad hoc developed Python script (see Supporting information), able to generate 3D graphs from the 2D images; by plotting correlating grey values of each (x,y) image's pixel coordinates. The 3D graphs can be sectioned and exported at different height levels. For each sample, three sections at different levels were extracted. At last, the data points contained in the sections were imported to OriginPro software (ver. 2021), where a circle fitting function was applied to determine the diameter of the particle.

3. Results

Our article aims to expand the knowledge concerning the characteristics and behavior of AAV8-based VLPs suspended in electrolyte solutions at different pH levels. nES GEMMA technology is used as main analytical technique. As previously introduced, nES GEMMA can be easily employed to analyze viruses and deliver robust, high-throughput, and cost-effective analyses. Moreover, orthogonal analytical techniques such as AFM in tapping mode and cryo-TEM have been applied to corroborate our results obtained from gas-phase electrophoresis.

As already demonstrated in our previous work [35], nES GEMMA can discriminate between VLPs either carrying or lacking genomic information for the two individual sample types (Figure 1A). The small, albeit critical, difference in EM diameter between the two VLP preparations originates from the presence, or absence, of an encapsulated cargo (i.e., the genomic material). The discrimination is possible due to the operating conditions of nES GEMMA analysis, which induce evaporation of the hydrodynamic layer surrounding the particles, thus yielding surface-dry analytes. However, the surface drying effect seemingly propagates across the proteinaceous layer of the capsid to its core, as the VLP EM diameter is affected by the content of its cargo. Two possible outcomes can be surmised: in the presence of a payload, the particles mostly retain their original shape thanks to stabilizing forces emerging from protein-genome interactions. In the absence of a genomic load instead, a higher degree of deformation is allowed, thus making the particles more flexible and detectable at smaller EM diameters.

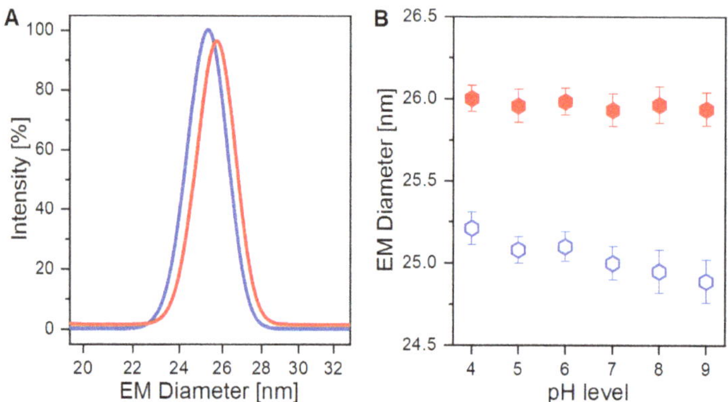

Figure 1. nES GEMMA analysis of empty and filled VLP preparations. (**A**) Gaussian fitting overlay of empty (blue trace) and filled (red trace) VLPs at pH 7. (Adapted from Zoratto et al., 2021 [35]) (**B**) EM diameter changes of empty (empty blue hexagons) and filled (filled red hexagons) VLPs according to pH levels ($n = 6$ analyses per pH level, average EM diameters and standard deviations are depicted, $p < 0.0001$).

A mandatory step for nES GEMMA includes the exchange of the original sample buffer to a volatile electrolyte solution. This drastically reduces non-volatile components, which hinder correct and accurate particle EM diameter determination. More details can be found in previous works [36,37]. To some extent, either for nES GEMMA or for ESI MS analysis, aqueous ammonium acetate is applied in that context. However, it has to be considered that ammonium acetate droplets might undergo drastic pH changes upon evaporation, as discussed by Konermann in detail [29]. Thus, we took an interest in investigating the effects of different pH levels on VLP preparations, either carrying a genomic cargo or not, and both suspended in aqueous ammonium acetate. Results are displayed in Figures 1B and 2A, respectively.

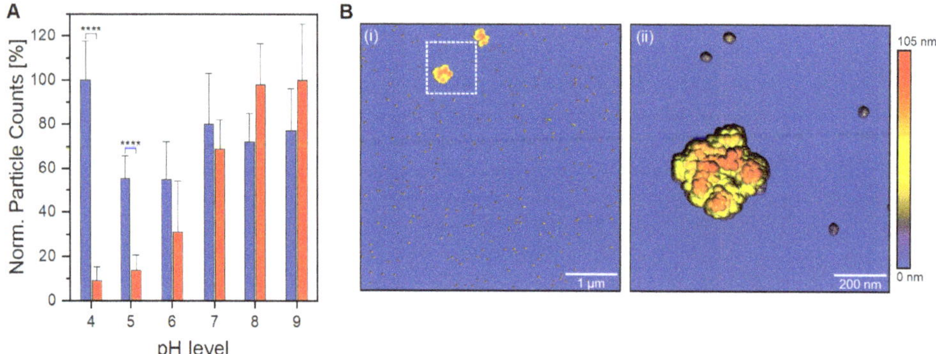

Figure 2. Behavior of VLP preparations at different pH levels. (**A**) Normalized particle counts detected via nES GEMMA for empty (blue columns) and filled (red columns) preparations at different pH levels ($n = 6$ measurements; average values and standard deviations are depicted; values at a corresponding pH are related to particle counts at pH 4 for empty VLPs and pH 9 for filled VLPs). For filled VLPs a decline in particle numbers with decreasing pH is observed in contrast to empty particles. (**B**) AFM analysis of filled VLPs at pH 5; (**i**) the analyzed area shows monomers and two aggregate clusters; (**ii**) magnification of the white dotted area in (**i**). **** $p < 0.0001$.

3.1. AAV8 VLPs in Aqueous Ammonium Acetate at Different pH Targeted by Gas-Phase Electrophoresis

Thanks to their distinct size differences, VLP preparations, either carrying or lacking cargo, can be easily discriminated ($p < 0.0001$, Figure 1B). In detail, filled VLPs have a constant higher EM diameter, and show a minimal downward trend towards a smaller EM diameter for alkaline pH. For empty VLPs instead, a sharper pH-dependent trend towards a smaller EM diameter is observed. Reasonably, pH-driven protein modification alters capsid's structural stiffness in both VLP preparations, but empty ones are again more affected.

In addition, below a physiological pH level, a significant contrast in the number of detected particles between filled and empty VLP preparation is noticed ($p < 0.0001$, Figure 2A). A high detection yield for empty VLPs confirms that nES GEMMA capabilities are not hindered in an acidic working environment; hence, the observed difference for VLP preparations is sample-dependent. The unique behavior demonstrated by filled VLPs is of great importance, both for VLP analysis and in the biopharmaceutical field during manufacturing steps. In vivo, viruses commonly hijack the cellular uptake mechanism to infect the cells. A number of them take advantage of the acidic environment of late endosomes and endolysosomes to shed their proteinaceous layer to deliver their naked viral load directly into the cytoplasm. Other viruses instead—AAV8 included—need to survive and accomplish endosomal escape intact in order to complete the infection [38,39]. Therefore, it is unlikely that bionanoparticles depletion is caused by viral uncoating. Moreover, capsid fragments would be detected at EM diameters below 25 nm; still, the nES GEMMA signals in this region are negligible and comparable to empty VLP preparations at comparable pH levels. Another viable option is aggregation, where filled VLPs at acidic pH generate clusters outside the operating range of nES GEMMA. To investigate this claim, filled VLPs in acidic conditions have been analyzed via AFM. In fact, VLP aggregates sized >200 nm are clearly visible in the AFM images (Figure 2B(i)). AFM images at higher magnification (Figure 2B(ii)) clearly show aggregate species stemming from monomeric particles. Longitudinal and medial sections of the imaged species match profiles fitting 25 nm monomeric particles (Supporting information, Figure S1). It is of note that particle aggregation yielding such large assemblies cannot be followed by the applied nES GEMMA setup. Furthermore, it has to be stressed that the resulting pH-dependant aggregation is partially reversible as adjusting the pH back to a neutral value restores the corresponding monomer peak in nES GEMMA measurements (Supporting information, Figure S2).

As demonstrated, the presence or absence of an encapsulated cargo significantly alters the capsid's structural characteristics. The novel behavior observed in filled VLPs additionally shows aggregation once exposed to acidic conditions. This pH-driven phenomenon is likely caused by a newly acquired feature linked to an increase in the capsid-to-capsid adherence with or without DNA mediation, and might be caused by fine capsid's structural changes. Although nES GEMMA can determine particles' diameter with both high accuracy and precision with high sample throughput times and good statistics, it can only partially infer the shape details of the sample. For this purpose, analytical techniques capable of imaging at near-atomic levels, such as cryo-TEM, are more fitted, although very time-consuming. Therefore, VLP preparations at both pH extremes (i.e., pH 4 and 9) have been analyzed with cryo-TEM technology to investigate if any distinct features would arise between samples (Figure 3).

3.2. Cryogenic Transmission Electron Microscopy (cryo-TEM) of AAV8 VLPs

Cryo-TEM RAW images have low signal-to-noise contrast between the investigated material and the background. Hence, averaging with software is necessary to increase contrast and highlight the sample's characteristics. In our work, we employed the Particle Estimation for Electron Tomography (PEET) open-source software to catalog, align, and average hundreds of particles per sample (Figure 3A). Originally designed as an averaging program for 3D tomograms, we used it for 2D averaging. To ensure a reliable, biased-free diameter determination, the averages generated by PEET were processed via a Python script written in-house (included in the Supporting information), where sections at different height were extracted. The data points included in these sections were later fitted with a circle function via the OriginPro (ver. 2021) software. This workflow, providing unbiased, reliable data for PEET-generated cryo-TEM images of VLPs, is summarized in Figure 4 and could be applied to images generated from different analytical techniques.

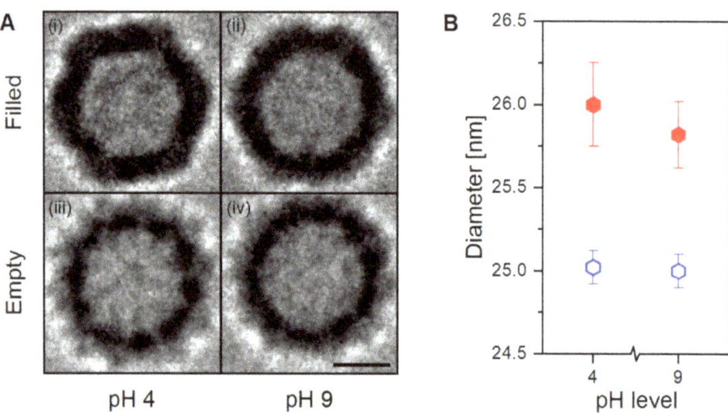

Figure 3. **Cryo-TEM analysis of empty and filled AAV8 VLPs.** (**A**) AAV8 VLPs filled at pH 4 (**i**) and pH 9 (**ii**), and empty at pH 4 (**iii**) and pH 9 (**iv**) after 2D averaging. (**B**) Particles' diameter determined via Python script from the aligned particles showed in (**A**). Black bar equals 10 nm.

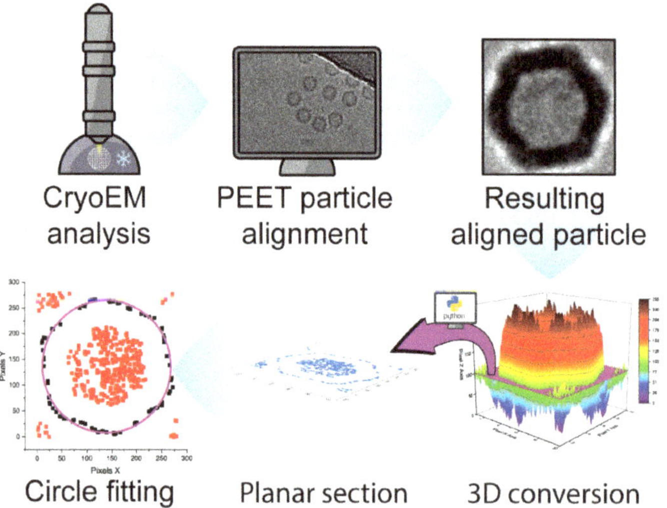

Figure 4. **Workflow scheme from cryo-TEM analysis to biased-free diameter determination.** In the last tile, the magenta circle represents the fitted function applied to the black data points. Red data points are instead excluded from the fitting process.

Visual comparison of the PEET-generated Cryo-TEM images show remarkable structural differences (Figure 3A). Within the filled VLP preparation, the sample at pH 4 (Figure 3A(i)) shows a well-defined icosahedral capsid's structure with regular, straight edges, and clear vertices; at pH 9 (Figure 3A(ii)) the same characteristics are instead less pronounced. Whereas in the empty VLP preparation, regardless of the pH value, the capsid's structure appears smoother, rounder, and slightly smaller (Figure 3A(iii,iv)). Anticipated by nES GEMMA results, and now corroborated by cryo-TEM images, the particle's shape is linked with the presence or absence of an encapsulated cargo, and this could affect the capsid's interactions at the macromolecular scale.

Diameter determination results in filled VLPs exhibiting larger diameters than empty ones, in agreement with nES GEMMA observations ($p < 0.001$, Figure 3B). Moreover, new information about the sample's characteristics can be inferred. In detail, empty VLPs report identical diameter values regardless of the solvent's pH, which is contrary to the downward trend observed with nES GEMMA. In hydrated conditions, regardless of the pH, empty VLPs maintain a constant size and shape. Filled VLPs instead show a small diameter variability between pH conditions, probably due to the sharper contours as reflected in the imaged particles (Figure 3A(i,ii)). Nonetheless, this slight visual and diameter difference does not provide any functional proof to explain the observed aggregation behavior.

4. Discussion

In this study, we demonstrated the feasibility of nES GEMMA to discriminate, detect, and characterize novel behaviors of AAV8-based VLP preparations analyzed with nES GEMMA-compatible electrolyte solution adjusted to different pHs. Moreover, to observe the VLP preparations in their hydrated native state, cryo-TEM analysis was performed.

Results from nES GEMMA of the two VLP preparations show peculiar, unique characteristics. Filled VLPs report limited EM diameter deviation across the different pHs, but detection efficiency is hindered in acidic conditions. Instead, the particle detection efficiency of empty VLPs is unperturbed by the electrolyte's pH, but pH-depended EM diameter variability is observed.

Results from PEET-based cryo-TEM images highlight the structural differences between the two VLP preparations. Moreover, the developed method for biased-free diameter determination via cryo-TEM confirms size-specific differences between empty and filled VLP preparations as observed with nES GEMMA. However, within VLP preparations, pH-dependent variation is observed, contrasting nES GEMMA results—specifically, a constant diameter for empty VLPs and a variable one for filled VLPs. One hypothesis to explain the discrepancy observed between nES GEMMA and cryo-TEM resides in the different working environments of these techniques. In nES GEMMA, the particles are sorted and classified in a solvent-free environment and are in a surface-dry state. Consequently, empty VLPs, lacking an encapsulated cargo, collapse due to structural stress caused by solvent evaporation. Likely, the solvent's pH alters protein conformation, which translates to different degrees of structural stability or stiffness. Therefore, labile structures are detected at smaller EM diameters. Instead, filled VLPs remain unaffected, thanks to the incompressible characteristics of their payload. On the contrary, in cryo-TEM, the VLPs are analyzed in a solubilized native state. Indeed, empty VLPs are not subjected to structural stress from solvent evaporation, and maintain their original form. Filled VLPs instead show a slight difference between the two pH extremes. The origin of this difference is as yet unclear. Putatively, and in addition to pH-induced changes at the protein level in acidic conditions, an osmotic pressure difference is generated at the semipermeable membrane level (i.e., proteinaceous capsid layer) caused by the negative charges of the genomic cargo. Hence, solvent intake is favored to equilibrate the osmotic pressure, causing the particle to swell.

In acidic conditions, nES GEMMA reveals a steep decrease in particle counts for filled VLPs, while empty ones are unaffected. Indeed, AFM analysis of filled VLPs in acidic conditions confirmed our hypothesis: the lower particle count was due to particle aggregation. Although nES GEMMA's working range of detection could be expanded, the heterogeneity of these clusters impedes their detection without an upstream enrichment technique, as observed in previous work [36]. Aggregation could be explained by the shape of the particles, highlighted by cryo-TEM-based PEET averaged images. The well-defined capsid faces of filled VLPs might provide the infrastructure for non-covalent capsid-to-capsid interaction, thus supporting aggregation. However, since no aggregate formation was observed in cryo-TEM averages, it suggests that the onset of aggregation requires further contributing factors. Once again, the critical factor might lie in the different working environments, dry in both nES GEMMA and AFM, and wet (i.e., plunge frozen in a solvent) for cryo-TEM. Putatively, if aggregation is promoted in a dry environment, it might be an evolutionary adaptation of AAV to increase viral infection transmission efficiency through airborne routes.

The results presented in this study demonstrate the feasibility and resilience of nES GEMMA as a method of analysis for AAV Serotype-8 VLP preparations across different pH levels. This is of great importance, especially in overseeing the different phases of a drug production process, where simple, reliable, and inexpensive control is often sought.

The results observed via AFM and cryo-TEM analysis provided complementary information to investigate the behavior observed with nES GEMMA when filled VLPs are exposed to acidic pH. AFM reveals the formation of heterogeneous aggregate products that could not be directly targeted with nES GEMMA but whose presence could be inferred by the significant drop in particle counts in the EM diameter range characteristic of VLPs in a monomeric state. Initially employed to investigate particle aggregation, cryo-TEM analysis did not confirm cluster formation observation. However, its extreme magnification power provided valuable information, and its application will propel future development toward the characterization of VLPs.

To conclude, the developed method and analytical approaches presented in this study provided new insight and possible applications for both the scientific and the biopharmaceutical field. The observations herein open several discussion points, encouraging further studies to investigate such valuable drug-delivery vectors.

Supplementary Materials: The following supporting information can be downloaded at: https://www.mdpi.com/article/10.3390/v15061361/s1, Figure S1: AFM analysis of filled VLPs at pH 5; Figure S2: nES GEMMA reversibility experiment of a filled VLP preparation; Python script to extract a planar section from the cryo-TEM 2D averaged images.

Author Contributions: Conceptualization, V.U.W.; Formal analysis, S.Z.; Investigation, S.Z., T.H. and G.F.; Resources, T.H., G.F., R.P. and M.M.-D.; Writing—original draft, S.Z. and V.U.W.; Writing—review & editing, S.Z., R.P., M.G. and V.U.W.; Supervision, M.M.-D. and V.U.W.; Project administration, M.M.-D. and V.U.W.; Funding acquisition, R.P., M.G. and M.M.-D. All authors have read and agreed to the published version of the manuscript.

Funding: This research was funded by Takeda and received financial support from TU Wien Bibliothek through its Open Access Funding programme. The Vienna BioCenter Core Facilities gratefully acknowledge funding from the Austrian Federal Ministry of Education, Science and Research and the city of Vienna.

Data Availability Statement: Data is available upon request.

Acknowledgments: We thank Takeda for providing the AAV8 samples and supervising this work, Thomas Heuser for the access to the cryo-TEM instrumentation and his support during the corresponding analysis, and Open Access Funding by TU Wien. We dedicate this manuscript to the memory of Günter Allmaier.

Conflicts of Interest: The authors declare no conflict of interest.

References

1. Nathwani, A.C. Gene therapy for hemophilia. *Hematol. Am. Soc. Hematol. Educ. Program.* **2022**, *2022*, 569–578. [CrossRef] [PubMed]
2. Mejia-Mendez, J.L.; Vazquez-Duhalt, R.; Hernandez, L.R.; Sanchez-Arreola, E.; Bach, H. Virus-like Particles: Fundamentals and Biomedical Applications. *Int. J. Mol. Sci.* **2022**, *23*, 8579. [CrossRef] [PubMed]
3. Sant'Anna, T.B.; Araujo, N.M. Adeno-associated virus infection and its impact in human health: An overview. *Virol. J.* **2022**, *19*, 173. [CrossRef] [PubMed]
4. Kaufman, S.L.; Skogen, J.W.; Dorman, F.D.; Zarrin, F.; Lewis, K.C. Macromolecule analysis based on electrophoretic mobility in air: Globular proteins. *Anal. Chem.* **1996**, *68*, 1895–1904. [CrossRef]
5. Wiedensohler, A.; Fissan, H.J. Aerosol charging in high purity gases. *J. Aerosol Sci.* **1988**, *19*, 867–870. [CrossRef]

6.	Kallinger, P.; Steiner, G.; Szymanski, W.W. Characterization of four different bipolar charging devices for nanoparticle charge conditioning. *J. Nanoparticle Res.* **2012**, *14*, 1–8. [CrossRef]

7.	Weiss, V.U.; Frank, J.; Piplits, K.; Szymanski, W.W.; Allmaier, G. Bipolar Corona Discharge-Based Charge Equilibration for Nano Electrospray Gas-Phase Electrophoretic Mobility Molecular Analysis of Bio- and Polymer Nanoparticles. *Anal. Chem.* **2020**, *92*, 8665–8669. [CrossRef]

8.	Allmaier, G.; Weiss, V.U.; Engel, N.Y.; Marchetti-Deschmann, M.; Szymanski, W.W. Soft X-ray Radiation Applied in the Analysis of Intact Viruses and Antibodies by Means of Nano Electrospray Differential Mobility Analysis. In *Molecular Technologies for Detection of Chemical and Biological Agents*; Banoub, J., Caprioli, R., Eds.; NATO Science for Peace and Security Series A: Chemistry and Biology; Springer: Dordrecht, The Netherlands, 2017; pp. 149–157. [CrossRef]

9.	Carazzone, C.; Raml, R.; Pergantis, S.A. Nanoelectrospray ion mobility spectrometry online with inductively coupled plasma-mass spectrometry for sizing large proteins, DNA, and nanoparticles. *Anal. Chem.* **2008**, *80*, 5812–5818. [CrossRef]

10.	Shah, V.B.; Orf, G.S.; Reisch, S.; Harrington, L.B.; Prado, M.; Blankenship, R.E.; Biswas, P. Characterization and deposition of various light-harvesting antenna complexes by electrospray atomization. *Anal. Bioanal. Chem.* **2012**, *404*, 2329–2338. [CrossRef]

11.	Weiss, V.U.; Bereszcazk, J.Z.; Havlik, M.; Kallinger, P.; Gosler, I.; Kumar, M.; Blaas, D.; Marchetti-Deschmann, M.; Heck, A.J.; Szymanski, W.W.; et al. Analysis of a common cold virus and its subviral particles by gas-phase electrophoretic mobility molecular analysis and native mass spectrometry. *Anal. Chem.* **2015**, *87*, 8709–8717. [CrossRef]

12.	Allmaier, G.; Blaas, D.; Bliem, C.; Dechat, T.; Fedosyuk, S.; Gosler, I.; Kowalski, H.; Weiss, V.U. Monolithic anion-exchange chromatography yields rhinovirus of high purity. *J. Virol. Methods* **2018**, *251*, 15–21. [CrossRef]

13.	Fernandez de la Mora, J. A singularly narrow 29 nm aerosol size standard based on the ΦX174 bacteriophage. *J. Aerosol Sci.* **2022**, *161*, 105949. [CrossRef]

14.	Guha, S.; Pease, L.F., 3rd; Brorson, K.A.; Tarlov, M.J.; Zachariah, M.R. Evaluation of electrospray differential mobility analysis for virus particle analysis: Potential applications for biomanufacturing. *J. Virol. Methods* **2011**, *178*, 201–208. [CrossRef]

15.	Bereszczak, J.Z.; Havlik, M.; Weiss, V.U.; Marchetti-Deschmann, M.; van Duijn, E.; Watts, N.R.; Wingfield, P.T.; Allmaier, G.; Steven, A.C.; Heck, A.J. Sizing up large protein complexes by electrospray ionisation-based electrophoretic mobility and native mass spectrometry: Morphology selective binding of Fabs to hepatitis B virus capsids. *Anal. Bioanal. Chem.* **2014**, *406*, 1437–1446. [CrossRef]

16.	Havlik, M.; Marchetti-Deschmann, M.; Friedbacher, G.; Messner, P.; Winkler, W.; Perez-Burgos, L.; Tauer, C.; Allmaier, G. Development of a bio-analytical strategy for characterization of vaccine particles combining SEC and nanoES GEMMA. *Analyst* **2014**, *139*, 1412–1419. [CrossRef]

17.	Weiss, V.U.; Pogan, R.; Zoratto, S.; Bond, K.M.; Boulanger, P.; Jarrold, M.F.; Lyktey, N.; Pahl, D.; Puffler, N.; Schelhaas, M.; et al. Virus-like particle size and molecular weight/mass determination applying gas-phase electrophoresis (native nES GEMMA). *Anal. Bioanal. Chem.* **2019**, *411*, 5951–5962. [CrossRef]

18.	Epstein, H.; Afergan, E.; Moise, T.; Richter, Y.; Rudich, Y.; Golomb, G. Number-concentration of nanoparticles in liposomal and polymeric multiparticulate preparations: Empirical and calculation methods. *Biomaterials* **2006**, *27*, 651–659. [CrossRef]

19.	Chattopadhyay, S.; Modesto-Lopez, L.B.; Venkataraman, C.; Biswas, P. Size Distribution and Morphology of Liposome Aerosols Generated By Two Methodologies. *Aerosol Sci. Technol.* **2010**, *44*, 972–982. [CrossRef]

20.	Allmaier, G.; Laschober, C.; Szymanski, W.W. Nano ES GEMMA and PDMA, new tools for the analysis of nanobioparticles-protein complexes, lipoparticles, and viruses. *J. Am. Soc. Mass. Spectrom.* **2008**, *19*, 1062–1068. [CrossRef]

21.	Weiss, V.U.; Urey, C.; Gondikas, A.; Golesne, M.; Friedbacher, G.; von der Kammer, F.; Hofmann, T.; Andersson, R.; Marko-Varga, G.; Marchetti-Deschmann, M.; et al. Nano electrospray gas-phase electrophoretic mobility molecular analysis (nES GEMMA) of liposomes: Applicability of the technique for nano vesicle batch control. *Analyst* **2016**, *141*, 6042–6050. [CrossRef]

22.	Weiss, V.U.; Wieland, K.; Schwaighofer, A.; Lendl, B.; Allmaier, G. Native Nano-electrospray Differential Mobility Analyzer (nES GEMMA) Enables Size Selection of Liposomal Nanocarriers Combined with Subsequent Direct Spectroscopic Analysis. *Anal. Chem.* **2019**, *91*, 3860–3868. [CrossRef] [PubMed]

23.	Chernyshev, V.S.; Rachamadugu, R.; Tseng, Y.H.; Belnap, D.M.; Jia, Y.; Branch, K.J.; Butterfield, A.E.; Pease, L.F., 3rd; Bernard, P.S.; Skliar, M. Size and shape characterization of hydrated and desiccated exosomes. *Anal. Bioanal. Chem.* **2015**, *407*, 3285–3301. [CrossRef] [PubMed]

24.	Steinberger, S.; Karuthedom George, S.; Laukova, L.; Weiss, R.; Tripisciano, C.; Birner-Gruenberger, R.; Weber, V.; Allmaier, G.; Weiss, V.U. A possible role of gas-phase electrophoretic mobility molecular analysis (nES GEMMA) in extracellular vesicle research. *Anal. Bioanal. Chem.* **2021**, *413*, 7341–7352. [CrossRef] [PubMed]

25.	Steinberger, S.; Karuthedom George, S.; Laukova, L.; Weiss, R.; Tripisciano, C.; Marchetti-Deschmann, M.; Weber, V.; Allmaier, G.; Weiss, V.U. Targeting the Structural Integrity of Extracellular Vesicles via Nano Electrospray Gas-Phase Electrophoretic Mobility Molecular Analysis (nES GEMMA). *Membranes* **2022**, *12*, 872. [CrossRef]

26.	Hinterwirth, H.; Wiedmer, S.K.; Moilanen, M.; Lehner, A.; Allmaier, G.; Waitz, T.; Lindner, W.; Lammerhofer, M. Comparative method evaluation for size and size-distribution analysis of gold nanoparticles. *J. Sep. Sci.* **2013**, *36*, 2952–2961. [CrossRef]

27.	Weiss, V.U.; Lehner, A.; Kerul, L.; Grombe, R.; Kratzmeier, M.; Marchetti-Deschmann, M.; Allmaier, G. Characterization of cross-linked gelatin nanoparticles by electrophoretic techniques in the liquid and the gas phase. *Electrophoresis* **2013**, *34*, 3267–3276. [CrossRef]

28. Elzey, S.; Tsai, D.H.; Yu, L.L.; Winchester, M.R.; Kelley, M.E.; Hackley, V.A. Real-time size discrimination and elemental analysis of gold nanoparticles using ES-DMA coupled to ICP-MS. *Anal. Bioanal. Chem.* **2013**, *405*, 2279–2288. [CrossRef]

29. Konermann, L. Addressing a Common Misconception: Ammonium Acetate as Neutral pH "Buffer" for Native Electrospray Mass Spectrometry. *J. Am. Soc. Mass. Spectrom.* **2017**, *28*, 1827–1835. [CrossRef]

30. Tycova, A.; Prikryl, J.; Foret, F. Reproducible preparation of nanospray tips for capillary electrophoresis coupled to mass spectrometry using 3D printed grinding device. *Electrophoresis* **2016**, *37*, 924–930. [CrossRef]

31. Mastronarde, D.N. Automated electron microscope tomography using robust prediction of specimen movements. *J. Struct. Biol.* **2005**, *152*, 36–51. [CrossRef]

32. Kremer, J.R.; Mastronarde, D.N.; McIntosh, J.R. Computer visualization of three-dimensional image data using IMOD. *J. Struct. Biol.* **1996**, *116*, 71–76. [CrossRef]

33. Nicastro, D.; Schwartz, C.; Pierson, J.; Gaudette, R.; Porter, M.E.; McIntosh, J.R. The molecular architecture of axonemes revealed by cryoelectron tomography. *Science* **2006**, *313*, 944–948. [CrossRef]

34. Heumann, J.M.; Hoenger, A.; Mastronarde, D.N. Clustering and variance maps for cryo-electron tomography using wedge-masked differences. *J. Struct. Biol.* **2011**, *175*, 288–299. [CrossRef]

35. Zoratto, S.; Weiss, V.U.; van der Horst, J.; Commandeur, J.; Buengener, C.; Foettinger-Vacha, A.; Pletzenauer, R.; Graninger, M.; Allmaier, G. Molecular weight determination of adeno-associate virus serotype 8 virus-like particle either carrying or lacking genome via native nES gas-phase electrophoretic molecular mobility analysis and nESI QRTOF mass spectrometry. *J. Mass. Spectrom.* **2021**, *56*, e4786. [CrossRef]

36. Zoratto, S.; Weiss, V.U.; Friedbacher, G.; Buengener, C.; Pletzenauer, R.; Foettinger-Vacha, A.; Graninger, M.; Allmaier, G. Adeno-associated Virus Virus-like Particle Characterization via Orthogonal Methods: Nanoelectrospray Differential Mobility Analysis, Asymmetric Flow Field-Flow Fractionation, and Atomic Force Microscopy. *Acs Omega* **2021**, *6*, 16428–16437. [CrossRef]

37. Weiss, V.U.; Kerul, L.; Kallinger, P.; Szymanski, W.W.; Marchetti-Deschmann, M.; Allmaier, G. Liquid phase separation of proteins based on electrophoretic effects in an electrospray setup during sample introduction into a gas-phase electrophoretic mobility molecular analyzer (CE-GEMMA/CE-ES-DMA). *Anal. Chim. Acta* **2014**, *841*, 91–98. [CrossRef]

38. Wang, D.; Tai, P.W.L.; Gao, G. Adeno-associated virus vector as a platform for gene therapy delivery. *Nat. Rev. Drug. Discov.* **2019**, *18*, 358–378. [CrossRef]

39. Dhungel, B.P.; Bailey, C.G.; Rasko, J.E.J. Journey to the Center of the Cell: Tracing the Path of AAV Transduction. *Trends Mol. Med.* **2021**, *27*, 172–184. [CrossRef]

Article

Modulation of Reoviral Cytolysis (I): Combination Therapeutics

Yoshinori Mori [1,2], Sandra G. Nishikawa [1], Andreea R. Fratiloiu [1], Mio Tsutsui [1], Hiromi Kataoka [2], Takashi Joh [2] and Randal N. Johnston [1,*]

[1] Cumming School of Medicine, University of Calgary, Calgary, AB T2N 1N4, Canada; ysnmori@yahoo.co.jp (Y.M.); snishika@ucalgary.ca (S.G.N.); arfratil@ucalgary.ca (A.R.F.)

[2] Graduate School of Medical Sciences, Nagoya City University, Nagoya 467-8601, Japan

* Correspondence: rnjohnst@ucalgary.ca

Abstract: Patients with stage IV gastric cancer suffer from dismal outcomes, a challenge especially in many Asian populations and for which new therapeutic options are needed. To explore this issue, we used oncolytic reovirus in combination with currently used chemotherapeutic drugs (irinotecan, paclitaxel, and docetaxel) for the treatment of gastric and other gastrointestinal cancer cells in vitro and in a mouse model. Cell viability in vitro was quantified by WST-1 assays in human cancer cell lines treated with reovirus and/or chemotherapeutic agents. The expression of reovirus protein and caspase activity was determined by flow cytometry. For in vivo studies, athymic mice received intratumoral injections of reovirus in combination with irinotecan or paclitaxel, after which tumor size was monitored. In contrast to expectations, we found that reoviral oncolysis was only poorly correlated with Ras pathway activation. Even so, the combination of reovirus with chemotherapeutic agents showed synergistic cytopathic effects in vitro, plus enhanced reovirus replication and apoptosis. In vivo experiments showed that reovirus alone can reduce tumor size and that the combination of reovirus with chemotherapeutic agents enhances this effect. Thus, we find that oncolytic reovirus therapy is effective against gastric cancer. Moreover, the combination of reovirus and chemotherapeutic agents synergistically enhanced cytotoxicity in human gastric cancer cell lines in vitro and in vivo. Our data support the use of reovirus in combination with chemotherapy in further clinical trials, and highlight the need for better biomarkers for reoviral oncolytic responsiveness.

Keywords: reovirus; oncolytic virus; oncolytic viral therapy; Ras pathway; chemotherapy

Citation: Mori, Y.; Nishikawa, S.G.; Fratiloiu, A.R.; Tsutsui, M.; Kataoka, H.; Joh, T.; Johnston, R.N. Modulation of Reoviral Cytolysis (I): Combination Therapeutics. *Viruses* **2023**, *15*, 1472. https://doi.org/10.3390/v15071472

Academic Editors: Ottmar Herchenröder and Brigitte Pützer

Received: 5 June 2023
Revised: 17 June 2023
Accepted: 20 June 2023
Published: 29 June 2023

1. Introduction

Nearly one million cases of gastric cancer are diagnosed worldwide each year, with the highest incidence occurring in eastern Asia, parts of South America and eastern Europe. Gastric cancer is the second most common cause of cancer-related death worldwide, with around 700,000 deaths a year. In the US, more than 20,000 new cases of gastric cancer and 10,000 deaths occur annually [1]. Survival of patients with gastric cancer is substantially worse than that of patients with most other solid malignancies. The only current treatment that offers potential cure is complete resection of the tumor [2]. However, after complete resection alone, patients who are shown to have extensive lymph node involvement on surgical pathology specimens have a 5-year survival of only 7% [3]. Recently, the use of chemotherapy such as cisplatin, irinotecan, paclitaxel, or docetaxel for gastric cancer has yielded some improvements in outcome, but the beneficial effects are still modest and there is a real need for new therapeutic strategies [4]. It is possible that oncolytic virus therapy may be useful in this context, and perhaps especially so for viruses that proliferate in the gastrointestinal tract.

Human reovirus is a ubiquitous, non-enveloped virus containing 10 segments of double-stranded RNA (dsRNA) as its genome, with infections that are generally mild, restricted to the upper respiratory and gastrointestinal tracts, and often asymptomatic [5]. Reovirus has innate oncolytic potential in a wide range of murine and human tumor cells, and this is at least partly dependent on the transformed state of the cell [6,7]. The precise mechanism of reoviral tropism and selective oncolysis in malignant cells is yet to be fully determined. In normal cells, the presence of an intact double-stranded RNA-activated protein kinase system limits the establishment of a productive reoviral infection. In malignant cells with an activated Ras pathway, it has been suggested that either directly through Ras mutation or indirectly via upregulation of epidermal growth factor receptor (EGFR) signaling or of other signaling components [5,8–11], this cellular antiviral response mechanism may be perturbed, viral replication enhanced, and subsequent lysis of the host cancer cell facilitated. The modulation of other pathways that regulate viral attachment, penetration, uncoating, assembly, and propagation further influence the efficiency of viral oncolysis (see below).

Thus, reovirus is considered a promising candidate for oncolytic virus therapy and Phase II and III clinical studies are underway (oncolyticsbiotech.com (accessed on 5 June 2023)) for multiple tumor categories, and an orphan drug designation has recently been approved by the FDA for the use of reovirus in the treatment of gastric and certain other cancers [12]. While monotherapy with oncolytic reovirus has been explored, most current efforts use reovirus or other oncolytic viruses such as herpes virus in combination with chemotherapy or radiotherapy in preclinical or clinical studies to potentially increase treatment efficacy for various malignancies [13–18]. For example, several studies have suggested that reovirus, in combination with paclitaxel against lung cancer and with cisplatin against melanoma, may have synergistic antitumoral effects [19,20]. However, so far there have been few reports of oncolytic reovirus and combination therapy against gastric cancer in preclinical animal models [21]. Here we show that oncolytic reovirus therapy is effective against gastric cancer xenografts in a mouse model, and that the combination of reovirus and chemotherapeutic agents paclitaxel and irinotecan has clear synergistic benefits.

2. Materials and Methods

2.1. Cells and Virus

Human gastric cancer cell lines (KATOIII, SNU16, AGS, NCI-N87, Hs746T) and human colon cancer cell lines (HCT116, HT-29) were obtained from the American Type Culture Collection (ATCC; Rockville, MD, USA). Other human gastric cancer cell lines (MKN1, MKN7, MKN45, MKN74, HGC-27, GCIY) were obtained from the Cell Bank, RIKEN BioResource Center (Kyoto, Japan). Another human gastric cancer cell line (FU97) was obtained from the Health Science Research Resources Bank (Japan), and ISt-1 was a gift from Dr. Masanori Terashima. All cell lines were tested and free of mycoplasma contamination. The Dearing strain of reovirus serotype 3 (a gift from P. Lee, Dalhousie University, Halifax, NS, Canada) was propagated in suspension cultures of L929 cells (from ATCC) and purified according to previously established methods [8,9] with the exception that β-mercaptoethanol was omitted from the extraction buffer. Viral titers were also established using L929 cells [9].

2.2. Chemotherapeutic Agents

Irinotecan (Mayne Pharma, Raleigh, NC, USA), Paclitaxel (Hospita, Canada), and Docetaxel (Taxotere®; Sanofi Aventis, Bridgewater, NJ, USA) were kindly provided by Dr. Aru Narendran, University of Calgary. These agents were diluted with the respective medium just before use for in vitro studies and with phosphate-buffered saline (PBS) for in vivo studies.

2.3. Cell Viability Assay

All cells were seeded in 96-well plates at a density of 2×10^3 cells/well with appropriate medium. GC cells were mock infected or infected with reovirus at a MOI of 1 or 10 and then treated with chemotherapeutic agents. Experiments were repeated three times and results presented as mean +/− standard deviation. Numbers of viable cells were evaluated by a colorimetric WST-1 assay at 3, 6, and 9 days post treatment. WST-1 (Roche), a tetrazolium salt, is cleaved to a colored formazan product by enzymes in metabolically active cells, and the reaction is quantitated with an automatic plate reader at 450 nm. The potential synergistic effect arising from the combination of reovirus with chemotherapy on cell proliferation was assessed by calculating combination index (CI) values using the method of Zhao et al. [22]. The CI provides a quantitative measure of the degree of interaction between two or more agents. A CI of 1 denotes an additive interaction, >1 represents antagonism, and <1 indicates synergy, with lower values indicating a higher degree of augmentation of effect of the two agents working together.

2.4. Ras Activation Assay

First, 85–90% confluent cells grown in 150 mm dishes were lysed with $1 \times Mg^{2+}$ lysis buffer (Ras activation assay kit; Millipore). To determine the level of activated Ras (Ras-GTP) in these cells, 1 mg of cell lysate was incubated with 10μL of Raf-1 Ras binding domain agarose conjugate at 4 °C for 45 min. The beads were then collected, washed, resuspended in 2× Laemmli buffer, and boiled for 5 min. This was then followed by SDS-PAGE and Western blotting with an anti-Ras antibody (clone RAS 10) according to the manufacturer's instructions. To determine the level of total Ras, cell lysates were directly subjected to SDS-PAGE and Western blotting with anti-Ras antibody. The membrane was incubated with horseradish peroxidase-conjugated goat antimouse antibody, and specific bands were detected with an ECL system (GE Healthcare).

2.5. FACS Analysis

After treatment with reovirus and/or chemotherapeutic agents, general caspase activity was assessed by the carboxyfluorescein caspase detection kit (Apologix; cat. no. FAM100-2; Cell Technology, Inc., Newport News, VI, USA), which is based on carboxyfluorescein-labeled fluoromethyl ketone (FMK)—peptide inhibitors of caspases. These inhibitors are cell permeable and noncytotoxic. Once inside the cell, the fluorescent inhibitor binds covalently to the active caspase. Primary rabbit antireovirus polyclonal antibody was made in our lab and detected by binding to PE goat anti-rabbit IgG (Cedarlane Laboratories Ltd., Burlington, ON, Canada). Fixed and permeabilized cells were analyzed by flow cytometry.

2.6. Subcutaneous Tumor Xenograft Model in Nude Mice

Six-week-old male CD-1 nude mice, purchased from Charles River, were kept under pathogen-free conditions according to a protocol approved by the University of Calgary Animal Care Committee. MKN45 cells (2×10^6) were implanted subcutaneously in the left flanks of mice under anesthesia. When the tumors reached a diameter of ~5 mm, the mice were randomly divided into four groups (5 mice/group), and a 50 μL solution containing reovirus (1×10^8 PFU/animal) or PBS was injected into the tumor (any excess injected fluid was distributed in surrounding tissues). Simultaneously, each mouse received an intraperitoneal injection of 100 μL paclitaxel at a dose of 10 mg/kg or irinotecan at a dose of 5 mg/kg. The tumor size was calculated by external caliper measurements every 2 or 3 days. Tumor volume was calculated using the following formula: tumor volume $(mm^3) = a \times b^2 \times 0.5$, where a is the longest diameter, b is the shortest diameter, and 0.5 is a constant to calculate the volume of an ellipsoid. Statistical differences among groups were assessed using the Mann–Whitney U test.

2.7. Immunodetection of Reoviral Replication

For histological analysis, tumors were fixed in 10% neutral buffered formalin, embedded in paraffin, and sectioned. Sections were then immersed in xylene, followed by rehydration in decreasing concentrations of ethanol. Endogenous peroxidase was inactivated in 3% hydrogen peroxide in PBS for 15 min. Sections were then incubated in primary rabbit antireovirus polyclonal antibody (1:1000 in PBS with 10% goat serum and 0.3% Triton

X-100) partially purified by ammonium sulfate precipitation. Slides were washed in PBS and then subjected to avidin-biotin horseradish peroxidase staining as recommended by the manufacturer (Vector, Burlington, ON, Canada) and counterstained in hematoxylin.

3. Results

3.1. Reovirus Cytotoxicity in Gastric Cancer Cell Lines

We first surveyed the cytotoxicity of reovirus against gastric cancer using the WST-1 assay in 14 different human gastric cancer cell lines. After 72 h exposure, reovirus alone showed moderate cytopathic effects (relative cell viability was between 0.2 and 0.8) in six gastric cancer cell lines (AGS, MKN1, NCI-N87, Hs746T, FU97, ISt-1) and low cytopathic effects (relative cell viability was more than 0.8) in seven gastric cancer cell lines (HGC-27, KATOIII, MKN7, MKN45, MKN74, NUGC4, SNU16). Only GCIY cells showed a high cytopathic effect (relative viability was less than 0.2; Figure 1a).

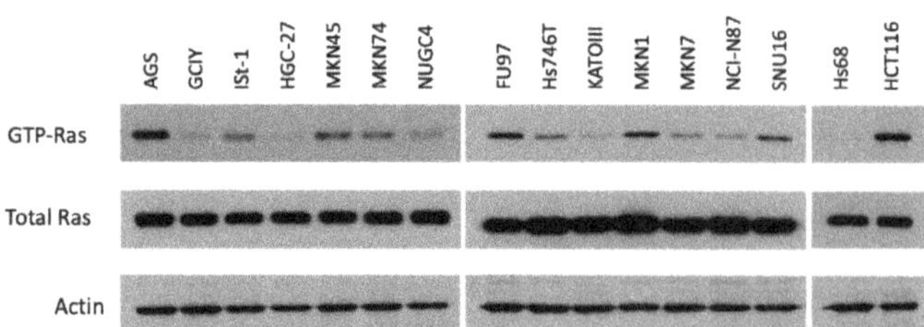

Figure 1. Response to reovirus is poorly correlated with levels of active Ras. (**a**) Gastric and colon cancer cell lines infected at a MOI of 10 were evaluated for viability by WST-1 assay at 72 h post infection. Bars, standard deviation (SD). (**b**) Ras activity in gastric cancer cell lines. The activated GTP-bound form of Ras was detected in most gastric cancer cell lines. Hs68 (normal human fibroblast cells) and HCT116 (colon cancer cells) were negative and positive controls, respectively. The levels of Ras activation correlated only poorly with cytolytic responses to reovirus infection.

To determine whether the variable cellular responses to reovirus might be explained by differing levels of Ras activation, we then measured the levels of GTP-Ras in the various gastric cancer cell lines. Activated Ras was detectable in most gastric cancer cell lines (Figure 1b) when compared with a negative control of normal human fibroblasts. However, we observed no obvious correlation between Ras activity levels and cytolytic effects—some gastric cancer cell lines with prominent Ras activation were relatively resistant to reovirus (e.g., AGS), whereas the most responsive cell line (GCIY) displayed only very modest activation of GTP-Ras. While variable cytolytic responses to reovirus may reflect multiple cellular features (abundance of receptor, efficiency of viral uncoating, etc.), it is clear that in these gastric cancer cell lines, variables beyond simple Ras pathway activation must be modulating cellular responses to reovirus infection.

3.2. Reovirus Cytotoxicity with Chemotherapeutic Agents in Gastric Cancer Cell Lines

We then chose four different gastric cancer cell lines (GCIY, AGS, NCI-N87, and MKN45, showing strong, medium, or minimal responses to reovirus) to examine cytotoxicity in more detail. We evaluated cell viability with combinations of reovirus and chemotherapeutic agents using WST-1 assays at days 3, 6, and 9 after treatment. We chose irinotecan, paclitaxel, and docetaxel as combination chemotherapeutic agents, as these are already commonly used in treatments of human patients. Although the various cell lines showed modest differences in responses to the chemotherapeutic drugs alone, the various agents showed clear enhancement of cell killing when supplemented with reovirus in MKN45 and AGS cells (Figure 2a,b).

Reovirus alone killed GCIY very well, so we could not evaluate synergy with chemotherapy in these cells, and NCI-N87 cells showed no enhancement in combination experiments (Supplementary Figure S1a,b). All Combination Indices for MKN45 and AGS cells were less than 1, which therefore showed synergy of reovirus with chemotherapy (Figure 2d); enhanced killing was also revealed in photomicrographs of these two cell populations (Figure 2c).

3.3. Combinations of Reovirus and Chemotherapeutic Agents Enhanced Reovirus Replication and Apoptosis

We then tested whether the administration of chemotherapeutic agents might enhance or diminish viral activity, while promoting cell death. For this purpose, we evaluated reoviral protein expression and caspase activity using FACS analysis of MKN45 and AGS cells treated in eight groups (control, reovirus alone, irinotecan alone, paclitaxel alone, docetaxel alone, reovirus and irinotecan, reovirus and paclitaxel, and reovirus and docetaxel). Nearly every combination of reovirus and chemotherapeutic agents enhanced reoviral protein synthesis in these cells, which as we have shown previously leads to the elevated release of infectious viral particles [11]. Reovirus or chemotherapy alone enhanced caspase activity to some extent, and this was enhanced further when reovirus and chemotherapy were combined (Figure 3).

In GCIY and NCI-N87 cells, reovirus alone enhanced caspase activity in both cell types (Supplementary Figure S2b). These results show that the combination of reovirus and chemotherapeutic agents can enhance reovirus protein synthesis in some cell lines and induces effects leading to cell death via apoptosis or possibly pyroptosis.

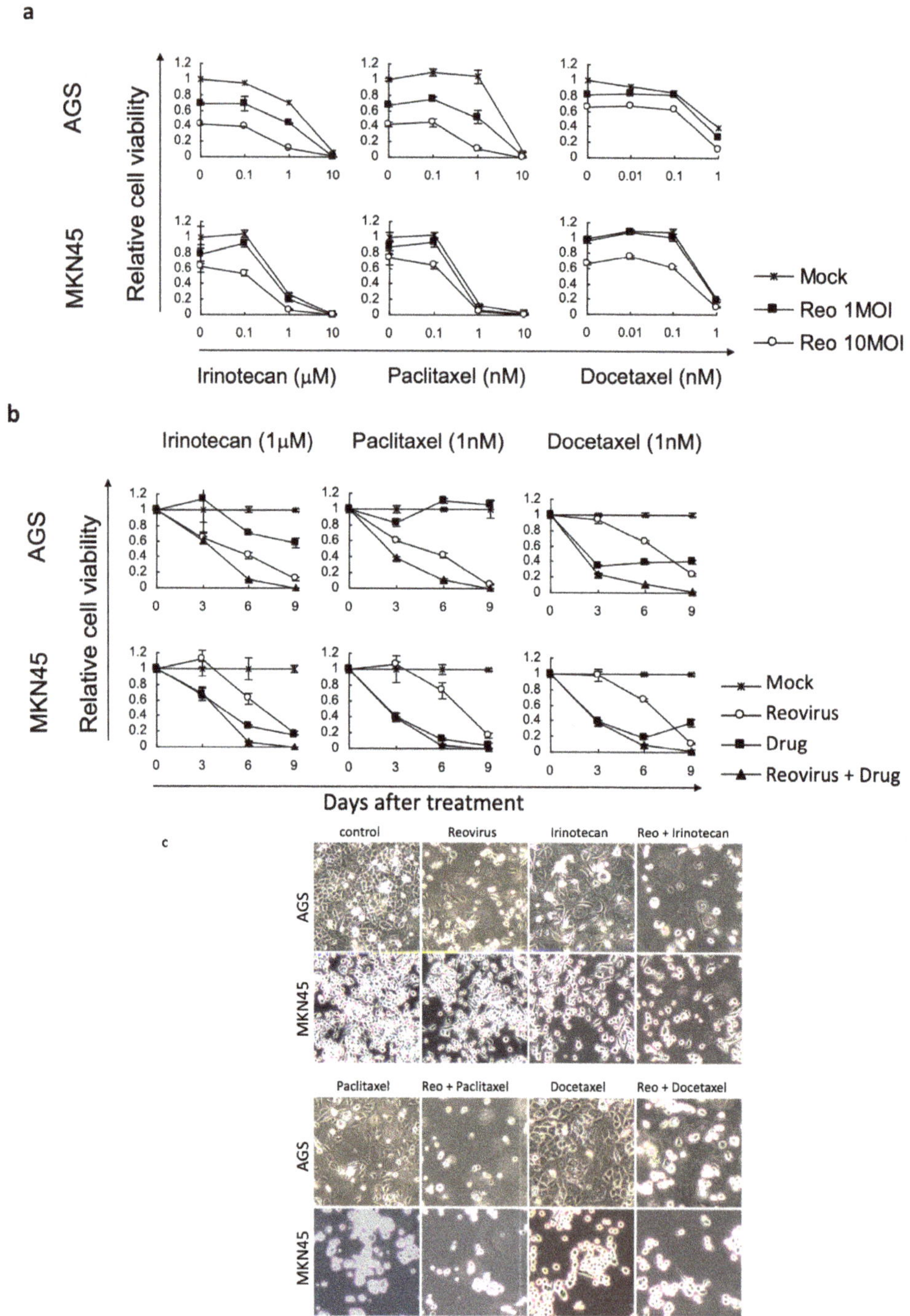

Figure 2. *Cont.*

d

Combination Index (CI)

	Irinotecan	Paclitaxel	Docetaxel
AGS	0.370	0.317	0.916
MKN45	0.669	0.851	0.809

Figure 2. Combination effects of reovirus and chemotherapeutic agents (irinotecan, paclitaxel, and docetaxel) on human gastric cancer cell lines (MKN45, AGS, GCIY, NCI-N87). (**a**) Cells were infected with 1 or 10 MOI of reovirus and exposed to chemotherapeutic agents at the indicated concentrations. Cell viability was assessed by WST-1 assay at 6 days after treatment. Experiments were repeated at least three times and results presented as means $+/-$ standard deviation (SD). (**b**) Time course of the combined effect of reovirus plus chemotherapeutic agents on gastric cancer cell lines. Cells were treated with 10 MOI of reovirus, chemotherapeutic agent (1 µM irinotecan, 1 nM paclitaxel, 1 nM docetaxel), or a combination of both, and cell killing efficacy was evaluated by WST-1 assay over 9 days. (**c**) Cytopathic effects of reovirus with chemotherapeutic agents. MKN45 and AGS were treated with reovirus, chemotherapeutic agents, or both (1 µM irinotecan, 1 nM paclitaxel, 1 nM docetaxel) according to the schedule described above, and photographed 5 days after treatment. $\times 100$ magnification. (**d**) Combination indices (CI) were calculated [22] for each combination after 6 days of treatment, when differences in effect were maximal. CI values are the means of three experiments, with levels below 0.9 indicating substantial synergy, whereas a value of 1 denotes an additive effect and values above 1 indicate antagonism between the agents.

3.4. Combined Reovirus and Chemotherapeutic Agents Enhanced Anti-Tumor Effects in a Murine Gastric Cancer Xenograft Model

We then assessed the therapeutic efficacy of reovirus in combination with chemotherapeutic agents against gastric cancer cells in vivo. We made two types of gastric cancer xenograft models, with CD-1 nude mice bearing either MKN45- or GCIY-based tumors. Then, we treated the former MKN45 model with combination therapy in four groups (control, reovirus alone, chemotherapeutic agent (irinotecan or paclitaxel) alone, and reovirus plus chemotherapeutic agent); the latter model was treated with monotherapy in two groups (control and reovirus alone). Administration of reovirus, irinotecan, or paclitaxel results in significant tumor growth suppression compared with the untreated control at 28 days after initiation of treatment. Importantly, the combination of reovirus plus irinotecan or reovirus plus paclitaxel produced a more profound inhibition of tumor growth compared with mice treated with either modality alone and control (Figure 4A–D).

In the GCIY model, reovirus monotherapy was effective and sufficient when compared with the untreated control (Supplementary Figure S3), similar to the effect observed in vitro. Extensive viral distribution in the MKN45 tumors was confirmed by immunohistochemical staining of reovirus protein (Figure 4E). There were no significant differences in the mean body weights among experimental groups, and no morbidity was attributable to therapy with reovirus, paclitaxel, irinotecan, or both in combination.

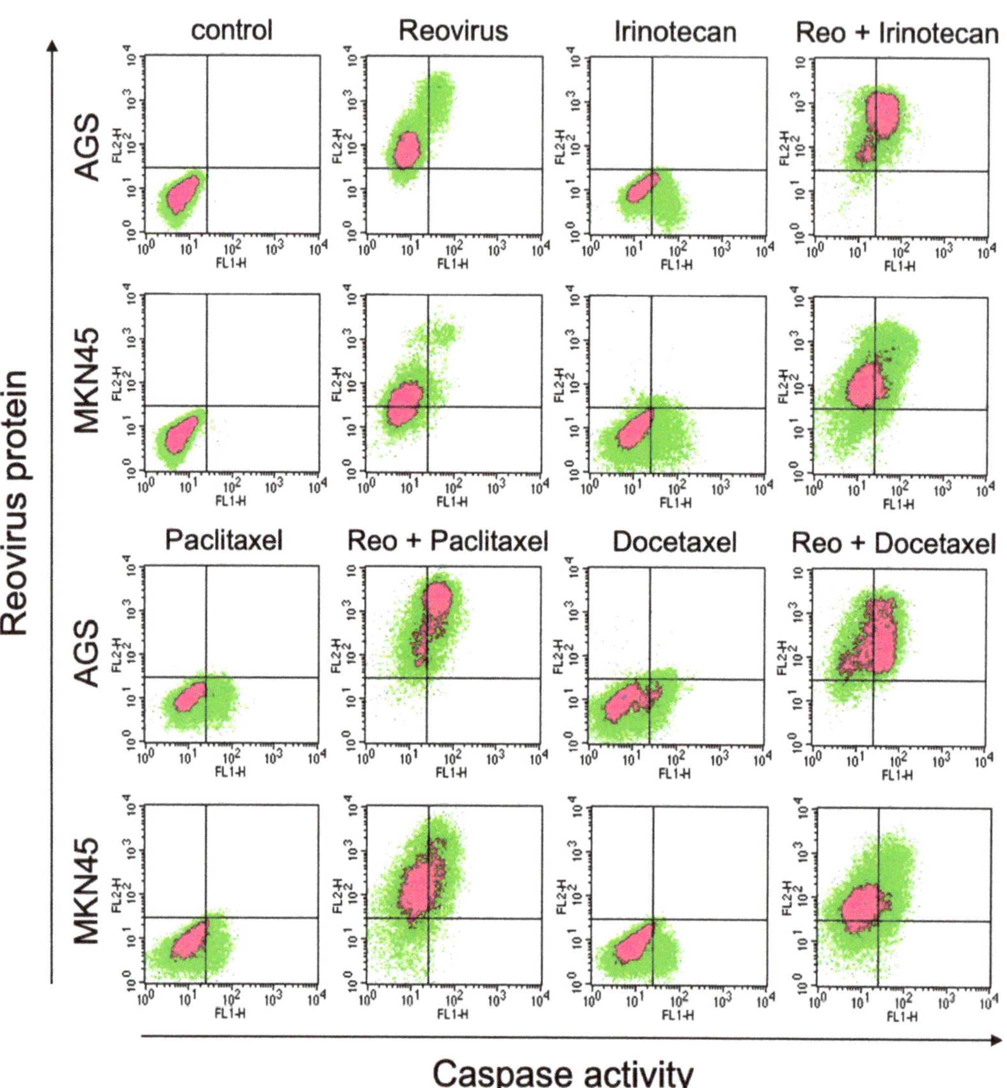

Figure 3. Caspase activation and reovirus protein expression. These were determined by flow cytometry in the gastric cancer cell lines MKN45 and AGS. Cells were treated with 10 MOI of reovirus, chemotherapeutic agent (1 μM irinotecan, 1 nM paclitaxel, 1 nM docetaxel), or a combination of both, and caspase activation and reovirus protein expression were evaluated by flow cytometry at 5 days post treatment. Three experiments under these conditions were performed with similar results and representative images are shown.

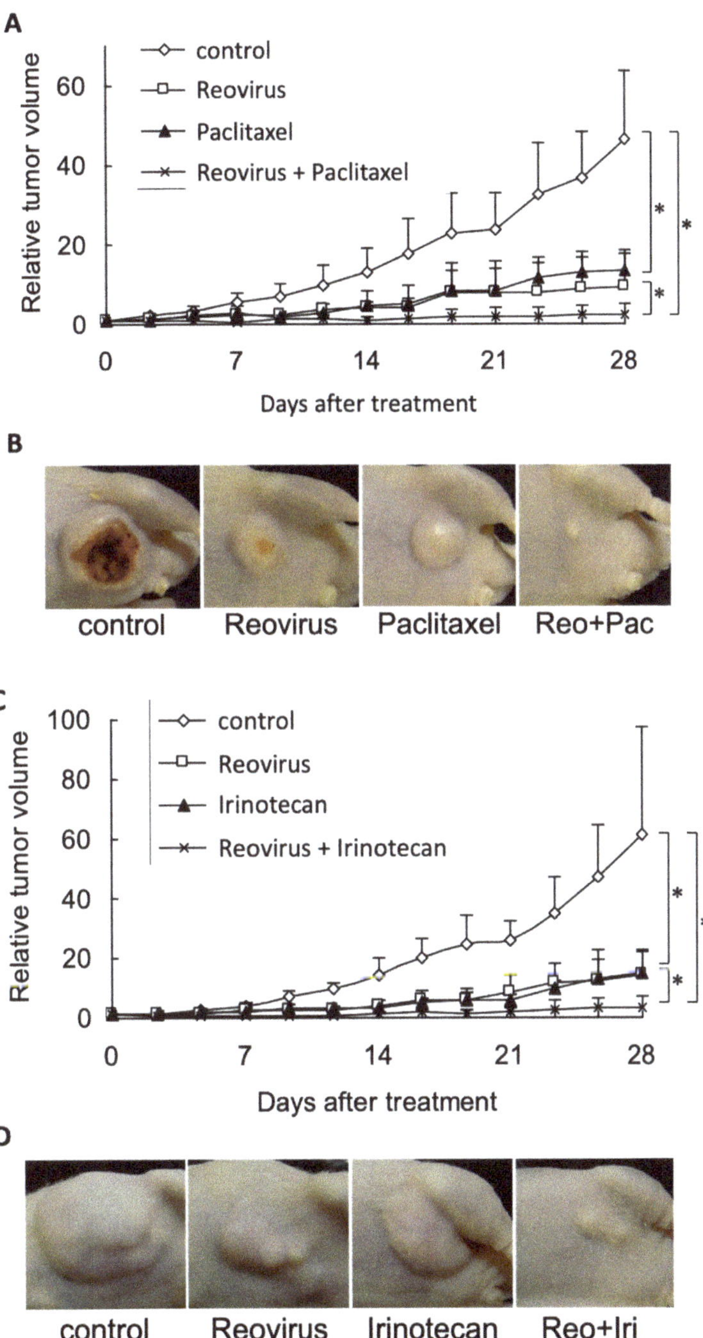

Figure 4. *Cont.*

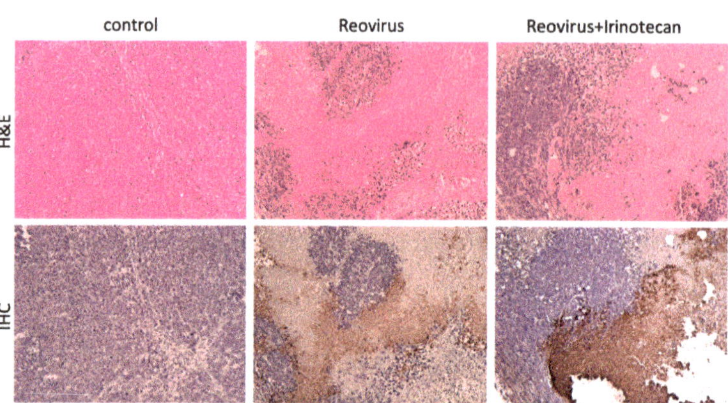

Figure 4. Antitumor effects of intratumorally injected reovirus and intraperitoneally administered irinotecan or paclitaxel against established flank MKN45 xenograft tumors in nu/nu mice. MKN45 cells (2×10^6 cells/each) were subcutaneously injected into the left flanks of mice. Reovirus (1×10^8 pfu)/body) and (**A,B**), irinotecan (5 mg/kg) or (**C,D**), paclitaxel (10 mg/kg) were administered intratumorally and intraperitoneally, respectively. Five mice were used for each group. Tumors were measured every 2 or 3 days, and tumor volume was expressed as tumor volume relative to volume at commencement of treatment. Statistical significance was defined as $p < 0.05$ (*) (Mann–Whitney *U* test). (**B,D**), representative tumors were photographed 28 days after treatment. (**E**) Representative histologic analysis in MKN45 xenografts treated with/without reovirus or irinotecan. Top, hematoxylin and eosin (H&E) staining. Bottom, immunohistochemistry (IHC) staining for reovirus protein. Magnification, ×200.

4. Discussion

Many oncolytic viruses have been developed for application in multiple kinds of cancer. However, there are relatively few reports about the potential use of oncolytic viruses with gastric cancer. For example, adenoviral vectors have been used for experimental treatment and gene therapy of various cancers because of their high transduction efficiency. However, adenoviral infectivity of gastrointestinal cancer cells is generally poor due to the limited expression of the coxsackie-adenovirus receptor [23]. Thus, we were interested in exploring the use of reovirus in gastric cancer, as it naturally replicates in the gastrointestinal tract and its cell surface receptor (JAM-A) is abundant.

One of our first questions was therefore to determine the level of activation of Ras signaling in our gastric cancer cells, as this has been reported in other systems to correlate with susceptibility to reoviral oncolysis [5,8]. Although oncogenic mutations of Ras are infrequent (2–7%) in gastric cancer [24]; we nevertheless found evidence for variably activated GTP-Ras in most of the gastric cancer cell lines we used in this experiment (Figure 1b), possibly due to upstream activation of receptors such as EGFR, which is often mutated in gastric cancer. Thus, it is plausible for this reason that reovirus can be used as a gastric cancer therapy, even though we were surprised by the poor correlation between the levels of activated GTP-Ras and cytolytic susceptibility to reovirus (compare Figure 1a,b). The weakness of this association has also been reported by others [25] and it is likely that in addition to the activation status of Ras-associated pathways [5,26], there are other molecular determinants of reovirus-sensitivity, such as the cellular abundance of putative reovirus receptors/coreceptors [27–29], intracellular virion uncoating or assembly processes [30,31], and viral release and propagation, all of which can affect reovirus infection and oncolytic efficiency. In addition, as we propose in our accompanying manuscript, it is possible that the degree of cellular stemness may vary among different cancers, and this may confer variable reoviral responsiveness.

The cytopathic effect of reovirus we observed with our gastric cancer cell lines was relatively modest when compared with other gastrointestinal cancers (colon, esophageal, liver, pancreas; in preparation). GCIY cells were very susceptible to reovirus, but most gastric cancer cell lines showed moderate or low cytopathic effects in vitro (Figure 1a). Thus, from this evidence, it would be difficult to justify the use of reovirus as monotherapy against gastric cancer. However, we wished to consider whether some combination of reovirus and chemotherapy might show synergistic effects in gastric cancer and perhaps expand the range of gastric cancers in which benefits could be achieved.

Many preclinical studies have provided experimental evidence for effective killing of cancer cells by oncolytic viruses [32–36]. In animal models, however, established xenograft tumors are rarely eliminated despite the existence of persistently high viral titers within the tumor, and it is possible that total elimination of solid tumors may require higher doses of oncolytic viruses that might prove toxic or lethal. In a report of a clinical trial of ONYX-015 adenovirus, no clinical benefit was noted in the majority of patients, despite encouraging biological activity [37]. Tumor progression was rapid in most patients, even though substantial necrosis was noted in the tumors after treatment [38,39]. Thus, we opted to evaluate chemovirotherapy, consisting of oncolytic virotherapy combined with low doses of a chemotherapeutic agent. We reasoned that sublethal doses of chemotherapy might damage cancer cell pathways and reduce, for example, innate anti-viral responses, thereby enhancing viral oncolysis while reducing the likelihood of adverse effects [40]. In this study, we chose to explore the co-administration of irinoctecan, paclitaxel, and docetaxel because these are often used for second-line chemotherapy in human patients and are therefore plausible choices for future clinical trials employing chemovirotherapy.

Indeed, several chemotherapy/oncolytic virus combinations have already been evaluated to date and have been shown to result in enhanced antitumor effects without compromising safety. For example, the adenovirus Onyx-015 enhanced clinical efficacy by combining intratumoral Onyx-015 with systemic cisplatin and 5-fluorouracil when compared with chemotherapy alone [41]. E1A-expressing adenoviral E3B mutants combined with cisplatin and paclitaxel [42] also showed synergistic activity in vitro and in vivo. Combinations using oncolytic herpesviruses, such as G207 with cisplatin and HSV-1716 with mitomycin C, resulted in synergistic activity in vitro and in vivo [43–45]. Even though the precise biochemical mechanisms by which these synergies are achieved remain unknown, their potential use in gastric cancer may provide new options for more successful treatment.

In the experiments described here, we found that irinotecan, paclitaxel, and docetaxel were all able to enhance reovirus replicative activity (Figure 3) in MKN45 and AGS cells. It is possible that these chemotherapeutic agents act by repressing the innate immune responses of cells, thereby enhancing virus replication. In any case, it appears that the elevated viral replication is linked to greater caspase activation and thus an acceleration in programmed cell death.

In our experiments with xenograft tumors in vivo, we did not expect reovirus alone to be very effective in repressing growth in the MKN45 gastric cancer model, simply because reovirus did not kill MKN45 cells very well in vitro. However, we found that even reovirus by itself was able to strongly repress (though not completely eliminate) growth of these tumor cells in vivo. Nevertheless, both combinations of reovirus plus paclitaxel or reovirus plus irinotecan, showed clear enhancement of cell killing in vivo (Figure 4), consistent with our results in vitro.

Thus, we conclude that our in vitro and in vivo data both encourage further studies of reovirus plus chemotherapy (and especially with irinotecan and paclitaxel, which were superior to docetaxel) as a viable therapeutic modality for gastric cancer, even though in some cases (Figure S3 and reference [46]), reovirus monotherapy may be partly or fully effective. Our results are obtained in immunodeficient mice, and thus studies with immunocompetent hosts and syngeneic tumors [47] could show further enhancement in tumor cell killing.

In other work [48], we found that trastuzumab is able to enhance reoviral oncolysis in gastric cancers that overexpress the Her2/neu oncogene. Thus, we argue that there is a clear cellular rationale for combining chemotherapeutic agents and oncolytic reovirus for the treatment of this disease, even though the precise mechanisms underlying synergy between reoviral oncolysis and specific chemotherapeutic drugs remain unclear. We also note the potential for different routes of viral administration, such as intraperitoneal or intravascular, or especially for tumors of the gastrointestinal tract, via direct administration orally or as we have shown anally [49]. As more is learned about the molecular pathways by which oncolytic reovirus kills many cancer cell types, while sparing normal cells and tissues, the rational combination of more potent agents or immunomodulators ([50–54]; also Kubota et al., submitted) plus the potential for engineered virus [55] with greater anticancer action and reduced side effects will become clearer. Finally, we return to the observation that we (Figure 1) and others have made, which is that the correlation between Ras activation and reoviral responsiveness is unexpectedly poor. Transfection of activated Ras genes into normal cells can indeed result in both cellular transformation and reoviral susceptibility, as originally reported by Lee's group [26], but many tumor cells with activated Ras may display reoviral resistance, whereas in other cases, those with low Ras pathway activity may still be sensitive to the virus. Ras pathway activation alone is therefore unlikely to be a reliable biomarker for reoviral responsiveness, which as we have argued is indeed subject to multiple cellular constraints [56]. In our accompanying manuscript, we propose that a previously unappreciated variable, that of cellular stemness in cancer (or embryonic) cells, may be a novel important factor in predicting reoviral responsiveness. Further work will be required to evaluate more fully the utility of this proposed relationship.

In our accompanying manuscript (Bourhill et al. [57]), we propose that a previously unappreciated variable, that of cellular stemness in cancer (or embryonic) cells, may be a novel important factor in predicting reoviral responsiveness. It is unlikely that stemness alone will be a definitive factor in responses to reovirus, but perhaps together with a panel of other relevant variables we may eventually be able to identify which patients will benefit most from therapy that includes this or other viruses. These variables may in turn provide novel targets for therapeutic intervention for better modulation of reoviral oncolysis, as we show here in part with chemotherapeutic agents and gastric cancers. Further work will be required to evaluate more fully the utility of these proposed relationships.

5. Conclusions

Reovirus shows synergistic benefit when used in combination with paclitaxel or irinotecan in the treatment of gastric cancer in a murine xenograft model system. This finding may facilitate the development of effective therapeutic strategies for treating gastric cancer in patients.

Supplementary Materials: The following supporting information can be downloaded at: https://www.mdpi.com/article/10.3390/v15071472/s1. Figure S1: Combination effects of reovirus and chemotherapeutic agents (irinotecan, paclitaxel, docetaxel) on human gastric cancer cell lines (GCIY, NCI-N87). Figure S2: (a) Cytopathic effects of reovirus in GCIY and NCI-N87. (b) Caspase activation and reovirus protein expression were determined by flow cytometry in gastric cancer cell lines, GCIY and NCI-N87. Figure S3: Antitumor effects of intratumourally injected reovirus against established flank GCIY xenograft tumours in nu/nu mice.

Author Contributions: Conceptualization, Y.M. and R.N.J.; methodology, Y.M., S.G.N., A.R.F. and M.T.; resources, R.N.J.; writing—original draft preparation, Y.M.; writing—review and editing, R.N.J.; supervision, H.K., T.J. and R.N.J.; project administration Y.M.; funding acquisition, R.N.J. All authors have read and agreed to the published version of the manuscript.

Funding: Supported in part by funds from the Canadian Breast Cancer Foundation (Funding Code: B34), the Cancer Research Society (Funding Code: RT757538), and the Canadian Institutes of Health Research (RNJ) (Funding Code: MOP-82850).

Viruses **2023**, *15*, 1472

Institutional Review Board Statement: The animal study protocol was approved by the Institutional Review Board of the University of Calgary (protocol code M389; July 2010).

Informed Consent Statement: Not applicable.

Data Availability Statement: Not applicable.

Conflicts of Interest: The authors declare no conflict of interest. The funders had no role in the design of the study; in the collection, analyses, or interpretation of data; in the writing of the manuscript, or in the decision to publish the results.

References

1. Winer, E.; Gralow, J.; Diller, L.; Karlan, B.; Loehrer, P.; Pierce, L.; Demetri, G.; Ganz, P.; Kramer, B.; Kris, M.; et al. Clinical cancer advances 2008: Major research advances in cancer treatment, prevention, and screening—A report from the American Society of Clinical Oncology. *J. Clin. Oncol.* **2009**, *27*, 812–826. [CrossRef] [PubMed]
2. Bozzetti, F.; Marubini, E.; Bonfanti, G.; Miceli, R.; Piano, C.; Gennari, L.; Italian Gastrointestinal Tumor Study Group. Subtotal versus total gastrectomy for gastric cancer: Five-year survival rates in a multicenter randomized Italian trial. *Ann. Surg.* **1999**, *230*, 170–178. [CrossRef] [PubMed]
3. Karpeh, M.S.; Leon, L.; Klimstra, D.; Brennan, M.F. Lymph node staging in gastric cancer: Is location more important than Number? An analysis of 1,038 patients. *Ann. Surg.* **2000**, *232*, 362–371. [CrossRef] [PubMed]
4. Wesolowski, R.; Lee, C.; Kim, R. Is there a role for second-line chemotherapy in advanced gastric cancer? *Lancet Oncol.* **2009**, *10*, 903–912. [CrossRef]
5. Strong, J.E.; Coffey, M.C.; Tang, D.; Sabinin, P.; Lee, P.W. The molecular basis of viral oncolysis: Usurpation of the Ras signaling pathway by reovirus. *EMBO J.* **1998**, *17*, 3351–3362. [CrossRef]
6. Duncan, M.R.; Stanish, S.M.; Cox, D.C. Differential sensitivity of normal and transformed human cells to reovirus infection. *J. Virol.* **1978**, *28*, 444–449. [CrossRef]
7. Hashiro, G.; Loh, P.C.; Yau, J.T. The preferential cytotoxicity of reovirus for certain transformed cell lines. *Arch. Virol.* **1977**, *54*, 307–315. [CrossRef]
8. Coffey, M.C.; Strong, J.E.; Forsyth, P.A.; Lee, P.W. Reovirus therapy of tumors with activated Ras pathway. *Science* **1998**, *282*, 1332–1334. [CrossRef]
9. Hirasawa, K.; Nishikawa, S.G.; Norman, K.L.; Alain, T.; Kossakowska, A.; Lee, P.W. Oncolytic reovirus against ovarian and colon cancer. *Cancer Res.* **2002**, *62*, 1696–1701.
10. Etoh, T.; Himeno, Y.; Matsumoto, T.; Aramaki, M.; Kawano, K.; Nishizono, A.; Kitano, S. Oncolytic viral therapy for human pancreatic cancer cells by reovirus. *Clin. Cancer Res.* **2003**, *9*, 1218–1223.
11. Kim, M.; Egan, C.; Alain, T.; Urbanski, S.J.; Lee, P.W.; Forsyth, P.A.; Johnston, R.N. Acquired resistance to reoviral oncolysis in Ras-transformed fibrosarcoma cells. *Oncogene* **2007**, *26*, 4124–4134. [CrossRef]
12. Oncolytics Biotech® Inc. Announces Receipt of Orphan Drug Designation from the U.S. FDA for Gastric Cancer. 2015. Available online: https://www.oncolyticsbiotech.com/press-releases (accessed on 5 June 2023).
13. Toyoizumi, T.; Mick, R.; Abbas, A.E.; Kang, E.H.; Kaiser, L.R.; Molnar-Kimber, K.L. Combined therapy with chemotherapeutic agents and herpes simplex virus type 1 ICP34.5 mutant (HSV-1716) in human non-small cell lung cancer. *Hum. Gene Ther.* **1999**, *10*, 3013–3029. [CrossRef] [PubMed]
14. Nawa, A.; Nozawa, N.; Goshima, F.; Nagasaka, T.; Kikkawa, F.; Niwa, Y.; Nakanishi, T.; Kuzuya, K.; Nishiyama, Y. Oncolytic viral therapy for human ovarian cancer using a novel replication-competent herpes simplex virus type I mutant in a mouse model. *Gynecol. Oncol.* **2003**, *91*, 81–88. [CrossRef] [PubMed]
15. Cinatl, J., Jr.; Cinatl, J.; Michaelis, M.; Kabickova, H.; Kotchetkov, R.; Vogel, J.U.; Doerr, H.W.; Klingebiel, T.; Driever, P.H. Potent oncolytic activity of multimutated herpes simplex virus G207 in combination with vincristine against human rhabdomyosarcoma. *Cancer Res.* **2003**, *63*, 1508–1514.
16. Petrowsky, H.; Roberts, G.D.; Kooby, D.A.; Burt, B.M.; Bennett, J.J.; Delman, K.A.; Stanziale, S.F.; Delohery, T.M.; Tong, W.P.; Federoff, H.J.; et al. Functional interaction between fluorodeoxyuridine-induced cellular alterations and replication of a ribonucleotide reductase-negative herpes simplex virus. *J. Virol.* **2001**, *75*, 7050–7058. [CrossRef]
17. Bennett, J.J.; Adusumilli, P.; Petrowsky, H.; Burt, B.M.; Roberts, G.; Delman, K.A.; Zager, J.S.; Chou, T.C.; Fong, Y. Up-regulation of GADD34 mediates the synergistic anticancer activity of mitomycin C and a $\gamma_1$34.5 deleted oncolytic herpes virus (G207). *FASEB J.* **2004**, *18*, 1001–1003. [CrossRef] [PubMed]
18. Aghi, M.; Rabkin, S.; Martuza, R.L. Effect of chemotherapy-induced DNA repair on oncolytic herpes simplex viral replication. *J. Natl. Cancer Inst.* **2006**, *98*, 38–50. [CrossRef]
19. Sei, S.; Mussio, J.K.; Yang, Q.E.; Nagashima, K.; Parchment, R.E.; Coffey, M.C.; Shoemaker, R.H.; Tomaszewski, J.E. Synergistic antitumor activity of oncolytic reovirus and chemotherapeutic agents in non-small cell lung cancer cells. *Mol. Cancer* **2009**, *8*, 47. [CrossRef]

20. Pandha, H.S.; Heinemann, L.; Simpson, G.R.; Melcher, A.; Prestwich, R.; Errington, F.; Coffey, M.; Harrington, K.J.; Morgan, R. Synergistic effects of oncolytic reovirus and cisplatin chemotherapy in murine malignant melanoma. *Clin. Cancer Res.* **2009**, *15*, 6158–6166. [CrossRef]

21. Cho, I.R.; Koh, S.S.; Min, H.J.; Park, E.H.; Srisuttee, R.; Jhun, B.H.; Kang, C.D.; Kim, M.; Johnston, R.N.; Chung, Y.H. Reovirus infection induces apoptosis of TRAIL-resistant gastric cancer cells by down-regulation of Akt activation. *Int. J. Oncol.* **2010**, *36*, 1023–1030.

22. Zhao, L.; Wientjes, M.G.; Au, J.L. Evaluation of combination chemotherapy: Integration of nonlinear regression, curve shift, isobologram, and combination index analyses. *Clin. Cancer Res.* **2004**, *10*, 7994–8004. [CrossRef] [PubMed]

23. Ono, H.A.; Davydova, J.G.; Adachi, Y.; Takayama, K.; Barker, S.D.; Reynolds, P.N.; Krasnykh, V.N.; Kunisaki, C.; Shimada, H.; Curiel, D.T.; et al. Promoter-controlled infectivity-enhanced conditionally replicative adenoviral vectors for the treatment of gastric cancer. *J. Gastroenterol.* **2005**, *40*, 31–42. [CrossRef] [PubMed]

24. Mita, H.; Toyota, M.; Aoki, F.; Akashi, H.; Maruyama, R.; Sasaki, Y.; Suzuki, H.; Idogawa, M.; Kashima, L.; Yanagihara, K.; et al. A novel method, digital genome scanning detects *KRAS* gene amplification in gastric cancers: Involvement of overexpressed wild-type KRAS in downstream signaling and cancer cell growth. *BMC Cancer* **2009**, *9*, 198. [CrossRef] [PubMed]

25. Van Houdt, W.J.; Smakman, N.; van Den Wollenberg, D.J.M.; Emmink, B.L.; Veenendaal, L.M.; Van Diest, P.J.; Hoeben, R.C.; Borel Rinkes, I.H.M.; Kranenburg, O. Transient infection of freshly isolated human colorectal tumor cells by reovirus T3D intermediate subviral particles. *Cancer Gene Ther.* **2008**, *15*, 284–292. [CrossRef]

26. Norman, K.L.; Hirasawa, K.; Yang, A.D.; Shields, M.A.; Lee, P.W. Reovirus oncolysis: The Ras/RalGEF/p38 pathway dictates host cell permissiveness to reovirus infection. *Proc. Natl. Acad. Sci. USA* **2004**, *101*, 11099–11104. [CrossRef]

27. Barton, E.S.; Forrest, J.C.; Connolly, J.L.; Chappell, J.D.; Liu, Y.; Schnell, F.J.; Nusrat, A.; Parkos, C.A.; Dermody, T.S. Junction adhesion molecule is a receptor for reovirus. *Cell* **2001**, *104*, 441–451. [CrossRef]

28. Strong, J.E.; Lee, P.W. The v-erbB oncogene confers enhanced cellular susceptibility to reovirus infection. *J. Virol.* **1996**, *70*, 612–616. [CrossRef]

29. Chappell, J.D.; Gunn, V.L.; Wetzel, J.D.; Baer, G.S.; Dermody, T.S. Mutations in type 3 reovirus that determine binding to sialic acid are contained in the fibrous tail domain of viral attachment protein sigma1. *J. Virol.* **1997**, *71*, 1834–1841. [CrossRef]

30. Golden, J.W.; Linke, J.; Schmechel, S.; Thoemke, K.; Schiff, L.A. Addition of exogenous protease facilitates reovirus infection in many restrictive cells. *J. Virol.* **2002**, *76*, 7430–7443. [CrossRef]

31. Alain, T.; Kim, T.S.; Lun, X.; Liacini, A.; Schiff, L.A.; Senger, D.L.; Forsyth, P.A. Proteolytic disassembly is a critical determinant for reovirus oncolysis. *Mol. Ther.* **2007**, *15*, 1512–1521. [CrossRef]

32. Bischoff, J.R.; Kirn, D.H.; Williams, A.; Heise, C.; Horn, S.; Muna, M.; Ng, L.; Nye, J.A.; Sampson-Johannes, A.; Fattaey, A.; et al. An adenovirus mutant that replicates selectively in p53-deficient human tumor cells. *Science* **1996**, *274*, 373–376. [CrossRef]

33. Rodriguez, R.; Schuur, E.R.; Lim, H.Y.; Henderson, G.A.; Simons, J.W.; Henderson, D.R. Prostate attenuated replication competent adenovirus (ARCA) CN706: A selective cytotoxic for prostate-specific antigen-positive prostate cancer cells. *Cancer Res.* **1997**, *57*, 2559–2563. [PubMed]

34. Tsukuda, K.; Wiewrodt, R.; Molnar-Kimber, K.; Jovanovic, V.P.; Amin, K.M. An E2F-responsive replication-selective adenovirus targeted to the defective cell cycle in cancer cells: Potent antitumoral efficacy but no toxicity to normal cell. *Cancer Res.* **2002**, *62*, 3438–3447. [PubMed]

35. Li, Y.; Yu, D.C.; Chen, Y.; Amin, P.; Zhang, H.; Nguyen, N.; Henderson, D.R. A hepatocellular carcinoma-specific adenovirus variant, CV890, eliminates distant human liver tumors in combination with doxorubicin. *Cancer Res.* **2001**, *61*, 6428–6436.

36. Kirn, D.; Martuza, R.L.; Zwiebel, J. Replication-selective virotherapy for cancer: Biological principles, risk management and future directions. *Nat. Med.* **2001**, *7*, 781–787. [CrossRef] [PubMed]

37. Nemunaitis, J.; Ganly, I.; Khuri, F.; Arseneau, J.; Kuhn, J.; McCarty, T.; Landers, S.; Maples, P.; Romel, L.; Randlev, B.; et al. Selective replication and oncolysis in p53 mutant tumors with ONYX-015, an E1B-55kD gene-deleted adenovirus, in patients with advanced head and neck cancer: A phase II trial. *Cancer Res.* **2000**, *60*, 6359–6366.

38. Jacobs, C.; Lyman, G.; Velez-García, E.; Sridhar, K.S.; Knight, W.; Hochster, H.; Goodnough, L.T.; Mortimer, J.E.; Einhorn, L.H.; Schacter, L. A phase III randomized study comparing cisplatin and fluorouracil as single agents and in combination for advanced squamous cell carcinoma of the head and neck. *J. Clin. Oncol.* **1992**, *10*, 257–263. [CrossRef]

39. Vokes, E.E. Chemotherapy and integrated treatment approaches in head and neck cancer. *Curr. Opin. Oncol.* **1991**, *3*, 529–534. [CrossRef]

40. Fujiwara, T.; Kagawa, S.; Kishimoto, H.; Endo, Y.; Hioki, M.; Ikeda, Y.; Sakai, R.; Urata, Y.; Tanaka, N.; Fujiwara, T. Enhanced antitumor efficacy of telomerase-selective oncolytic adenoviral agent OBP-401 with docetaxel: Preclinical evaluation of chemovirotherapy. *Int. J. Cancer* **2006**, *119*, 432–440. [CrossRef]

41. Khuri, F.R.; Nemunaitis, J.; Ganly, I.; Arseneau, J.; Tannock, I.F.; Romel, L.; Gore, M.; Ironside, J.; MacDougall, R.H.; Heise, C.; et al. A controlled trial of intratumoral ONYX-015, a selectively-replicating adenovirus, in combination with cisplatin and 5-fluorouracil in patients with recurrent head and neck cancer. *Nat. Med.* **2000**, *6*, 879–885. [CrossRef]

42. Cheong, S.C.; Wang, Y.; Meng, J.H.; Hill, R.; Sweeney, K.; Kirn, D.; Lemoine, N.R.; Hallden, G. E1A-expressing adenoviral E3B mutants act synergistically with chemotherapeutics in immunocompetent tumor models. *Cancer Gene Ther.* **2008**, *15*, 40–50. [CrossRef] [PubMed]

43. Eisenberg, D.P.; Adusumilli, P.S.; Hendershott, K.J.; Yu, Z.; Mullerad, M.; Chan, M.K.; Chou, T.C.; Fong, Y. 5-fluorouracil and gemcitabine potentiate the efficacy of oncolytic herpes viral gene therapy in the treatment of pancreatic cancer. *J. Gastrointest. Surg.* **2005**, *9*, 1068–1077. [CrossRef] [PubMed]

44. Post, D.E.; Fulci, G.; Chiocca, E.A.; Van Meir, E.G. Replicative oncolytic herpes simplex viruses in combination cancer therapies. *Curr. Gene Ther.* **2004**, *4*, 41–51. [CrossRef] [PubMed]

45. Lin, S.F.; Gao, S.P.; Price, D.L.; Li, S.; Chou, T.C.; Singh, P.; Huang, Y.Y.; Fong, Y.; Wong, R.J. Synergy of a herpes oncolytic virus and paclitaxel for anaplastic thyroid cancer. *Clin. Cancer Res.* **2008**, *14*, 1519–1528. [CrossRef]

46. Kawaguchi, K.; Etoh, T.; Suzuki, K.; Mitui, M.T.; Nishizono, A.; Shiraishi, N.; Kitano, S. Efficacy of oncolytic reovirus against human gastric cancer with peritoneal metastasis in experimental animal model. *Int. J. Oncol.* **2010**, *37*, 1433–1438. [CrossRef] [PubMed]

47. Hsu, H.P.; Wang, C.Y.; Hsieh, P.Y.; Fang, J.H.; Chen, Y.L. Knockdown of serine/threonine-protein kinase 24 promotes tumorigenesis and myeloid-derived suppressor cell expansion in an orthotopic immunocompetent gastric cancer animal model. *J. Cancer* **2020**, *11*, 213–228. [CrossRef]

48. Hamano, S.; Mori, Y.; Kataoka, H.; Tanaka, M.; Ebi, M.; Kubota, E.; Mizoshita, T.; Tanida, S.; Johnston, R.N.; Asai, K.; et al. Oncolytic reovirus combined with trastuzumab enhances antitumor efficacy through TRAIL signaling in human HER2-positive gastric cancer cells. *Cancer Lett.* **2015**, *356*, 846–854. [CrossRef]

49. Alain, T.; Wong, J.F.; Urbanski, S.J.; Lee, P.W.; Muruve, D.A.; Johnston, R.N.; Forsyth, P.A.; Beck, P.L. Reovirus decreases azoxymethane-induced aberrant crypt foci and colon cancer in a rodent model. *Cancer Gene Ther.* **2007**, *14*, 867–872. [CrossRef]

50. Mahalingam, D.; Wilkinson, G.A.; Eng, K.H.; Fields, P.; Raber, P.; Moseley, J.L.; Cheetham, K.; Coffey, M.; Nuovo, G.; Kalinski, P.; et al. Pembrolizumab in Combination with the Oncolytic Virus Pelareorep and Chemotherapy in Patients with Advanced Pancreatic Adenocarcinoma: A Phase Ib Study. *Clin. Cancer Res.* **2020**, *26*, 71–81. [CrossRef]

51. Mahalingam, D.; Goel, S.; Aparo, S.; Patel Arora, S.; Noronha, N.; Tran, H.; Chakrabarty, R.; Selvaggi, G.; Gutierrez, A.; Coffey, M.; et al. A Phase II Study of Pelareorep (REOLYSIN®) in Combination with Gemcitabine for Patients with Advanced Pancreatic Adenocarcinoma. *Cancers* **2018**, *10*, 160. [CrossRef]

52. Shao, S.; Yang, X.; Zhang, Y.N.; Wang, X.J.; Li, K.; Zhao, Y.L.; Mou, X.Z.; Hu, P.Y. Oncolytic Virotherapy in Peritoneal Metastasis Gastric Cancer: The Challenges and Achievements. *Front. Mol. Biosci.* **2022**, *28*, 835300. [CrossRef]

53. Mohamed, A.; Lin, Q.F.; Eaton, H.E.; Shmulevitz, M. p38 Mitogen-Activated Protein Kinase Signaling Enhances Reovirus Replication by Facilitating Efficient Virus Entry, Capsid Uncoating, and Postuncoating Steps. *J. Virol.* **2023**, *97*, 9–23. [CrossRef] [PubMed]

54. Zamarin, D.; Holmgaard, R.B.; Subudhi, S.K.; Park, J.S.; Mansour, M.; Palese, P.; Merghoub, T.; Wolchok, J.D.; Allison, J.P. Localized oncolytic virotherapy overcomes systemic tumor resistance to immune checkpoint blockade immunotherapy. *Sci. Trans. Med.* **2014**, *6*, 226. [CrossRef] [PubMed]

55. Mohamed, A.; Johnston, R.N.; Shmulevitz, M. 2015. Potential for improving potency and specificity of reovirus oncolysis with next-generation reovirus variants. *Viruses* **2015**, *7*, 6251–6278. [CrossRef]

56. Bourhill, T.; Mori, Y.; Rancourt, D.E.; Shmulevitz, M.; Johnston, R.N. Going (reo) viral: Factors promoting successful reoviral oncolytic infection. *Viruses* **2018**, *10*, 421. [CrossRef]

57. Bourhill, T.; Rohani, L.; Kumar, M.; Bose, P.; Rancourt, D.; Johnston, R.N. Modulation of Reoviral Cytolysis (II): Cellular Stemness. *Viruses* **2023**.

Article

Modulation of Reoviral Cytolysis (II): Cellular Stemness

Tarryn Bourhill, Leili Rohani, Mehul Kumar, Pinaki Bose, Derrick Rancourt and Randal N. Johnston *

Department of Biochemistry and Molecular Biology, Cumming School of Medicine, University of Calgary, Calgary, AB T2N 4N1, Canada; tarryn.bourhill@ucalgary.ca (T.B.)
* Correspondence: rnjohnst@ucalgary.ca

Abstract: Oncolytic viruses (OVs) are an emerging cancer therapeutic that are intended to act by selectively targeting and lysing cancerous cells and by stimulating anti-tumour immune responses, while leaving normal cells mainly unaffected. Reovirus is a well-studied OV that is undergoing advanced clinical trials and has received FDA approval in selected circumstances. However, the mechanisms governing reoviral selectivity are not well characterised despite many years of effort, including those in our accompanying paper where we characterize pathways that do not consistently modulate reoviral cytolysis. We have earlier shown that reovirus is capable of infecting and lysing both certain types of cancer cells and also cancer stem cells, and here we demonstrate its ability to also infect and kill healthy pluripotent stem cells (PSCs). This led us to hypothesize that pathways responsible for stemness may constitute a novel route for the modulation of reoviral tropism. We find that reovirus is capable of killing both murine and human embryonic and induced pluripotent stem cells. Differentiation of PSCs alters the cells' reoviral-permissive state to a resistant one. In a breast cancer cell line that was resistant to reoviral oncolysis, induction of pluripotency programming rendered the cells permissive to cytolysis. Bioinformatic analysis indicates that expression of the Yamanaka pluripotency factors may be associated with regulating reoviral selectivity. Mechanistic insights from these studies will be useful for the advancement of reoviral oncolytic therapy.

Keywords: reovirus; oncolytic virus; oncolytic viral therapy; stemness; pluripotency; cytolysis; reprogramming

Citation: Bourhill, T.; Rohani, L.; Kumar, M.; Bose, P.; Rancourt, D.; Johnston, R.N. Modulation of Reoviral Cytolysis (II): Cellular Stemness. *Viruses* **2023**, *15*, 1473. https://doi.org/10.3390/v15071473

Academic Editors: Ottmar Herchenröder and Brigitte Pützer

Received: 5 June 2023
Revised: 21 June 2023
Accepted: 25 June 2023
Published: 29 June 2023

1. Introduction

OVs are an exciting possible augmentation to current therapies for the treatment of cancer. This novel therapeutic consists of viruses that are either engineered or that naturally target and kill cancer cells while leaving healthy tissues mainly unharmed. The lytic replication of OVs allows for the generation and spread of progeny virus while killing the host infected cells. These viruses not only kill cancer cells directly but can also stimulate an anti-tumour immune response and thus have the potential to allow for the development of lasting immunity against particular cancers. As the therapy is targeted, it potentially allows for a greater therapeutic window and the reduction of potential side effects [1–3]. Reovirus is a promising OV as it is not linked to serious disease, has natural tumour tropism and is well-studied. It is undergoing phase III clinical trials and has received FDA approval for its use as an orphan drug for certain cancers [4–12].

The name reovirus is derived as a partial acronym for Respiratory Enteric Orphan virus [13]. The natural endemic virus preferentially infects respiratory and intestinal tissues but does not elicit severe pathology and is therefore designated an orphan virus. Reovirus is a highly stable non-enveloped virus that consists of two protein coats, an outer capsid and the inner core that surround ten dsRNA genome segments [14]. Most OVs in use have been engineered to preferentially replicate in cancers by taking advantage of the aberrant signalling that establishes the hallmarks of cancer. A fascinating property of reovirus is that it has an inherent tropism for many types of transformed cells. This unique ability to preferentially replicate in transformed cells was first demonstrated in 1977 by

Hashiro et al. [15]. One year later Duncan et al. corroborated this finding by showing that reovirus would preferentially lyse transformed WI-38 lung fibroblast cells [16]. Almost 20 years later, Lee's group demonstrated the first use of reovirus as an oncolytic agent in vivo for Ras-transformed cancer cells and showed that it could reduce tumour burden in mice [7].

Since then, reovirus has emerged as one of the most clinically advanced oncolytic therapies under development, with over thirty clinical trials (phase I-III) in numerous cancer types including colorectal, bladder, skin, head and neck, brain, non-small-cell lung cancer, ovarian, gastric, pancreatic prostate and breast [17,18]. In 2015, reovirus received orphan drug status from the FDA and European Medicine Agency (EMA) for the treatment of gastric, pancreatic and ovarian cancers [4,5,19]. Reolysin® has also received fast track designation and a special protocol assessment agreement from the FDA for the treatment of metastatic breast cancer [20,21]. Even so, responses to reoviral monotherapy in advanced human cancers have in most cases been modest and short-lived, despite the promising preclinical data in animals [22], and much effort is currently focused on the testing of combination therapies. Indeed, the clinical responses to OV therapies in general have been disappointing [6,23,24], and this has led to the development of next generation OVs that are more potent, specific and safe, plus efforts to identify patients with cancers that will respond best to these therapies.

There are numerous barriers to effective clinical application of oncolytic viral therapies, some of which include inefficient delivery of viral vectors, antiviral immunity, limited vector spread within tumours and inefficient infection of tumour cells [25]. There are many strategies being employed to design the next generation of oncolytic viral vectors that overcome these barriers. The first is through arming OVs to improve viral potency. This is done by engineering viruses to express pro-drug convertases (suicide genes) or therapeutic proteins. An additional strategy for advancing oncolytic vectors is to shield them from neutralising antibodies that can inactivate viruses before they reach the tumour, through the use of chemical protective barriers, serotype switiching or the use of carrier cells. The final approach to enhancing oncolytic vectors through genetic engineering has been to improve viral specificity for tumours by the insertion of tumour or tissue specific enhancers and promoters that limit the expression of genes essential for viral replication. This approach can also be used to regulate the expression of suicide genes and therapeutic proteins [26–29].

Although attempts have been made to engineer reovirus in similar ways, it has a dsRNA segmented genome, making it challenging to engineer the virus and generate viable mutants. The first entirely plasmid based system for engineering reovirus was developed by Dermody's group in 2007 [30]. One of the first replication competent engineered reoviruses to be produced using this system was developed by Hoeben's group. They attached a fluorescent protein (improved Light, Oxygen or Voltage sensing domain from *Arabidopsis thaliana*, iLOV) into the binding region of σ1 [31]. Despite these recent advances in reoviral engineering, none of the isolated mutants has yet to progress to the clinic as an oncolytic vector. Mechanistic insights into what regulates reoviral tropism for transformed cells is critical for improving reoviral therapeutic strategies [32]. In addition, an understanding of the fundermental interactions that govern reoviral tropism will aid in the development of a panel of biomarkers that could indicate whether or not patients are likely to respond to reoviral treatment. An additional benefit of understanding reoviral selectivity would be the potential for developing rational drug combinations for cancer therapy as reovirus has so few off target effects.

However, the molecular mechanisms for reoviral selective replication in transformed cell lines are not completely understood. An early breakthrough was achieved by Lee's group, who argued that reoviral selectivity is dependent on aberrant RAS activation and subsequent down regulation of protein kinase R (PKR) [7,12,32–41]. In healthy cells, PKR functions as part of an antiviral cellular defence system. When two of these proteins bind to dsRNA, transphosphorylation activates these kinases, which then prevent the formation of the translation initiation complexes and thereby preventing protein translation [42,43]. When the presence of dsRNA viral genomes is detected by PKR, protein synthesis is halted, which causes rapid apoptosis and viral clearance. PKR has been shown to be inactive in RAS transformed cells [36]. In cancer cells where PKR is dysfunctional and is no longer able to detect the presence of dsRNA and activate appropriate immune responses, reoviral replication is in theory uninhibited, leading to cytolysis. However, tests of this model yielded conflicting results and the exact mechanisms co-ordinating RAS activation and PKR signalling in promoting reoviral replication [35] and the general control of protein synthesis [44,45] remain unclear. Indeed, numerous groups have shown that RAS and PKR do not fully account for reovirus' tumourigenic selectivity [46]. Contrary to the model proposed by Lee, Terasawa and colleagues were able to show that some cancers with low RAS activity are unexpectedly susceptible to reoviral oncolysis, while others with high RAS activation were resistant [46] (see also our accompanying paper). Knockdown of RAS did not prevent reoviral replication in colorectal cancer cell lines [47]. Smakman et al. subsequently showed that deletion of KRAS through homologous recombination did not prevent reoviral replication [48]. In accordance with these findings, work by van Holt et al. showed that cell lines or tissue fragments taken from colorectal tumours were non-permissive to reovirus regardless of KRAS activation [49]. Similarly, in work focusing on PKR activation Zhang et al. demonstrated that siRNA knockdown of PKR expression did not increase reoviral lysis in tumour cells [50]. Work presented by Twigger et al. also showed that inhibiting PKR activity in resistant cell lines did not cause an increase in reoviral response [51]. Schiff's group presented data entirely contrary to Lee's work showing that the presence of either RNAse L or PKR modulated reoviral replication [52]. This team subsequently verified their findings by showing that reovirus replication was more efficient in PKR knockout cells [53]. The debate surrounding the RAS/PKR pathway clearly indicates viral tropism is a complex process involving numerous integrated signalling networks.

The interactions between host and viral signaling that determine successful reoviral oncolysis are likely to be complex and varied. Efficient receptor targeting and binding, viral replication, type of virally induced programmed cell death (autophagy, apoptosis, necrosis, pyroptosis) and the susceptibility of a cancer to different cell death programs, as well as host cell immune responses, all influence the efficiency of oncolysis [23,32,54–56]. As our understanding of what governs reoviral selectivity in transformed cells is incomplete, there remain unique opportunities to uncover these mechanisms and apply this knowledge to enhance reoviral oncolytic efficiency.

Remarkably, reovirus not only has anti-tumour activity but its selective cytolysis extends to cancer stem cells (CSCs) [57,58]. This is intriguing as both bulk tumour cells and CSCs exist in a de-differentiated state relative to their healthy counterparts. What is also striking about reoviral infection is that is capable of lysing healthy, non-cancerous stem cells. An earlier observation from our lab demonstrated that murine embryonic stem cells are sensitive to reovirus both in vitro and in vivo. We found that wild type reovirus could reduce in SCID mice the size of teratomas that were derived from non-cancerous murine embryonic stem cells [59]. Reovirus can therefore selectively lyse bulk cancer cells, cancer stem cells and healthy murine embryonic stem cells, all of which have stem-like characteristics. In addition, healthy differentiated tissues are mostly refractory to infection. These observations led us to speculate that pathways responsible for stemness may govern some aspects of reoviral selectivity. To test this hypothesis we began by characterizing the capabilities of reovirus to lyse non-cancerous stem cells. We then investigated whether inducing pluripotency and differentiation could alter the cellular responses to reoviral infection. Finally, we performed bioinformatic analysis to understand the potential role of Yamanaka stem cell factors in regulating reoviral selectivity.

An appreciation for the signalling pathways responsible for governing reoviral tropism will be a foundation for improving the efficacy of this oncolytic agent. Findings from this study could help establish which cancer backgrounds are amenable to cytolysis, information that will be critical for developing effective biomarkers to screen patients. Additionally, knowledge of reoviral selectivity can aid in the development of synergistic therapeutic combinations enhancing potency. Mechanistic insights on the signalling pathways regulating viral tropism will enable the development of next generation reoviral oncolytic therapies.

2. Materials and Methods

2.1. Cancer Cell Lines and Culture Conditions

Cancer cells were grown in high glucose (25 mM) Dulbecco's Modified Eagle Medium (DMEM) supplemented with 10% (*v/v*) foetal bovine serum (FBS) (Thermo Fisher Scientific, Waltham, MA, USA) and 100 units/mL penicillin and 100 mg/mL streptomycin (Thermo Fisher Scientific). Cell lines cultured in this medium formulation included: L929 (murine fibroblasts, American Type Culture Collection, ATCC, Manassas, VA, USA), Hela (human cervical adenocarcinoma, ATCC), ZR-30–75 (human mammary ductal carcinoma, ATCC), MDA-MB-231 (human mammary adenocarcinoma, ATCC), MDA-MB-468 (human mammary adenocarcinoma, ATCC), MCF7 (human mammary adenocarcinoma, ATCC) and EMT6 (murine mammary carcinoma, ATCC). All cell lines used gave broadly similar responses, but only experiments performed with cells in triplicate or greater are presented in the Results. The N_2O_2 (mammary tumour cells from MMTV-Neu mice) cell line was generated by DR [60]. Cells were grown in monolayers and incubated in a humidified atmosphere of 5% CO2 with 95% humidity at 37 °C. When cultures reached 70–80% confluency they were passaged using Trypsin-ethylenediaminetetraacetic acid (EDTA), (Thermo Fisher Scientific; 0.25% Trypsin, 0.913 mM EDTA). Trypsinized cells were incubated with fresh medium and counted using Trypan blue (Thermo Fisher Scientific) and a haemocytometer. Cells were then placed in fresh medium and seeded onto appropriate culture dishes. Stocks of these cell lines were maintained in liquid nitrogen for long term storage in a 90% FBS 10% Dimethyl sulfoxide (DMSO, Sigma, St. Louis, MO, USA) solution.

Induced pluripotent cancer stem cells (14ex-11, Yulia Shulga, Rancourt lab) were grown in high glucose (25 mM) DMEM medium supplemented with 15% (*v/v*) embryonic stem cell FBS (Thermo Fisher Scientific), 1 mM sodium pyruvate (Thermo Fisher Scientific), 0.1 mM non-essential amino acids (Thermo Fisher Scientific), 0.1 mM 2-mercaptoethanol (Thermo Fisher Scientific) and 10 ng/mL leukaemia inhibitory factor (LIF, Sigma). These cells were grown on a feeder cell layer of inactivated murine embryonic fibroblasts

Induced pluripotent cancer stem cells were generated by reprogramming N_2O_2 tumour cell lines with PiggyBac transposon vectors that expressed cDNA for OCT4, c-MYC, KLF4 and SOX2 [61]. Reprogramming was confirmed through qRT-PCR and immunofluorescence imaging for stem cell markers. Pluripotency of this reprogrammed cell line was additionally confirmed via in vivo teratoma assay.

2.2. Stem Cell Culture Conditions

2.2.1. Murine Cell Lines

Murine embryonic (D3) and induced pluripotent iPS-3, stem cells were cultured in high glucose (25 mM)) DMEM that was supplemented with; 15% (*v/v*) embryonic stem cell FBS, 1 mM sodium pyruvate, 0.1 mM non-essential amino acid, 0.1 mM 2-mercaptoethanol and 10 ng/mL LIF. Both cell lines were provided by the Rancourt lab [62]. Stem cells were co-cultured on inactivated murine embryonic fibroblast cells, as described next.

Murine embryonic fibroblast (MEF) cells were cultured on 0.1% gelatin (Sigma) coated tissue culture plates. These cells were grown in high glucose (25 mM) DMEM medium with 10% (*v/v*) FBS, 0.1 mM non-essential amino acids, 50 units/mL penicillin and 50 mg/mL streptomycin. MEFs were used for experimental purposes and as a feeder layer for stem-like cells. MEFs were prepared as a feeder layer and when they reached 100% confluency the cells were treated with 10 μg/mL mitomycin C (Sigma) for 2 h at 37 C. The cells were then washed with PBS and incubated in fresh medium overnight. Inactivated MEFs were used as feeder cells the following day [63]. When stem cells were required for experiments they were removed from MEFs. Once the stem cells reach 70–80% confluency on the gelatin plates the cells were dissociated and counted using a haemocytometer and Trypan blue.

The cells were then aliquoted in appropriate amounts for subsequent experiments. Stocks of these cell lines were maintained in liquid nitrogen for long term storage in a 90% FBS 10% DMSO solution.

2.2.2. Human Stem Cells

Human embryonic stem cell lines H9 (Wicell, Madison, WI, USA), H1 (Wicell, WA01) and induced pluripotent human stem cells BJ-EOS clone 4YA (provided by the Ellis lab) [64] were cultured in mTeSR complete medium (Stem Cell Technologies, Vancouver, Canada) supplemented with 50 units/mL penicillin and 50 mg/mL streptomycin. These cells were cultured on Matrigel (Corning, New York, NY, USA) treated 60 mm plates as per the manufacturers specifications [64].

Cells were passaged by washing the cultures with Dulbecco's phosphate-buffered saline without calcium and magnesium (DPBS, Stem Cell Technologies) twice followed by accutase (Stem Cell Technologies) treatment. Once the cells were in a single cell suspension fresh medium was used to inhibit accutase activity. Cells were then counted and aliquoted for subsequent experiments. Cell were maintained in medium containing 10 µM Rho-associated kinase (ROCK) inhibitor (Y-27632; Stem Cell Technologies) for 24 h after passaging and then returned to their regular medium.

Human foreskin fibroblast cells (HFF) [64] were cultured on gelatin coated plates in high glucose (25 mM) DMEM medium supplemented with; 15% (*v/v*) embryonic stem cell FBS, 1.0 mM sodium pyruvate, 0.1 mM non-essential amino acid and 0.1 mM 2-mercaptoethanol.

2.3. Differentiation Protocols
2.3.1. Murine Cell Lines

Murine embryonic and induced stem cells were differentiated through the hanging drop technique [63,65,66]. Trypsinized stem cells were resuspended in medium that did not contain LIF [high glucose (25 mM) DMEM that was supplemented with 15% (*v/v*) embryonic stem cell FBS, 1.0 mM sodium pyruvate, 0.1 mM non-essential amino acids and 0.1 mM 2-mercaptoethanol]. Cells were counted and diluted to a concentration of 25,000 cells/mL (or 500 cells per 20 µL drop). Drops of 20 µL were plated onto the lid of a 90-mm bacterial culture plate. The lid was then placed on top of a plate that contained 10 mL of PBS. Hanging drops were incubated at 37 C in 5% CO_2, with 95% humidity for a period of 3–4 days [63]. Embryoid bodies were then transferred into 24, 48 or 96 well cell culture plates (Greiner Bio-One, #662160, Kremsmünster, Austria) that were treated with 0.1% gelatin. The embryoid bodies were either infected with reovirus directly as the cells were being transferred into the multiwell plates or after 1 or 10 days of differentiation. Spontaneous differentiation was confirmed by the presence of beating cardiomyocytes in cultures. Beating cells were identified through the use of an Olympus CKX41 inverted phase contrast microscope.

2.3.2. Human Cell Lines

Human embryonic stem cell (hESC) line H1 (WiCell) and human induced pluripotent stem cell (hiPSC) line 4YA (generated from BJ cell lines in Dr. James Ellis' laboratory at the University of Toronto) were differentiated into liver organoids using a modified previously published protocol [67]. The differentiation process was performed through three stepwise stages: definitive endoderm (DE) formation, hepatoblast, and mature hepatocytes.

Briefly, human pluripotent stem cells (hPSCs) were seeded on matrigel-coated 60-mm plates at a density of 4×10^5 cells per plate, and when the cells reached to 90% confluency the directed differentiation toward definitive endoderm was started for three days. The first step of differentiation (DE) was done on 2D culture plates. The cells were treated with RPMI 1640 medium (Thermo Fisher scientific) containing 1% BSA (Sigma-Aldrich, St. Louis, MO, USA), 1% B27 (Thermo Fisher Scientific). On day 1 this medium was supplemented with 6 mM CHIR99021 (Stem Cell Technologies), and following this the medium was supplemented with 10 ng/mL activin A (R&D Systems, Minneapolis, MN, USA) on days 2 and 3 of the differentiation protocol.

Stage two of differentiation (hepatoblast) began with aggregating putative endoderm cells into aggregates using AggreWell 400 24-well plates. Aggregates containing 100 cells were used for spheroid formation. To calculate the number of cells needed in each well of AggreWell 400 24-well plate, the following formula was used:

$$1200 \text{ microwells per well (24 wp)} = 1200 \text{ aggregate} \times \# \text{ cells/aggregate}$$

The endodermal cells were dissociated using Accutase and collected into fresh stage two medium containing DMEM-F12 medium (Thermo Fisher Scientific) supplemented with 2% knockout serum replacement (KOSR, Thermo Fisher Scientific), 10 ng/mL hepatocyte growth factor, (HGF, R&D Systems), and 10 ng/mL fibroblast growth factor-4 (FGF4, R&D Systems). Cells were aggregated by adding 600 μL cell suspension to each well of a 24-well plate containing 10 μM ROCK inhibitor (Y27632, Stem Cell Technologies) and placed in the incubator for 24 h. The next day, aggregates were placed into fresh stage two medium in 6-well plates (non-TC treated) on a shaker (f < 60 rpm). Aggregates were then treated with stage two differentiation medium for a period of 7 days on the shaker.

Formed organoids were then subjected to the third and final stage of differentiation (mature hepatocytes) where they were treated for a period of 10 days using HCMTM Hepatocyte Culture Medium BulletKitTM (Lonza, Basel, Switzerland) supplemented with 0.1 mM DMSO, 1 mM non-essential amino acids, 1 mM GlutaMAX and 10 ng/mL oncostatin M (R&D Systems) [67]. Stage 3 hepatocyte-like differentiation liver organoids were then subject to reoviral infection. Differentiation of the liver organoids was confirmed through immunofluorescence imaging of liver specific markers KRT7, CYP3A4, HNF4-alpha and Sox9.

2.3.3. Reoviral Infection

Reovirus type 3 Dearing strain was used for all infections and was obtained from Dr. Patrick Lee (Dalhousie University, Canada). To determine concentration and stability of reovirus, all batches were extensively titred by plaque assay in L929 cells [33].

2.4. Viability Assays

To determine cell survival after reoviral infection, crystal violet staining was used to quantify viable cells that adhered to the multi-well plate surface [68,69]. For viability assays reoviral infection took place one day after adherent cells were seeded into 24 well plates. Reoviral infection of murine embryoid bodies took place immediately as embryoid bodies were seeded into plates, or 10 days after seeding. In the case of human liver organoids, these were infected immediately or one day after seeding. A standard range of MOIs (multiplicity of infection or the ratio of plaque forming units, viral particles, per cell) from 1–100 was used. Reovirus was added to wells in a desired final volume and cells were left on a rocking platform at room temperature for 30 min. Infected cells were then placed in an incubator (at 37 °C with 5% CO_2 with 95% humidity) and left until the experimental endpoint was reached. Viability assays were conducted for a period of three days.

After reoviral infection medium was removed from the cells, and they were then fixed to the plate through the addition of 4% paraformaldehyde (PFA, Sigma) solution. The fixative solution was removed and plates allowed to air dry. Subsequently a 1% crystal violet (Sigma) solution was added to each well and plates incubated at room temperature for 10 min. The plates were then washed with water and allowed to air dry. The crystal violet stains were dissolved in a 1% sodium dodecyl sulfate (SDS, Sigma) solution. The optical density of each well was measured at 570 nm using a microplate reader. Experiments were conducted in replicates of three, and data were normalised to an untreated control that had 100% survival. This method resulted in low levels of residual staining of cellular debris even when all cells were killed, as verified by microscopy.

2.5. Reovirus Protein Synthesis
Immunostaining and Widefield Microscopy

Reoviral protein abundance and distribution were monitored through the use of a polyclonal rabbit anti-reovirus (serotype 3 Dearing) antibody serum [70]. This primary antibody was used in both Western Blot analysis and immunostaining. Adherent cell lines (cancer and stem cell) were cultured on chambered coverslips (Ibidi μ-Slide 8 Well). Cells were seeded at 100,000 cells per chamber and subsequently infected (MOI 1-100) one day later. Cells were usually fixed one day post-infection. In the case of embryoid bodies and liver organoids, cells were fixed two and three days post infection, respectively. All cells were washed with DPBS and then fixed by adding 200 μL of 4% PFA to each well. The plates were incubated at room temperature for 15 min. The cells were then washed with DPBS and permeabilised with 100% methanol and left to incubate for 8 min at −20 C. All slides were rinsed with DPBS twice and then incubated in a 5% bovine serum albumin (BSA, Sigma) solution in DPBS for 45 min at room temperature on a rocking platform. Primary antibody (polyclonal rabbit anti-reovirus, 1 in 5000 dilution) was added to the cells for a period of 2 h. Cells were then rinsed three times in DPBS. Following this a secondary antibody (donkey anti-rabbit-Alex647, Thermo Fisher Scientific 1:5000) solution was added to the cells for one hour at room temperature. Plates were covered in foil from this point onwards. Plates were then washed with DPBS and stained with 4′,6-diamidino-2-phenylindole (DAPI; Thermo Fisher Scientific) solution. Plates were incubated for 5 min on a rocking platform and rinsed with DPBS twice. Cells were stored in the dark at 4 °C until imaged on an Olympus IX71 Evolve wide-field microscope [71]. Image capture and analysis was performed using Volocity version 6.5.1.

2.6. Western Blots

After cells had been infected (MOI 1-10) they were lysed and proteins collected for Western blot analysis. Cells were lysed using 120 μL lysis buffer per well (12-well plate, 4×10^5 cells). Lysis buffer consisted of 1 M Tris-base (VWR), 8.75% glycerol, 10% SDS, 150 mM sodium chloride (Sigma) and 1% TritonX100 (Sigma). Protease inhibitor complex (Sigma) and 2 mM phenylmethylsulfonyl fluoride (PMSF, Sigma) was added fresh for every lysis reaction. Lysis buffer was added to the adherent cells and plates were placed on ice for 5 min while spent medium were centrifuged at $1000 \times g$ for 5 min. Lysates from the plates were collected and placed on pellets from the centrifuged medium. The combined cell lysates were then heated at 90 °C for 10 min.

Protein samples were then quantified using Detergent Compatible (DC) Protein Assay Kit (Bio-Rad). BSA was used for the standard curve and reactions conducted in 96-well format. A microplate reader was used to quantify the optical density of each sample at 750 nm.

Protein samples were then separated using sodium dodecyl sulphate-poly acrylamide gel electrophoresis (SDS-PAGE) with stacking gels at 5% while resolving gels were 12%. Proteins were transferred to nitrocellulose membranes and were incubated with primary antibodies (Supplementary Materials) overnight on a rocker at 4 C. Membranes were rinsed with TBST three times and incubated with appropriate (species specific) secondary antibodies that were conjugated to horseradish peroxidase (HRP). Membranes were washed four times in TBST and Amersham ECL Prime Western Blotting Detection Reagent (GE Healthcare, Chicago, IL, USA) was added to each membrane. Proteins were detected using the Bio-Rad ChemiDoc Touch Imaging system. Images of the membranes were captured and analysed using Image Lab version 5.2.

2.7. Bioinformatic Analysis
2.7.1. Analysis of Reoviral Sensitive vs. Resistant Cell Lines

To determine the extent of Yamanaka factor expression within reoviral resistant vs. sensitive cancer cell lines, a literature search was conducted in PubMed to compile a dataset of relevant cell lines. Cell lines were designated sensitive if cytopathic effects were noted after infection with an MOI 20 or lower. Robust multichip average (RMA) normalised expression data for 27 sensitive and 12 resistant cell lines from the Genomics of Drug Sensitivity in Cancer database were available for analysis [72]. These data were then collated and analysed using the R programming language.

A linear model for microarray data analysis (Limma) was performed to compare resistant and sensitive cell lines to determine differential gene expression profiles [73]. Four genes (SNAP25, PON2, PRCD, C9orf16) were determined to be differentially expressed (false discovery rate 0.1). Literature review revealed PON2 was implicated as a having a role in regulating stemness and as a result genes involved in regulating PON2 expression were investigated. Spearman correlation analysis was used to investigate the relationships between the continuous gene expression of PON2 and STAT5/KLF4. Spearman correlations were visualised using scatter plots.

2.7.2. Analysis of Gene Expression in Tumours

To further our analysis of the Yamanaka factors possibly involved in tumours, expression datasets from nearly 10,000 samples from 30 different cancers from The Cancer Genome Atlas (TCGA) were analysed [74]. Expression of the Yamanaka factors within the breast cancer subtype (BRCA) in the TCGA were log 2 transformed and Z-scores normalised to plot heat maps of expression. Hierarchical clustering was performed using Euclidean distancing. Spearman correlation coefficients were calculated for expression of OCT4, KLF4, SOX2, and MYC. A similar analysis was conducted in only breast cancer tumours with 1092 samples from the TCGA.

2.8. Statistical Analysis

Experimental raw data were normalised and maintained in Microsoft Excel. For viability assays infected cells (at various MOIs) were only compared to the untreated controls to determine whether a significant difference in the means was present. Two-tailed unpaired student's t-tests were used to compare treated groups to untreated controls, with p values of significance set at 0.05. Data presented show standard error of the mean. Inkscape version 0.92 was used to produce graphs and visualisations of the data.

3. Results

In our earlier studies of an attenuated reoviral strain, we noted that wild type reovirus (Dearing strain) was surprisingly capable of infecting and killing healthy (non-cancerous) murine embryonic stem cells (mESCs), despite previous assumptions that reovirus causes little or no harm to non cancerous cells [59]. We then showed that wild type reovirus was capable of inhibiting the growth in SCID mice of teratomas that normally arise after subcutaneous mESC transplantation [59]. The mechanism for why reovirus can infect some cancers and not others remains undefined and we speculated that an investigation into the similarities between stem cells and cancer cells provides a new avenue of investigation into what regulates this phenomenon. Although sensitivity of adult stem cells to reovirus has been investigated previously, particularly as groups are interested in using mesenchymal stem cells as carrier cells to deliver virus [75,76], no studies have investigated the sensitivity of pluripotent stem cells to reovirus. The first objective of this work was therefore to investigate both murine and human stem cell responses to reoviral infection.

3.1. Characterisation of Reoviral Cytolysis in Embryonic Stem Cells

To determine the extent to which murine ESCs were sensitive to reoviral infection, D3 embryonic stem cells were infected with reovirus with various multiplicities of infection (MOIs). L929 cells were used as a positive control in these experiments as these cells are known to be susceptible to reoviral infection and lysis (Figure 1A). It is clear that mESCs are exquisitely sensitive to reoviral infection and cytolysis when infected with MOIs of one (Figure 1A). To investigate whether these observations were restricted to murine cell types or if they were applicable to the human cells, similar experiments were performed in human embryonic stem cells lines H9 (hESCs). HeLa cells were used as a positive control. It was clear that human embryonic stem cells are also highly sensitive to reoviral infection (Figures 1B and S1). To establish if viral replication was occurring within mESCs and hESCs, Western blot analysis and immunofluorescence imaging were performed. Antibodies directed towards the capsid proteins of the virus were utilized to determine whether viral proteins were being produced, which is indicative of viral replication [70]. Western blots demonstrate that reovirus can effectively replicate in mESCs and hESCs (Figure 1C,D). These results were in accordance with the immunofluorescence imaging which showed that mESCs and hESCs support active production of reoviral particles (Figure 1E,F).

It is clear from these results that healthy, non-cancerous embryonic stem cells are highly susceptible to reoviral infection and cytolysis. Both mouse and human cell lines appear to be highly sensitive to infection with low concentrations of virus producing massive cell death within the population. Likewise, both murine and human embryonic stem cells are capable of supporting reoviral replication. Cellular machinery is recruited to produce viral capsid proteins, which is indicative of active viral replication. The observation that reovirus can infect and kill non-cancerous stem cells is fascinating as it has implications for its mechanism of selectivity. Cancerous cells are thought to either arise from stem cells that have undergone oncogenic mutation and transformation or from differentiated cells that have undergone a process of reversion to a more stem-like state. We infer these responses reflect similarities between cancer cells and embryonic cells, especially in responses to cytotoxic agents such as viruses. The differentiation status of the cell may influence embryonic and cancer cells' permissiveness to infection.

3.2. Characterisation of Reoviral Cytolysis in Induced Pluripotent Stem Cells

iPSCs offer a unique model in which to investigate the role that pluripotency plays in regulating cellular susceptibility to reoviral infection. These cells begin in a differentiated state and progress towards pluripotency upon introduction of the Yamanaka factors. This provides the opportunity to investigate whether a cell that is initially resistant to reovirus can be made susceptible by inducing a pluripotent state within the cell.

Murine embryonic fibroblast (MEF) cells are a mature cell type that is derived from connective tissue of the skeletal muscle of new-born mice and are frequently used to generate murine induced pluripotent stem cells (miPSC). In our lab, MEFs were transduced with lentiviral vectors carrying a payload of cDNA that encoded the four Yamanaka factors to produce a stable iPSC cell line [62]. Similarly, human induced pluripotent stem cells (hiPSC, 4YA cell line) were derived from human foreskin fibroblast (HFF) cells and were transduced with a lentiviral vector that stably integrates into genomic DNA and overexpresses the Yamanaka factors [64,77]. To determine whether the differentiation status influences reoviral selectivity, the differentiated cells (MEF and HFF) as well as the established induced pluripotent stem cells (miPSC and 4YA) were infected with reovirus. Sensitivity to the virus was subsequently determined through cell viability assays, Western blots and fluorescent imaging (Figure 2). Both murine and human induced pluripotent cells in both murine and human models are sensitive to infection and subsequent cell death (Figure 2A,B). The sensitivity was similar to that seen in embryonic stem cells. The more mature ancestral cell types however were resistant to cytolytic effects from the virus. Both Western blot analysis and immunofluorescence imaging confirmed the production of new viral capsid proteins one day after infection (MOI 10) in the pluripotent cells while differentiated cell types did not produce viral capsids proteins, indicating a lack of viral replication (Figure 2C–F).

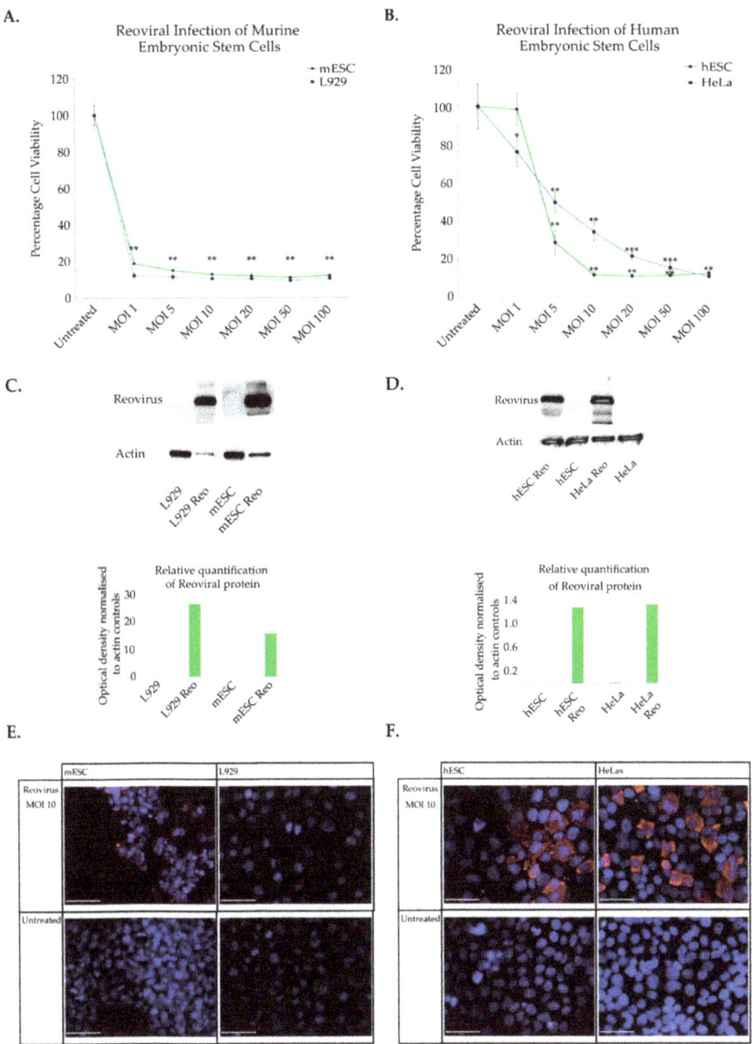

Figure 1. Characterisation of reoviral cytolysis and replication in cancer and embryonic stem cells. (**A**) mESCs (D3) and (**B**) hESCs (H9) cells were infected with wild type reovirus (T3D) twenty-four hours after seeding. Three days after infection cell viability was determined using crystal violet colorimetric analysis (residual staining represents remaining cell debris; all cells were killed). Absorbance was read at 570 nm and normalised to untreated controls. The assays were performed in triplicate; error bars represent standard deviation. Two-tailed paired student's *t*-tests were used to compare treated groups to untreated controls, * $p < 0.05$; ** $p < 0.01$ and *** $p < 0.001$. Viral protein synthesis in mESCs (**C**) and hESC (**D**) was initially confirmed via Western blot. Twenty-four hours after infection with MOI 10 cells were lysed and Western blots performed with antibodies directed toward viral capsids; experiments were performed in triplicates. Relative protein quantification was measured by normalizing capsid protein expression to actin controls. Immunofluorescence imaging was also performed to assess reoviral replication. mESCs (**E**) and hESCs (**F**) were infected at an MOI 10. One day post infection cells were fixed, permeabilised and stained with antibodies directed toward viral capsids. Red—reovirus, Blue—DAPI. Scale bar is 50 μm. L929 and HeLa cells were used as positive controls. Experiments were performed in triplicates with similar results, and representative images are shown.

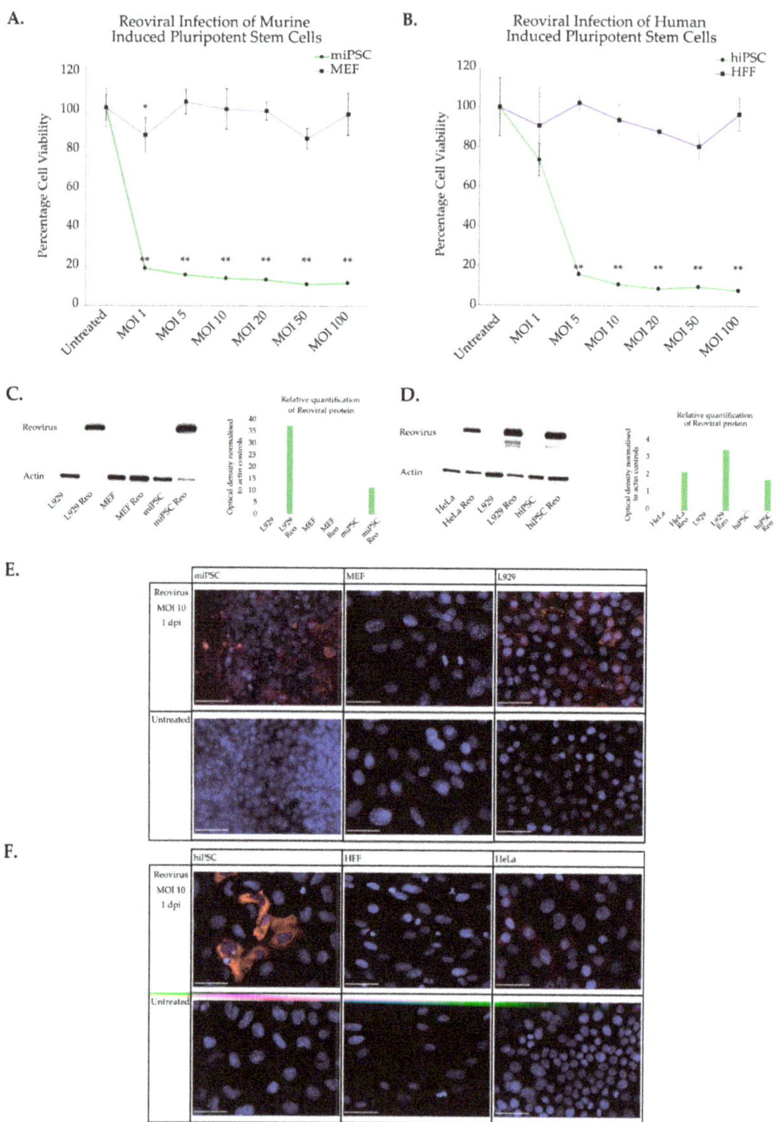

Figure 2. Induced pluripotent stem cell sensitivity to reoviral cytolysis and replication. (**A**) Murine differentiated cells (MEFs) and miPSCs were infected with reovirus and three days post infection were subject to crystal violet colorimetric analysis. (**B**) Similarly human foreskin fibroblasts and hiPSCs (4YA) cells were infected with wild type reovirus (T3D) twenty-four hours after seeding. Three days after infection cell viability was determined. Absorbance was read at 570 nm and normalised to untreated control. Residual staining reflects cellular debris. The assay was performed in triplicate (n = 3). Error bars represent standard deviation. Two-tailed paired student's *t*-tests were used to compare treated groups to untreated controls, * $p < 0.05$; ** $p < 0.01$. Viral replication in miPSCs (**C**) and hiPSC (**D**) was confirmed via Western blot one day after infection (MOI 10). Relative protein quantification was measured by normalizing capsid protein expression to actin controls. Immunofluorescence imaging was performed to assess reoviral replication. One day post infection (MOI 10) miPSCs (**E**) and hiPSCs (**F**) were fixed, permeabilised and stained. Red—reovirus, Blue—DAPI. Scale bar is 50 μm. L929 and HeLa cells were used as positive controls.

What is particularly interesting about the results presented here is that upon induction of pluripotency programs there is a shift in reoviral sensitivity from a resistant phenotype to that of a sensitive one. The sensitivity of the induced pluripotent cells is similar to that of their embryonic counter parts with murine cells being the most sensitive while human cells are slightly less so. The fact that cells become susceptible to infection upon reprogramming is a novel finding with intriguing implications, not only for reoviral biology but also for oncogenesis. Reprogramming appears to promote reoviral susceptibility, implicating the Yamanaka factors or their downstream targets in determining cellular permissiveness to reoviral replication. We may infer that pluripotency induction mimics certain aspects of oncogenic progression. The Yamanaka factors that are required for reprogramming are often upregulated or are overexpressed in a number of cancers [78]. A good example of this is the Yamanaka factor MYC, which is a well-known oncogene and accounts for a great deal of transcriptional similarity between iPSCs and cancer cells. It is intriguing then to speculate on what role, if any, the Yamanaka factors may have on reoviral replication.

3.3. Reoviral Cytolysis in Differentiated Cells

It is evident that induction of pluripotency can promote characteristics of a cell that make it susceptible to reovirus. It would therefore be important to know if stem cells (induced or embryonic) could be made refractory to reoviral infection through the loss of pluripotency by promoting differentiation.

In the murine PSC context, spontaneous differentiation can be induced through the removal of leukaemia inhibitory factor (LIF) from culture media. To promote differentiation, a hanging drop technique was used to form 3D spheres known as embryoid bodies, which are reflective of spontaneous cell differentiation [65,66]. Murine pluripotent stem cells (induced and embryonic) were induced to differentiate in hanging drops for a period of three days. Following this embryoid bodies were transferred to gelatin coated plates and allowed to differentiate. Time course analysis was performed and differentiated cells were infected immediately after their hanging drop treatment as they were transferred to gelatin plates (Figure 3A), or they were allowed to differentiate for a period of five or ten days following seeding. Beating cardiomyocytes after five days indicated that differentiation had occurred in some of the cells. Cells were infected with reovirus and crystal violet viability assays performed. Cells that had undergone differentiation were refractory to reovirus as there was very little change in their cell viability particularly when compared to their stem cell counter parts (Figures 1A and 2A). Surprisingly, this switch from susceptible cell type to refractory occurred relatively soon in the differentiation process where refractory cells were noted only three days after removal of LIF and suspension in hanging drops. Differentiated murine cells were then subject to immunofluorescence imaging and Western blot to determine if reoviral replication occurred within these cell types. Reovirus cann produce capsid protein as evidenced through Western blot analysis (Figure 3C). The amount of protein produced however is quite low relative to the susceptible L929 control cell lines. This trend was mirrored in the immunofluorescence images: some cells within the population appear to support reoviral replication, while the large majority of cells appear to be resistant to infection and concomitant production of capsid proteins (Figure 3G). The amount of infection seen in these cells is lower than in their pluripotent counterparts, particularly as the immunofluorescence in the differentiated cells was measured two days post infection. It appears that cells along the periphery of the embryoid bodies are infected and produce actively replicating virus. In general however, the differentiated cell populations produced much lower levels of reoviral protein when compared to susceptible controls and their stem cell counterparts.

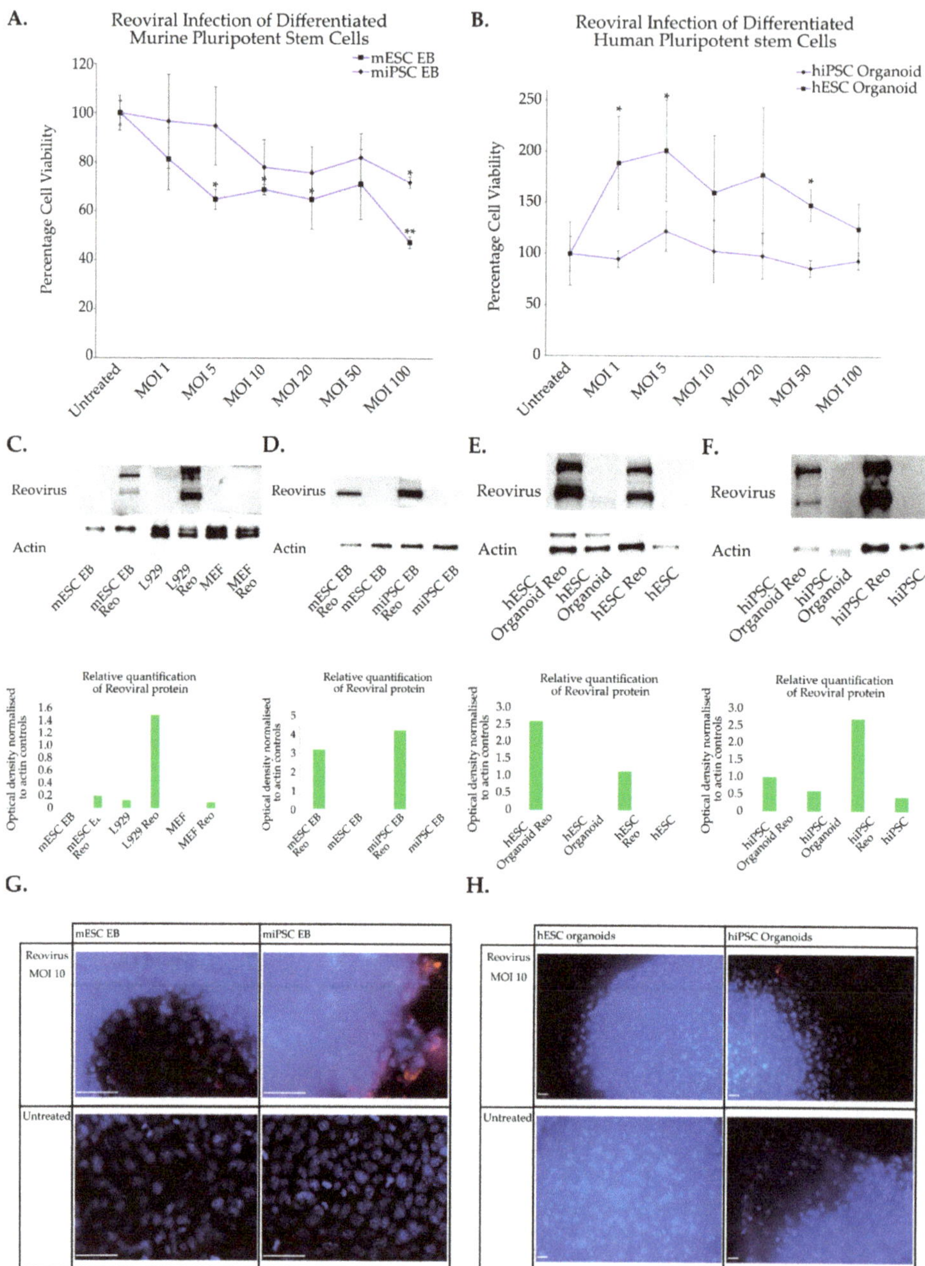

Figure 3. Reoviral sensitivity in differentiated stem cells. (**A**) Murine embryonic and induced pluripotent stem cells were induced to differentiate in hanging drops for three days to form embryoid bodies (EB). The cells were infected with reovirus and three days post infection crystal violet analysis was used to determine cell viability. Absorbance was read at 570 nm and normalised to untreated control. The assays were performed in triplicate (n = 3). Error bars represent standard deviation. Two-tailed paired student's *t*-test was used to compare treated groups to untreated controls, * $p < 0.05$;

** $p < 0.01$. (**B**) Human embryonic and induced pluripotent stem cells were differentiated into liver organoids. Three days post infection cell viability was determined using crystal violet viability assay. Replicates of six (n = 6) were used for hESCs organoids (H1 organoids) and replicates of nine (n = 9) for hiPSCs organoids (4YA), * $p < 0.05$. (**C**). Error bars represent standard deviation. Two-tailed paired student's *t*-test was used to compare treated groups to untreated controls. No biological replicates were performed for these experiments. (**C,D**) Murine EBs were then subjected to reoviral infection (MOI 10) for a period of 3 days and proteins were harvested and Western blots performed. (**E,F**) Human liver organoids were infected with reovirus (MOI 10) and three days post infection proteins was harvested for Western blot analysis. Stem cell controls were infected with MOI 10 and left for twenty-four hours before reoviral proteins were harvested. Relative protein quantification was measured by normalizing capsid protein expression to actin controls for all Western blots (**G,H**) Immunofluorescence imaging was performed on murine EBs and human liver organoids that had been subjected to reoviral infection (MOI 10) for two days and three days in murine and human samples respectively. Red—reovirus, Blue—DAPI. Scale bar is 50 μm.

In the case of human stem cells a method for directed differentiation was used. Cells were again encouraged to form 3D spheres in the Aggrewell system. These aggregates were then exposed to a variety of growth factors such as Activin A, Hepactocyte Growth Factor and Fibroblast Growth Factor 4 to promote endodermal differentiation [67]. Differentiation in human cells was allowed to proceed to stage three of the differentiation protocol where liver organoids are produced. Organoids were characterised using immunofluorescence imaging and the expression of liver-specific markers such as KRT7, CYP3A4, HNF4-α and Sox9 was confirmed (Figure S2). hESCs (H1) cells and hiPSCs (4YA) organoids were infected with reovirus and three days post-infection cell viability was determined. The cell viability was not reduced upon infection (Figure 3B) and was much higher when compared to stem cell populations (Figures 1B, 2B and S1). As organoids can vary in size and shape this creates variability in the data. To reconcile this difficulty additional replicates (H1 n = 6 and 4YA n = 9) were included in these experiments. To determine if reovirus could replicate within the organoids Western blotting and immunofluorescence imaging were employed. Western blots of the hESC organoids showed a relatively high production of reoviral proteins, especially when compared to embryonic stem cells (Figure 3E). This was unexpected and the result was not echoed in the immunofluorescence images captured from organoids where very little active reoviral replication was seen (Figure 3H). In differentiated hiPSCs organoids some reoviral protein was detectable on the Western blot, although it was much less than the protein production seen in hiPSCs (Figure 3F). Fluorescent imaging showed a similar situation where some of the peripheral cells support reoviral replication but the vast majority appeared refractory (Figure 3H).

In these experiments, murine and human stem cells become refractory to reoviral infection after differentiation has taken place. In the case of murine cells this switch from susceptible to refractory occurs relatively soon in the differentiation process, occurring three days after LIF removal. These differentiated cells appear to be resistant to reoviral lysis and there is a reduction in the amount of reoviral replication that is supported by the cells. Reoviral capsid proteins are still observed in cells after differentiation occurs. The production of reoviral proteins in some cells within differentiated cell populations is indicative of replication. This is perhaps not surprising given that differentiation is not a synchronous process and produces a population of heterogeneous cell types. The cells that sustain reoviral replication may be slower to differentiate and more stem-like in character. This may explain why only a few cells within the population support replication. This trend is similar in the differentiated hiPSCs (4YA) where cytolysis and reoviral replication are not supported after differentiation. This is in contrast to the differentiated hESCs (H1) where Western blots indicate that reoviral replication occurs after differentiation has occurred but the immunofluorescence images indicate that virus replication within this cell population is limited. From these results it appears that pluripotency can modulate reoviral sensitivity within healthy cells. This leads to a bigger question of whether cellular stemness varies within cancers such that the de-differentiated state within some cancers and not others contributes to reoviral selectivity.

3.4. Induced Pluripotent Cancer Stem Cell Sensitivity to Reovirus

To investigate whether pluripotency programming in cancer cells could modulate reoviral susceptibility, resistant mammary tumour cells were induced to become pluripotent through the use of PiggyBac transposon vectors carrying the Yamanaka factors. Once these induced pluripotent cancer stem cells were fully characterised, they were subjected to reoviral infection and their sensitivity was investigated.

To investigate the role of pluripotency in regulating reoviral selectivity in the context of a cancer, induced pluripotent cancer stem cells (ipCSC) were generated. A murine cell line (N_2O_2) derived from mammary tumours in transgenic female mice of the MMTV-Neu strain was tested for its sensitivity to reoviral cytolysis [60,79]. It was found that these cancerous cells were resistant to reoviral lysis (Figure 4A). The N_2O_2 cells were reprogrammed using a PiggyBac vector that expressed all four Yamanaka factors. Resulting ipCSCs (cell line 14ex-11) were characterised using immunofluorescence staining to detect OCT4, Nanog and SSEA-1 expression (Figure S3A). Teratoma assays were used to determine the differentiation potential of these cells and histochemical analysis revealed the formation of all three germ layers indicating pluripotent potential in these cells (Figure S3B). IpCSCs were subjected to reoviral infection and subsequent cell viability was determined (Figure 4A). The induction of pluripotency within cancer cells enhanced reoviral susceptibility and transitioned the cells from a resistant phenotype into a sensitive one. Western blot and immunofluorescence imaging assays were conducted to determine the extent to which reoviral replication occurs within parental cancer cells (N_2O_2) and ipCSCs. The ipCSCs and sensitive control L929 both produce reoviral capsid after initial infection while the N_2O_2 tumour cells were unable to support active replication (Figure 4B). Fluorescent imaging confirmed this finding showing that ipCSCs produced large amounts of reoviral protein while N_2O_2 cells did not (Figure 4C).

These results show that reoviral sensitivity can be introduced into an initially resistant breast cancer cell line through the induction of pluripotency programming. This is a compelling finding as reoviral sensitivity has only been induced in cancers previously through the activation of the RAS pathway. It is evident that pluripotency programming creates a cellular environment that is conducive to reoviral infection, replication and lysis. This strongly implicates reprogramming and pluripotency in the regulation of reoviral tropism and particularly, it points to the involvement of the Yamanaka factors in the mechanisms for regulating reoviral selectivity.

3.5. Bioinformatic Analysis of the Yamanaka Factors and Their Potential Role in Regulating Reoviral Sensitivity

To investigate the pathways responsible for cellular susceptibility to reovirus, analysis of the Yamanaka factors and the pathways they regulate was conducted, as one or a combination of these factors may influence cellular responsiveness to virus.

To begin the bioinformatic investigation into the possible pathways involved in regulating reoviral selection, a dataset with information regarding the sensitivity of cancer cell lines to reovirus was created. Cell lines were identified by review of published analyses of reoviral responses and deemed sensitive to reovirus if cytopathic effects were observed after infection at MOI 20 or lower. A data set of 106 cell lines confirmed as sensitive or resistant to virus was compiled, and among these normalised expression (RMA) data were found for 27 sensitive and 10 resistant cell lines in the Genomics of Drug Sensitivity in Cancer database.

A.

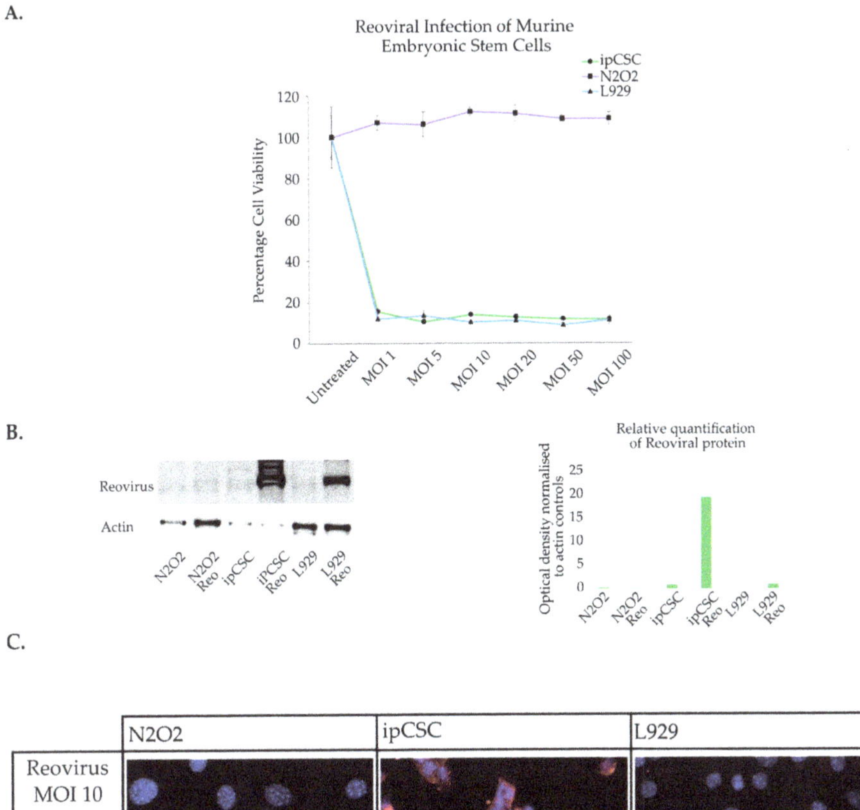

B.

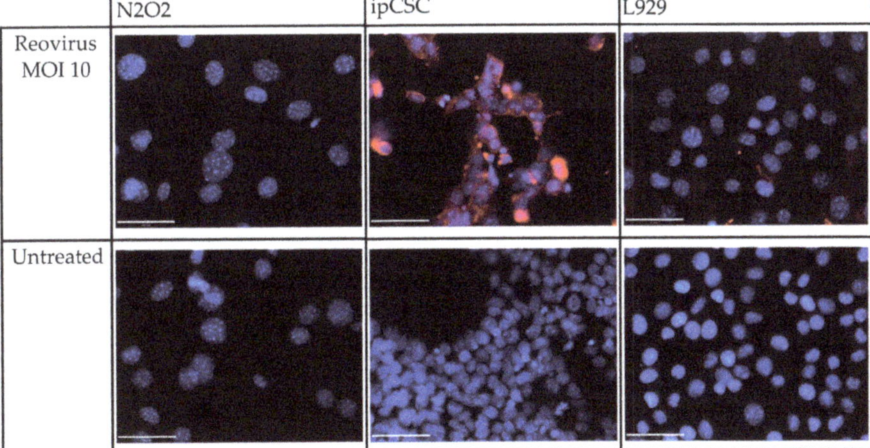

C.

<table>
<tr><td></td><td>N2O2</td><td>ipCSC</td><td>L929</td></tr>
<tr><td>Reovirus MOI 10</td><td></td><td></td><td></td></tr>
<tr><td>Untreated</td><td></td><td></td><td></td></tr>
</table>

Figure 4. Induced pluripotent cancer stem cell sensitivity to reovirus. (**A**) Cancer cell line N_2O_2 derived from mammary tumours in mice and cognate ipCSC (14ex-11) were infected with reovirus at various MOIs. L929 cells were used as a positive control. Cell viability was determined using crystal violet assays three days after infection. The assay was performed in triplicate (n = 3). Absorbance was read at 570 nm and normalised to untreated control. Error bars represent standard deviation. (**B**) Western blot was conducted on proteins extracted from N_2O_2 and ipCSC cells three days post infection (MOI 10). Relative protein quantification was measured from the Western blots by normalising capsid protein expression to actin controls. (**C**) Immunofluorescence imaging was also performed on these cell lines one day post infection (MOI 10). Red—reovirus, Blue—DAPI. Scale bar is 50 μm.

The analysis began by assessing differentially expressed genes in sensitive and resistant cell types. Differential expression was analysed using linear models for microarray data (limma) analysis to compare the resistant and sensitive cell lines. Four candidate genes were identified, namely *SNAP25, PON2, PRCD* and *C9orf16*. *PON2* showed promise as a candidate for further investigation, as lower expression levels of this gene are associated with differentiation. In *PON2* knockouts, mice have an increase in the fraction of differentiated stem cells [80]. *PON2* gene expression was significantly upregulated in reoviral sensitive cells (Figure 5A) which may be indicative of the stem-like nature of these cell types. STAT5B directly modulates PON2 activity and there is a significant negative association between PON2 and STAT5B expression in our cell lines of interest (Figure 5B) [81]. Interestingly, there was a trend toward down regulated STAT5B expression within reoviral sensitive cells lines (Figure 5C). This led to an investigation into the relationship between STAT5B, PON2 and the Yamanaka factors within the cell lines of interest. Of the four factors investigated KLF4 showed a significant negative correlation with STAT5B expression and a positive one with PON2 expression (Figure 5D,E). These associations provide intriguing new leads for further research and require further experimental validation.

To further the analysis of the potential involvement of Yamanaka factors in reoviral selectivity, expression data from TCGA were analysed to determine patterns of expression in cancers derived from patient samples. A heat map generated using data from TCGA (BRCA- breast cancers) showed that OCT4 and MYC were overexpressed in basal-like breast cancers while SOX2 and KLF4 were down regulated in this subtype (Figure 5F). A multivariate linear analysis was performed and indicated that OCT4, MYC and KLF4 were independently and significantly associated with the basal subtype. While OCT4 ($t = 27.175$; $p < 2 \times 10^{-16}$) and MYC ($t = 7.807$; $p = 1.38 \times 10^{-14}$) had a positive association, KLF4 ($t = -6.698$; $p = 3.39 \times 10^{-11}$) had a negative association.

The four Yamanaka factors do not function in isolation and they have specific patterns of expression that regulate reprogramming and induction of pluripotency. To determine if there was a correlation among the expression levels of the four Yamanaka factors in different cancers, a Spearman correlation coefficient was calculated for the expression of OCT4, KLF4, SOX2, MYC, in a pan-cancer analysis that included all cancer types (Figure 5G) as well as a targeted breast cancers analysis (Figure 5H). MYC and KLF4 expression showed the most highly correlated expression profile ($r = 0.35$, $p < 0.05$) in the pan-cancer analysis while OCT4 and MYC expression was most positively correlated in the breast cancer analysis ($r = 0.29$, $p < 0.05$). These mild associations may provide clues for further investigation into pluripotency regulation within cancers and how this may be involved in contributing to susceptibility to reoviral cytolysis.

This bioinformatic analysis of resistant and sensitive cell lines brought novel leads for mechanistic insight into view. The role of PON2 in reoviral replication and cytolysis has yet to be determined. The interactions between KLF4, STAT5B and PON2 in reoviral tropism present an exciting opportunity for further assessment and may have a novel role in viral replication that has yet to be elucidated. Expression of OCT4 and MYC present an interesting avenue of investigation, particularly in basal breast cancer subtypes and may yield interesting insights into the mechanisms regulating reoviral selectivity in breast cancers. It is interesting to speculate about the relevance of these associations, as basal breast cancer are more stem-like in their characteristics, which implies these tumour types may be more responsive to reoviral therapy. Additionally, there are numerous redundant pathways that regulate pluripotency such as sonic hedgehog, notch, WNT, bone morphogenic protein, and Janus family kinase (JFK) pathways that may also contribute to reoviral selectivity but were not investigated in this study. The fact that embryonic and induced pluripotent stem cells are exquisitely sensitive to infection while their differentiated counterparts remain refractory are strong indicators that either the Yamanaka factors or additional dependent pluripotency programs are involved.

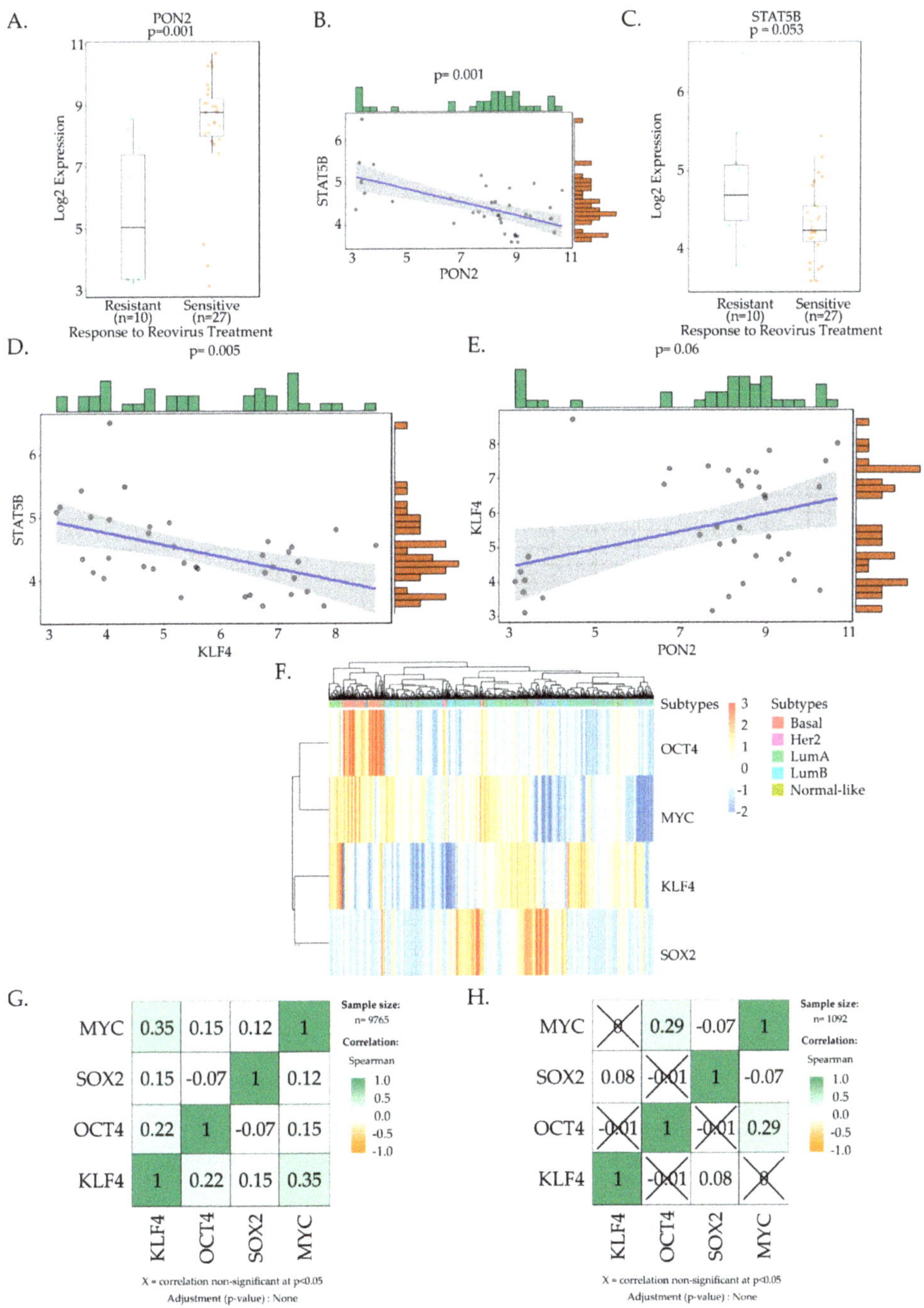

Figure 5. Bioinformatic analysis of Yamanaka factor expression and reoviral selectivity. (**A**) A Wilcoxon rank sum test was used to compare *PON2* expression in resistant and sensitive cell types while box plots indicate the distribution of *PON2* expression data within this dataset, $p = 0.001$. (**B**) Spearman correlation analysis was used to investigate the correlation between *PON2* and *STAT5B* expression.

STAT5B is negatively associated with *PON2* expression in cancer cell lines $p < 0.001$. (**C**) STAT5B expression is down regulated in cancer cells that are sensitive to reoviral cytolysis. A Wilcoxon rank sum test was used to compare *STAT5B* expression in resistant and sensitive cancer cell types, $p = 0.053$. Box plots indicate the distribution of *STAT5B* expression. (**D**) To investigate the relationship between *KLF4* and *STAT5B* expression Spearman correlation analysis was performed. *KLF4* expression negatively correlates with *STAT5B* expression r $= -0.45$ and $p = 0.005$. (**E**) *KLF4* expression is positively associated with *PON2* expression in cancer cell lines. Spearman correlation analysis was performed and r $= 0.31$, $p = 0.06$. (**F**) The Heat map demonstrates the relationship between the expression levels of each of the Yamanaka factors in differing breast cancer subtypes. Expression values were log2 transformed and Z-scores normalised. Hierarchical clustering was performed using Euclidean distancing. Additional annotation tracks were included for the breast cancer subtypes. (**G**) Associations between the Yamanaka factor expression levels were analysed in a group of 30 different cancer types from the TCGA. Expression data from nearly 10 000 samples were evaluated and Spearman correlation coefficients were calculated for *OCT4, KLF4, SOX2, MYC*. $p < 0.05$. Mild associations between *KLF4* and *OCT4* as well as *KLF4* and *MYC* expression were identified. (**H**) Expression data for breast cancers were taken from the TCGA database and assessed to determine the presence of associations between Yamanaka factor expression levels. Gene expression data from 1092 samples was considered while performing Spearman correlation calculations *for OCT4, KLF4, SOX2* and *MYC*. $p < 0.05$. Mild associations between *MYC* and *OCT4* expression were noted in this cancer subtype.

4. Discussion

Oncolytic viral therapies are an emerging anti-cancer strategy. Reovirus is a well-studied OV that is the only wild type virus to gain FDA approval. However, despite the initial pre-clinical promise of reovirus, results from the clinic have shown benefits for only a fraction of patients, and consequently there have been efforts to improve therapeutic efficacy [23,82–85]. The mechanisms governing reoviral tropism in cancers are not well defined and are a topic of debate. Reovirus is capable of infecting and lysing cancer cells and cancer stem cells, and, as shown here, healthy ESCs. Intriguingly, adult (mesenchymal and hematopoietic) stem cells appear to be refractory to infection [76,86]. In light of the recent view that pluripotency induction shares similar pathways with carcinogenesis, it is possible to suggest that the same subset of genes that are activated in some cancers may also be up-regulated in pluripotent cells, thereby rendering them permissive to infection. This led us to speculate that the pathways responsible for stemness in cancer and in embryonic cells also modulate reoviral susceptibility. Here we characterised the ability of reovirus to infect and kill both murine and human embryonic stem cells. We further demonstrated that reovirus can lyse induced pluripotent stem cells while their original, non-reprogrammed counterparts remained refractory. Differentiation of both embryonic and induced pluripotent stem cells altered the cells' reoviral sensitive state to a resistant one. Additionally, murine mammary cancer cells that were non-responsive to reoviral lysis were reprogrammed using the Yamanaka factors. The newly established ipCSCs were strikingly susceptible to infection and oncolysis. Finally, we briefly investigated the potential role of the Yamanaka factors in regulating this switch through bioinformatic analysis. These data strongly suggested that stemness and the differentiation status of a cell may contribute to regulating the successful oncolytic infection of reovirus.

Viruses such as reovirus with inherent tropism for transformed cells are potentially a powerful tool that can be used to understand mechanisms of oncolytic selection, which may contribute to the development of novel cancer specific therapies. The outcome of successful viral replication and lysis is thought to be largely dependent on cellular signalling. Innumerable changes within the cell are required for oncogenesis, making it difficult to pinpoint the essential targets for successful therapies. The improvement of viruses as oncolytic agents hinges on understanding the unique set of conditions that provide a beneficial niche for viral replication. Elucidating the pathways governing reoviral selectivity will aid the development of biomarker panels for patient screening and enable the rational design of drug combination strategies for enhancing reoviral efficacy. Additionally, an understanding of the basis for tumour-selective replication could result in the discovery of new targets for small molecules and aid in the development of novel engineering strategies for OVs.

Precision medicine strategies attempt to improve the application of anti-cancer therapies by using biomarkers to predict a patient's response to a therapy. This approach takes into account the variability of the tumour's genetic landscape, patient characteristics and cellular environment to determine the most effective therapeutic approach to treating a specific cancer [87,88]. Establishing a panel of predictive markers for response to oncolytic viruses could help maximise clinical response and minimize off-target adverse events [88]. Oncolytics Biotech Inc. has already begun screening patients in their clinical trials based on KRAS mutation, BRAF mutation, and EGFR mutational status and amplification (NCT00861627, NCT01274624) [24]. Even so, despite intense research into reoviral replication there is no single biomarker that unambiguously indicates whether cytolytic infections will occur. This is perhaps not surprising as a multitude of signalling events must coincide to enable for productive infections. Delivery, attachment, entry, replication and apoptosis induction are all collectively required for successful oncolytic replication. Based upon the research described here, stemness may in the future be added to the list of requirements for successful infection and lysis. Indeed, Dick's group developed a 17 gene stemness score to predict patient responses in acute leukaemia. It is therefore feasible that this type of biomarker panel can also be used to predict responses to reoviral treatment in the future [89]. Factors such as JAM-A, L and B cathepsin and β-integrin expression, RAS activation, and Yamanaka factor expression may all contribute to permissive phenotypes but not be permissive for replication in isolation [32,36,90–92]. A better understanding of the molecular mechanisms governing selective reoviral replication (for example, might Yamanaka factor expression modulate other pathways, such as JAM-A, cathepsins, etc.?) will aid in the development of a panel of biomarkers that can accurately and effectively be used to predict oncolytic success in patients, thereby optimising treatments.

The complexity and heterogeneity of tumours as well as their ability to develop resistance to single treatments (monotherapy) has fueled investigations into combination therapy for cancers. While oncolytic vectors are multimodal in their approach to eliminating cancers, their combination with standard treatment options and other viruses are topics of considerable research in a bid to improve their efficacy in the clinic [24,25]. Reovirus is an ideal candidate for combination therapeutics as it has minimal side effects in patients. In an attempt to rapidly translate reovirus from bench to bedside it has already been tested in combination with numerous chemotherapeutics as well as radiation therapy, thereby enhancing efficacy of both [8,18,23]. Exploration and discovery of the pathways and mechanisms responsible for reoviral selective infection will permit a more rational approach to drug combinations, allowing for the development of perhaps unexpected yet synergistic anti-cancer strategies. This may be particularly evident when considering stemness as a factor modulating reoviral tropism.

The next step in furthering the investigation into the role of pluripotency and stemness in regulating reoviral selectivity would be to infect teratomas in murine models and employ histochemical analysis and single cell sequencing to investigate the expression profile of cells within a tumour that is readily infected by reovirus. This would be fascinating in the context of a teratoma model as these tumours can produce differentiated tissue types from all three germ layers (mesoderm, ectoderm and endoderm). We earlier showed that wild type reovirus is capable of reducing the size of teratomas in NOD/SCID mice, although we did not investigate which cells within these tumours were readily infected [59]. It would be interesting to determine if differentiated tissues within the teratomas are susceptible or whether only stem-like cells within a tumour support infection and lysis. A similar approach could be taken with investigations into patient biopsies. Further investigations into the expression profiles of human cancer stem cells derived from breast tumours that have been shown to be responsive to reoviral infection may also help with this analysis [57]. The extent to which the Yamanaka factors are expressed in these cells may yield further insights into their role in regulating reoviral selection.

Although our bioinformatic analysis did not definitively elucidate the mechanism responsible for reoviral selectivity, it has provide novel avenues for investigation. In cell line analysis KLF4/STAT5B and PON2 signalling may contribute to regulating reoviral permissiveness. As KLF4 acts in concert with SOX2 and OCT4 it would be interesting to determine the combined roles of these factors in establishing reoviral susceptibility. The bioinformatic analysis of TCGA suggests that combinations of MYC and KLF4 as well as OCT4 and MYC were important for permissive cancers. It would also be interesting to determine the role of c-MYC in reoviral replication and whether it is an essential factor in regulating sensitivity. Reprogramming cells using only OCT4, SOX2 and KLF4 would be an interesting approach to testing this, as these factors can reprogram cells with reduced efficacy. Additionally, it would be intriguing to determine what role if any Nanog plays in establishing a suitable niche for reoviral replication, as stemness appears to be a key factor influencing replication. It may be of value to reprogram cells using entirely separate factors such as Lin28, Oestrogen related receptor β, Nanog, nuclear receptor subfamily 5 group A member 2 and TCL1A, all of which can promote reprograming by replacing OSKM [93]. This may be another approach to further investigate the extent to which pluripotency and stemness regulate reoviral selectivity.

Finally, if our proposal that cellular stemness modulates reoviral oncolysis is correct, we are left with the challenge of reconciling this explanation with that of Lee's group, which earlier posited a critical role for Ras pathway activation in promoting reoviral oncolysis [6–35]. As we and others have observed (see our accompanying manuscript [94]), the correlation between Ras activation and reoviral activity is good but imperfect, with many exceptions. Thus we are led to speculate that cells able to adapt to chronic Ras-mediated signalling may become cancerous, in part, by sometimes activating stemness pathways that also confer reoviral susceptibility. We are therefore intrigued by the recent observation that Ras expression can modulate miRNA signalling, opening new possibilities for altering multiple downstream pathways, including cellular stemness [95]. In addition, an investigation of primitive gut stem cells or induced stem cells in the digestive tract that are possible natural hosts for viral infection in humans and other animals could yield further insights into mechanisms of viral cytolysis [96]. Clearly much further work would be needed to explore this or other mechanisms that may link variable Ras activation, cellular stemness, oncogenic progression and reoviral susceptibility.

5. Conclusions

Many factors may contribute to reoviral selectivity in cancer cells. For the first time, this study demonstrates that reprogramming differentiated cells into more stem-like progeny can make them permissive to reoviral cytolysis. Induction of stemness phenotypes promotes reoviral infection in previously resistant cell types in both healthy and cancerous cell lines. The regulatory networks involved in reprogramming have not yet been fully investigated for their involvement in regulating reoviral tropism. It is evident that the stem cell properties of cancers contribute to the most dangerous and clinically relevant aspects of cancer biology. Understanding the similarities and differences in the responses to reoviral infection could lead to the discovery of novel strategies for cancer therapeutic intervention. Further investigation into the regulatory networks within the cell that facilitate reoviral tropism are essential for advancing reovirus as an oncolytic agent. This fundamental knowledge will enable biomarker selection, rational drug combinations and ultimately contribute to enhancing the safety, specificity and potency of reoviral therapy.

Supplementary Materials: The following supporting information can be downloaded at: https://www.mdpi.com/article/10.3390/v15071473/s1, File S1; Figure S1: Reovirus mediated cytolytic killing in human embryonic stem cells; Figure S2: Confirmation of differentiated Human liver organoids; Figure S3: Characterisation of ipCSC.

Author Contributions: Conceptualization, T.B., D.R. and R.N.J.; methodology, T.B. and L.R.; software, M.K. and P.B.; formal analysis, T.B. and M.K.; investigation, T.B. and L.R.; resources, D.R. and R.N.J.; writing—original draft preparation, T.B.; writing—review and editing, T.B., D.R. and R.N.J.; visualization, T.B.; supervision, D.R. and R.N.J.; project administration T.B.; funding acquisition, D.R. and R.N.J. All authors have read and agreed to the published version of the manuscript.

Funding: Supported in part by funds from the Canadian Breast Cancer Foundation (RNJ) (funding code:B34) and the Canadian Institutes of Health Research (RNJ and DER) (funding code: MOP-82850, RMF-82497).

Data Availability Statement: Data presented here may be obtained by contacting TB or RNJ.

Acknowledgments: We would like to thank and acknowledge the following people for the contribution of materials to this project: James Ellis' laboratory at the University of Toronto for the donation of BJ cell lines. Patrick Lee, Dalhousie University, Halifax, Nova Scotia for the donation of wild type Reovirus (Type 3 Dearing). We also thank Y.V. Shulga for expert assistance in the conduct of experiments and the provision of reagents.

Conflicts of Interest: The authors declare no conflict of interest. The funders had no role in the design of the study; in the collection, analyses, or interpretation of data; in the writing of the manuscript, or in the decision to publish the results.

References

1. Bischoff, J.R.; Kirn, D.H.; Williams, A.; Heise, C.; Horn, S.; Muna, M.; Ng, L.; Nye, J.A.; Sampson-Johannes, A.; Fattaey, A.; et al. An adenovirus mutant that replicates selectively in p53-deficient human tumor cells. *Science* **1996**, *274*, 373–376. [CrossRef] [PubMed]
2. Andtbacka, R.H.I.; Collichio, F.A.; Amatruda, T.; Senzer, N.N.; Chesney, J.; Delman, K.A.; Spitler, L.E.; Puzanov, I.; Doleman, S.; Ye, Y.; et al. OPTiM: A randomized phase III trial of talimogene laherparepvec (T-VEC) versus subcutaneous (SC) granulocyte-macrophage colony-stimulating factor (GM-CSF) for the treatment (tx) of unresected stage IIIB/C and IV melanoma. *J. Clin. Oncol.* **2013**, *31* (Suppl. S18), LBA9008. [CrossRef]
3. Senzer, N.N.; Kaufman, H.L.; Amatruda, T.; Nemunaitis, M.; Reid, T.; Daniels, G.; Gonzalez, R.; Glaspy, J.; Whitman, E.; Harrington, K.; et al. Phase II clinical trial of a granulocyte-macrophage colony-stimulating factor-encoding, second-generation oncolytic herpesvirus in patients with unresectable metastatic melanoma. *J. Clin. Oncol.* **2009**, *27*, 5763–5771. [CrossRef]
4. Oncolytics Biotech®Inc. Oncolytics Biotech®Inc. Announces Receipt of Orphan Drug Status from the EMA for Pancreatic Cancer. Available online: http//www.oncolyticsbiotech.com/news/press-release-details/2015/Oncolytics-Biotech-Inc-Announces-Receipt-of-Orphan-Drug-Status-from-the-EMA-for-Pancreatic-Cancer/default.aspx (accessed on 9 October 2015).
5. Oncolytics Biotech®Inc. Oncolytics Biotech®Inc. Announces Receipt of Orphan Drug Designation from the U.S. FDA for Gastric Cancer. Available online: http://www.oncolyticsbiotech.com/news/press-release-details/2015/Oncolytics-Biotech-Inc-Announces-Receipt-of-Orphan-Drug-Designation-from-the-US-FDA-for-Gastric-Cancer/default.aspx (accessed on 9 October 2015).
6. Galanis, E.; Markovic, S.N.; Suman, V.J.; Nuovo, G.J.; Vile, R.G.; Kottke, T.J.; Nevala, W.K.; Thompson, M.A.; Lewis, J.E.; Rumilla, K.M.; et al. Phase II trial of intravenous administration of Reolysin((R)) (Reovirus Serotype-3-dearing Strain) in patients with metastatic melanoma. *Mol. Ther.* **2012**, *20*, 1998–2003. [CrossRef]
7. Coffey, M.C.; Strong, J.E.; Forsyth, P.A.; Lee, P.W. Reovirus therapy of tumors with activated Ras pathway. *Science* **1998**, *282*, 1332–1334. [CrossRef]
8. Harrington, K.J.; Vile, R.G.; Melcher, A.; Chester, J.; Pandha, H.S. Clinical trials with oncolytic reovirus: Moving beyond phase I into combinations with standard therapeutics. *Cytokine Growth Factor Rev.* **2010**, *21*, 91–98. [CrossRef]
9. Wilcox, M.E.; Yang, W.; Senger, D.; Rewcastle, N.B.; Morris, D.G.; Brasher, P.M.; Shi, Z.Q.; Johnston, R.N.; Nishikawa, S.; Lee, P.W.; et al. Reovirus as an oncolytic agent against experimental human malignant gliomas. *J. Natl. Cancer Inst.* **2001**, *93*, 903–912. [CrossRef]
10. Morris, D.G.; Feng, X.; DiFrancesco, L.M.; Fonseca, K.; Forsyth, P.A.; Paterson, A.H.; Coffey, M.C.; Thompson, B. REO-001: A phase I trial of percutaneous intralesional administration of reovirus type 3 dearing (Reolysin(R)) in patients with advanced solid tumors. *Investig. New Drugs* **2013**, *31*, 696–706. [CrossRef]
11. Forsyth, P.; Roldan, G.; George, D.; Wallace, C.; Palmer, C.A.; Morris, D.; Cairncross, G.; Matthews, M.V.; Markert, J.; Gillespie, Y.; et al. A phase I trial of intratumoral administration of reovirus in patients with histologically confirmed recurrent malignant gliomas. *Mol. Ther.* **2008**, *16*, 627–632. [CrossRef]

12. Fukuhara, H.; Ino, Y.; Todo, T. Oncolytic virus therapy: A new era of cancer treatment at dawn. *Cancer Sci.* **2016**, *107*, 1373–1379. [CrossRef] [PubMed]

13. Fields, B.N.; Knipe, D.M.; Howley, P.M. *Fields Virology*; Wolters Kluwer Health/Lippincott Williams & Wilkins: Philadelphia, PA, USA, 2013.

14. Shatkin, A.J.; Sipe, J.D.; Loh, P. Separation of ten reovirus genome segments by polyacrylamide gel electrophoresis. *J. Virol.* **1968**, *2*, 986–991. [CrossRef]

15. Hashiro, G.; Loh, P.C.; Yau, J.T. The preferential cytotoxicity of reovirus for certain transformed cell lines. *Arch. Virol.* **1977**, *54*, 307–315. [CrossRef] [PubMed]

16. Duncan, M.R.; Stanish, S.M.; Cox, D.C. Differential sensitivity of normal and transformed human cells to reovirus infectin. *J. Virol.* **1978**, *28*, 444–449. [CrossRef]

17. Gong, J.; Sachdev, E.; Mita, A.C.; Mita, M.M. Clinical development of reovirus for cancer therapy: An oncolytic virus with immune-mediated antitumor activity. *World J. Methodol.* **2016**, *6*, 25–42. [CrossRef]

18. Clements, D.; Helson, E.; Gujar, S.A.; Lee, P.W. Reovirus in cancer therapy: An evidence-based review. *Oncolytic Virotherapy* **2014**, *3*, 69–82.

19. Oncolytics Biotech®Inc. Oncolytics Biotech®Inc. Announces Receipt of Orphan Drug Status from the EMA for Ovarian Cancer. Available online: https://www.oncolyticsbiotech.com/press-releases/detail/340/oncolytics-biotech-inc-announces-receipt-of-orphan-drug (accessed on 30 July 2018).

20. Oncolytics Biotech®Inc. Oncolytics Biotech®Receives Special Protocol Assessment Agreement from FDA for Phase 3 Clinical Trial of Pelareorep in Metastatic Breast Cancer. Available online: https://www.oncolyticsbiotech.com/press-releases/detail/412/oncolytics-biotech-receives-special-protocol-assessment (accessed on 9 October 2015).

21. Oncolytics Biotech®Inc. Oncolytics Biotech®Inc. Announces FDA Fast Track Designation for REOLYSIN®in Metastatic Breast Cancer. Available online: https://www.oncolyticsbiotech.com/press-releases/detail/39/oncolytics-biotech-inc-announces-fda-fast-track (accessed on 9 October 2015).

22. Joklik, W.K. Recent progress in reovirus research. *Annu. Rev. Genet.* **1985**, *19*, 537–575. [CrossRef] [PubMed]

23. Masemann, D.; Boergeling, Y.; Ludwig, S. Employing RNA viruses to fight cancer: Novel insights into oncolytic virotherapy. *Biol. Chem.* **2017**, *398*, 891–909. [CrossRef] [PubMed]

24. Zhao, X.; Chester, C.; Rajasekaran, N.; He, Z.; Kohrt, H.E. Strategic combinations: The future of oncolytic virotherapy with reovirus. *Mol. Cancer Ther.* **2016**, *15*, 767–773. [CrossRef]

25. Marchini, A.; Scott, E.; Rommelaere, J. Overcoming Barriers in Oncolytic Virotherapy with HDAC Inhibitors and Immune Checkpoint Blockade. *Viruses* **2016**, *8*, 9. [CrossRef]

26. Cattaneo, R.; Miest, T.; Shashkova, E.V.; Barry, M.A. Reprogrammed viruses as cancer therapeutics: Targeted, armed and shielded. *Nat. Rev. Microbiol.* **2008**, *6*, 529–540. [CrossRef] [PubMed]

27. Chaurasiya, S.; Hew, P.; Crosley, P.; Sharon, D.; Potts, K.; Agopsowicz, K.; Long, M.; Shi, C.; Hitt, M. Breast cancer gene therapy using an adenovirus encoding human IL-2 under control of mammaglobin promoter/enhancer sequences. *Cancer Gene Ther.* **2016**, *23*, 178–187. [CrossRef]

28. Harrington, K.; Freeman, D.J.; Kelly, B.; Harper, J.; Soria, J.-C. Optimizing oncolytic virotherapy in cancer treatment. *Nat. Rev. Drug Discov.* **2019**, *18*, 689–706. [CrossRef] [PubMed]

29. Miest, T.S.; Cattaneo, R. New viruses for cancer therapy: Meeting clinical needs. *Nat. Rev. Microbiol.* **2014**, *12*, 23–34. [CrossRef] [PubMed]

30. Kobayashi, T.; Antar, A.A.; Boehme, K.W.; Danthi, P.; Eby, E.A.; Guglielmi, K.M.; Holm, G.H.; Johnson, E.M.; Maginnis, M.S.; Naik, S. A plasmid-based reverse genetics system for animal double-stranded RNA viruses. *Cell Host Microbe* **2007**, *1*, 147–157. [CrossRef] [PubMed]

31. Van den Wollenberg, D.J.; Dautzenberg, I.J.; Ros, W.; Lipińska, A.D.; van den Hengel, S.K.; Hoeben, R.C. Replicating reoviruses with a transgene replacing the codons for the head domain of the viral spike. *Gene Ther.* **2015**, *22*, 267–279. [CrossRef]

32. Bourhill, T.; Mori, Y.; Rancourt, D.E.; Shmulevitz, M.; Johnston, R.N. Going (reo) viral: Factors promoting successful reoviral oncolytic infection. *Viruses* **2018**, *10*, 421. [CrossRef]

33. Strong, J.E.; Tang, D.; Lee, P.W. Evidence that the epidermal growth factor receptor on host cells confers reovirus infection efficiency. *Virology* **1993**, *197*, 405–411. [CrossRef]

34. Strong, J.E.; Lee, P. The v-erbB oncogene confers enhanced cellular susceptibility to reovirus infection. *J. Virol.* **1996**, *70*, 612–616. [CrossRef]

35. Norman, K.L.; Lee, P.W. Reovirus as a novel oncolytic agent. *J. Clin. Investig.* **2000**, *105*, 1035–1038. [CrossRef]

36. Strong, J.E.; Coffey, M.C.; Tang, D.; Sabinin, P.; Lee, P.W. The molecular basis of viral oncolysis: Usurpation of the Ras signaling pathway by reovirus. *EMBO J.* **1998**, *17*, 3351–3362. [CrossRef]

37. Norman, K.L.; Hirasawa, K.; Yang, A.-D.; Shields, M.A.; Lee, P.W. Reovirus oncolysis: The Ras/RalGEF/p38 pathway dictates host cell permissiveness to reovirus infection. *Proc. Natl. Acad. Sci. USA* **2004**, *101*, 11099–11104. [CrossRef]

38. Norman, K.L.; Farassati, F.; Lee, P.W.K. Oncolytic viruses and cancer therapy. *Cytokine Growth Factor. Rev.* **2001**, *12*, 271–282. [CrossRef] [PubMed]

39. Gong, J.; Mita, M.M. Activated ras signaling pathways and reovirus oncolysis: An update on the mechanism of preferential reovirus replication in cancer cells. *Front. Oncol.* **2014**, *4*, 167. [CrossRef] [PubMed]
40. Shmulevitz, M.; Marcato, P.; Lee, P.W. Unshackling the links between reovirus oncolysis, Ras signaling, translational control and cancer. *Oncogene* **2005**, *24*, 7720–7728. [CrossRef]
41. Van Den Wollenberg, D.J.; Van Den Hengel, S.K.; Dautzenberg, I.J.; Kranenburg, O.; Hoeben, R.C. Modification of mammalian reoviruses for use as oncolytic agents. *Expert. Opin. Biol. Ther.* **2009**, *9*, 1509–1520. [CrossRef] [PubMed]
42. Thomis, D.C.; Samuel, C.E. Mechanism of interferon action: Evidence for intermolecular autophosphorylation and autoactivation of the interferon-induced, RNA-dependent protein kinase PKR. *J. Virol.* **1993**, *67*, 7695–7700. [CrossRef]
43. Panniers, R.; Henshaw, E.C. A GDP/GTP exchange factor essential for eukaryotic initiation factor 2 cycling in Ehrlich ascites tumor cells and its regulation by eukaryotic initiation factor 2 phosphorylation. *J. Biol. Chem.* **1983**, *258*, 7928–7934. [CrossRef]
44. Bhat, M.; Robichaud, N.; Hulea, L.; Sonenberg, N.; Pelletier, J.; Topisirovic, I. Targeting the translation machinery in cancer. *Nat. Rev. Drug. Discov.* **2015**, *14*, 261–278. [CrossRef]
45. Krishnamoorthy, T.; Pavitt, G.D.; Zhang, F.; Dever, T.E.; Hinnebusch, A.G. Tight binding of the phosphorylated alpha subunit of initiation factor 2 (eIF2alpha) to the regulatory subunits of guanine nucleotide exchange factor eIF2B is required for inhibition of translation initiation. *Mol. Cell Biol.* **2001**, *21*, 5018–5030. [CrossRef]
46. Terasawa, Y.; Hotani, T.; Katayama, Y.; Tachibana, M.; Mizuguchi, H.; Sakurai, F. Activity levels of cathepsins B and L in tumor cells are a biomarker for efficacy of reovirus-mediated tumor cell killing. *Cancer Gene Ther.* **2015**, *22*, 188–197. [CrossRef]
47. Smakman, N.; van den Wollenberg, D.J.; Borel Rinkes, I.H.; Hoeben, R.C.; Kranenburg, O. Sensitization to apoptosis underlies KrasD12-dependent oncolysis of murine C26 colorectal carcinoma cells by reovirus T3D. *J. Virol.* **2005**, *79*, 14981–14985. [CrossRef]
48. Smakman, N.; van den Wollenberg, D.J.; Elias, S.G.; Sasazuki, T.; Shirasawa, S.; Hoeben, R.C.; Borel Rinkes, I.H.; Kranenburg, O. KRAS(D13) Promotes apoptosis of human colorectal tumor cells by ReovirusT3D and oxaliplatin but not by tumor necrosis factor-related apoptosis-inducing ligand. *Cancer Res.* **2006**, *66*, 5403–5408. [CrossRef]
49. Van Houdt, W.J.; Smakman, N.; van den Wollenberg, D.J.; Emmink, B.L.; Veenendaal, L.M.; van Diest, P.J.; Hoeben, R.C.; Borel Rinkes, I.H.; Kranenburg, O. Transient infection of freshly isolated human colorectal tumor cells by reovirus T3D intermediate subviral particles. *Cancer Gene Ther.* **2008**, *15*, 284–292. [CrossRef] [PubMed]
50. Zhang, P.; Samuel, C.E. Protein kinase PKR plays a stimulus- and virus-dependent role in apoptotic death and virus multiplication in human cells. *J. Virol.* **2007**, *81*, 8192–8200. [CrossRef] [PubMed]
51. Twigger, K.; Roulstone, V.; Kyula, J.; Karapanagiotou, E.M.; Syrigos, K.N.; Morgan, R.; White, C.; Bhide, S.; Nuovo, G.; Coffey, M.; et al. Reovirus exerts potent oncolytic effects in head and neck cancer cell lines that are independent of signalling in the EGFR pathway. *BMC Cancer* **2012**, *12*, 368. [CrossRef] [PubMed]
52. Smith, J.A.; Schmechel, S.C.; Williams, B.R.; Silverman, R.H.; Schiff, L.A. Involvement of the interferon-regulated antiviral proteins PKR and RNase L in reovirus-induced shutoff of cellular translation. *J. Virol.* **2005**, *79*, 2240–2250. [CrossRef] [PubMed]
53. Smith, J.A.; Schmechel, S.C.; Raghavan, A.; Abelson, M.; Reilly, C.; Katze, M.G.; Kaufman, R.J.; Bohjanen, P.R.; Schiff, L.A. Reovirus induces and benefits from an integrated cellular stress response. *J. Virol.* **2006**, *80*, 2019–2033. [CrossRef]
54. Kaufman, H.L.; Kohlhapp, F.J.; Zloza, A. Oncolytic viruses: A new class of immunotherapy drugs. *Nat. Rev. Drug Discov.* **2015**, *14*, 642–662. [CrossRef]
55. Vähä-Koskela, M.J.; Heikkilä, J.E.; Hinkkanen, A.E. Oncolytic viruses in cancer therapy. *Cancer Lett.* **2007**, *254*, 178–216. [CrossRef]
56. Bierman, H.R.; Crile, D.M.; Dod, K.S.; Kelly, K.H.; Petrakis, N.L.; White, L.P.; Shimkin, M.B. Remissions in leukemia of childhood following acute infectious disease: Staphylococcus and streptococcus, varicella, and feline panleukopenia. *Cancer* **1953**, *6*, 591–605. [CrossRef]
57. Marcato, P.; Dean, C.A.; Giacomantonio, C.A.; Lee, P.W. Oncolytic reovirus effectively targets breast cancer stem cells. *Mol. Ther.* **2009**, *17*, 972–979. [CrossRef] [PubMed]
58. Cripe, T.P.; Wang, P.Y.; Marcato, P.; Mahller, Y.Y.; Lee, P.W. Targeting cancer-initiating cells with oncolytic viruses. *Mol. Ther.* **2009**, *17*, 1677–1682. [CrossRef] [PubMed]
59. Kim, M.; Garant, K.A.; zur Nieden, N.I.; Alain, T.; Loken, S.D.; Urbanski, S.J.; Forsyth, P.A.; Rancourt, D.E.; Lee, P.W.; Johnston, R.N. Attenuated reovirus displays oncolysis with reduced host toxicity. *Br. J. Cancer* **2011**, *104*, 290–299. [CrossRef]
60. Shepherd, T.G.; Kockeritz, L.; Szrajber, M.R.; Muller, W.J.; Hassell, J.A. The pea3 subfamily ets genes are required for HER2/Neu-mediated mammary oncogenesis. *Curr. Biol.* **2001**, *11*, 1739–1748. [CrossRef] [PubMed]
61. Woltjen, K.; Michael, I.P.; Mohseni, P.; Desai, R.; Mileikovsky, M.; Hämäläinen, R.; Cowling, R.; Wang, W.; Liu, P.; Gertsenstein, M. piggyBac transposition reprograms fibroblasts to induced pluripotent stem cells. *Nature* **2009**, *458*, 766–770. [CrossRef] [PubMed]
62. Shafa, M.; Sjonnesen, K.; Yamashita, A.; Liu, S.; Michalak, M.; Kallos, M.S.; Rancourt, D.E. Expansion and long-term maintenance of induced pluripotent stem cells in stirred suspension bioreactors. *J. Tissue Eng. Regen. Med.* **2012**, *6*, 462–472. [CrossRef]
63. Conley, B.J.; Denham, M.; Gulluyan, L.; Olsson, F.; Cole, T.J.; Mollard, R. Mouse Embryonic Stem Cell Derivation, and Mouse and Human Embryonic Stem Cell Culture and Differentiation as Embryoid Bodies. *Curr. Protoc. Cell Biol.* **2005**, *28*, 23.2.1–23.2.22. [CrossRef]
64. Meng, G.; Liu, S.; Rancourt, D.E. Rapid isolation of undifferentiated human pluripotent stem cells from extremely differentiated colonies. *Stem Cells Dev.* **2011**, *20*, 583–591. [CrossRef]

65. Kurosawa, H. Methods for inducing embryoid body formation: In vitro differentiation system of embryonic stem cells. *J. Biosci. Bioeng.* **2007**, *103*, 389–398. [CrossRef]
66. Keller, G.M. In vitro differentiation of embryonic stem cells. *Curr. Opin. Cell Biol.* **1995**, *7*, 862–869. [CrossRef]
67. Heidariyan, Z.; Ghanian, M.H.; Ashjari, M.; Farzaneh, Z.; Najarasl, M.; Rezaei Larijani, M.; Piryaei, A.; Vosough, M.; Baharvand, H. Efficient and cost-effective generation of hepatocyte-like cells through microparticle-mediated delivery of growth factors in a 3D culture of human pluripotent stem cells. *Biomaterials* **2018**, *159*, 174–188. [CrossRef]
68. Vega-Avila, E.; Pugsley, M.K. An overview of colorimetric assay methods used to assess survival or proliferation of mammalian cells. *Proc. West. Pharmacol. Soc.* **2011**, *54*, 10–14. [PubMed]
69. Kueng, W.; Silber, E.; Eppenberger, U. Quantification of cells cultured on 96-well plates. *Anal. Biochem.* **1989**, *182*, 16–19. [CrossRef]
70. Alain, T.; Kim, M.; Johnston, R.; Urbanski, S.; Kossakowska, A.; Forsyth, P.; Lee, P. The oncolytic effect in vivo of reovirus on tumour cells that have survived reovirus cell killing in vitro. *Br. J. Cancer* **2006**, *95*, 1020–1027. [CrossRef]
71. Lichtman, J.W.; Conchello, J.-A. Fluorescence microscopy. *Nat. Methods* **2005**, *2*, 910–919. [CrossRef] [PubMed]
72. Yang, W.; Soares, J.; Greninger, P.; Edelman, E.J.; Lightfoot, H.; Forbes, S.; Bindal, N.; Beare, D.; Smith, J.A.; Thompson, I.R. Genomics of Drug Sensitivity in Cancer (GDSC): A resource for therapeutic biomarker discovery in cancer cells. *Nucleic Acids Res.* **2012**, *41*, D955–D961. [CrossRef] [PubMed]
73. Ritchie, M.E.; Phipson, B.; Wu, D.; Hu, Y.; Law, C.W.; Shi, W.; Smyth, G.K. limma powers differential expression analyses for RNA-sequencing and microarray studies. *Nucleic Acids Res.* **2015**, *43*, e47. [CrossRef]
74. Weinstein, J.N.; Collisson, E.A.; Mills, G.B.; Shaw, K.R.M.; Ozenberger, B.A.; Ellrott, K.; Shmulevich, I.; Sander, C.; Stuart, J.M.; Network, C.G.A.R. The cancer genome atlas pan-cancer analysis project. *Nat. Genet.* **2013**, *45*, 1113–1120. [CrossRef]
75. Banijamali, R.S.; Soleimanjahi, H.; Soudi, S.; Karimi, H.; Abdoli, A.; Khorrami, S.M.S.; Zandi, K. Kinetics of Oncolytic Reovirus T3D Replication and Growth Pattern in Mesenchymal Stem Cells. *Cell J.* **2020**, *22*, 283–292.
76. Park, J.-S.; Kim, M. Reovirus safety study for proliferation and differentiation of human adipose-derived mesenchymal stem cells. *J. Microbiol.* **2017**, *55*, 75–79. [CrossRef]
77. Hotta, A.; Cheung, A.Y.; Farra, N.; Vijayaragavan, K.; Séguin, C.A.; Draper, J.S.; Pasceri, P.; Maksakova, I.A.; Mager, D.L.; Rossant, J. Isolation of human iPS cells using EOS lentiviral vectors to select for pluripotency. *Nat. Methods* **2009**, *6*, 370–376. [CrossRef] [PubMed]
78. Schoenhals, M.; Kassambara, A.; De Vos, J.; Hose, D.; Moreaux, J.; Klein, B. Embryonic stem cell markers expression in cancers. *Biochem. Biophys. Res. Commun.* **2009**, *383*, 157–162. [CrossRef] [PubMed]
79. Guy, C.T.; Webster, M.A.; Schaller, M.; Parsons, T.J.; Cardiff, R.D.; Muller, W.J. Expression of the neu protooncogene in the mammary epithelium of transgenic mice induces metastatic disease. *Proc. Natl. Acad. Sci. USA* **1992**, *89*, 10578–10582. [CrossRef] [PubMed]
80. Spiecker, L.; Kindler, T.; Horke, S.; Witte, I. Paraoxonase-2 alters hematopoietic stem cell differentiation through redox signalling. *Exp. Hematol.* **2017**, *53*, S65. [CrossRef]
81. Yuan, J.; Devarajan, A.; Moya-Castro, R.; Zhang, M.; Evans, S.; Bourquard, N.; Dias, P.; Lacout, C.; Vainchenker, W.; Reddy, S.T. Putative innate immunity of antiatherogenic paraoxanase-2 via STAT5 signal transduction in HIV-1 infection of hematopoietic TF-1 cells and in SCID-hu mice. *J. Stem Cells* **2010**, *5*, 43–48.
82. Thirukkumaran, C.; Morris, D.G. Oncolytic viral therapy using reovirus. In *Gene Therapy of Solid Cancers*; Springer: Berlin/Heidelberg, Germany, 2015; pp. 187–223.
83. Choi, A.H.; O'Leary, M.P.; Fong, Y.; Chen, N.G. From benchtop to bedside: A review of oncolytic virotherapy. *Biomedicines* **2016**, *4*, 18. [CrossRef]
84. Kemp, V.; Hoeben, R.C.; Van den Wollenberg, D.J. Exploring reovirus plasticity for improving its use as oncolytic virus. *Viruses* **2016**, *8*, 4. [CrossRef]
85. Mohamed, A.; Johnston, R.N.; Shmulevitz, M. Potential for improving potency and specificity of reovirus oncolysis with next-generation reovirus variants. *Viruses* **2015**, *7*, 6251–6278. [CrossRef]
86. Thirukkumaran, C.M.; Luider, J.M.; Stewart, D.A.; Cheng, T.; Lupichuk, S.M.; Nodwell, M.J.; Russell, J.A.; Auer, I.A.; Morris, D.G. Reovirus oncolysis as a novel purging strategy for autologous stem cell transplantation. *Blood* **2003**, *102*, 377–387. [CrossRef]
87. Collins, D.C.; Sundar, R.; Lim, J.S.; Yap, T.A. Towards precision medicine in the clinic: From biomarker discovery to novel therapeutics. *Trends Pharmacol. Sci.* **2017**, *38*, 25–40. [CrossRef]
88. Miao, D.; Van Allen, E.M. Genomic determinants of cancer immunotherapy. *Curr. Opin. Immunol.* **2016**, *41*, 32–38. [CrossRef] [PubMed]
89. Ng, S.W.; Mitchell, A.; Kennedy, J.A.; Chen, W.C.; McLeod, J.; Ibrahimova, N.; Arruda, A.; Popescu, A.; Gupta, V.; Schimmer, A.D.; et al. A 17-gene stemness score for rapid determination of risk in acute leukaemia. *Nature* **2016**, *540*, 433–437. [CrossRef]
90. Alain, T.; Kim, T.S.; Lun, X.; Liacini, A.; Schiff, L.A.; Senger, D.L.; Forsyth, P.A. Proteolytic disassembly is a critical determinant for reovirus oncolysis. *Mol. Ther.* **2007**, *15*, 1512–1521. [CrossRef] [PubMed]
91. Barton, E.S.; Forrest, J.C.; Connolly, J.L.; Chappell, J.D.; Liu, Y.; Schnell, F.J.; Nusrat, A.; Parkos, C.A.; Dermody, T.S. Junction adhesion molecule is a receptor for reovirus. *Cell* **2001**, *104*, 441–451. [CrossRef] [PubMed]
92. Maginnis, M.S.; Forrest, J.C.; Kopecky-Bromberg, S.A.; Dickeson, S.K.; Santoro, S.A.; Zutter, M.M.; Nemerow, G.R.; Bergelson, J.M.; Dermody, T.S. Beta1 integrin mediates internalization of mammalian reovirus. *J. Virol.* **2006**, *80*, 2760–2770. [CrossRef]

93. Theunissen, T.W.; Jaenisch, R. Molecular control of induced pluripotency. *Cell Stem Cell* **2014**, *14*, 720–734. [CrossRef]
94. Mori, Y.; Nishikawa, S.G.; Fratiloiu, A.R.; Tsutsui, M.; Kataoka, H.; Joh, T.; Johnston, R.N. Modulation of Reoviral Cytolysis (I): Combination Therapeutics. *Viruses* **2023**, *15*, 1472. [CrossRef]
95. Bortoletto, A.S.; Parchem, R.J. KRAS Hijacks the miRNA Regulatory Pathway in Cancer. *Cancer Res.* **2023**, *83*, 1563–1572. [CrossRef]
96. Inoue, C.; Negoro, R.; Takayama, K.; Mizuguchi, H.; Sakurai, F. Asymmetric profiles of infection and innate immunological responses in human iPS cell-derived small intestinal epithelial-like cell monolayers following infection with mammalian reovirus. *Virus Res.* **2021**, *296*, 198–334. [CrossRef]

Article

The Optimized γ-Globin Lentiviral Vector GGHI-mB-3D Leads to Nearly Therapeutic HbF Levels In Vitro in CD34⁺ Cells from Sickle Cell Disease Patients

Ekati Drakopoulou [1,2,†], Maria Georgomanoli [1,2,†], Carsten W. Lederer [3], Fottes Panetsos [4], Marina Kleanthous [3], Ersi Voskaridou [5], Dimitrios Valakos [6], Eleni Papanikolaou [2] and Nicholas P. Anagnou [1,2,*]

[1] Laboratory of Cell and Gene Therapy, Centre of Basic Research, Biomedical Research Foundation of the Academy of Athens (BRFAA), 11527 Athens, Greece
[2] Laboratory of Biology, School of Medicine, National and Kapodistrian University of Athens, 11527 Athens, Greece
[3] The Molecular Genetics Thalassemia Department, The Cyprus Institute of Neurology and Genetics, 2371 Nicosia, Cyprus
[4] Bioiatriki SA Health Group Company, 11526 Athens, Greece
[5] Thalassemia and Sickle Cell Disease Centre, Laiko General Hospital, 11527 Athens, Greece
[6] Laboratory of Molecular Biology, Centre of Basic Research, Biomedical Research Foundation of the Academy of Athens (BRFAA), 11527 Athens, Greece
*   Correspondence: anagnou@med.uoa.gr
†   These authors contributed equally to this work.

Citation: Drakopoulou, E.; Georgomanoli, M.; Lederer, C.W.; Panetsos, F.; Kleanthous, M.; Voskaridou, E.; Valakos, D.; Papanikolaou, E.; Anagnou, N.P. The Optimized γ-Globin Lentiviral Vector GGHI-mB-3D Leads to Nearly Therapeutic HbF Levels In Vitro in CD34⁺ Cells from Sickle Cell Disease Patients. *Viruses* 2022, 14, 2716. https://doi.org/10.3390/v14122716

Academic Editors: Ottmar Herchenröder and Brigitte Pützer

Received: 20 September 2022
Accepted: 30 November 2022
Published: 5 December 2022

Publisher's Note: MDPI stays neutral with regard to jurisdictional claims in published maps and institutional affiliations.

Abstract: We have previously demonstrated that both the original γ-globin lentiviral vector (LV) GGHI and the optimized GGHI-mB-3D LV, carrying the novel regulatory elements of the 3D HPFH-1 enhancer and the 3′ β-globin UTR, can significantly increase HbF production in thalassemic CD34⁺ cells and ameliorate the disease phenotype in vitro. In the present study, we investigated whether the GGHI-mB-3D vector can also exhibit an equally therapeutic effect, following the transduction of sickle cell disease (SCD) CD34⁺ cells at MOI 100, leading to HbF increase coupled with HbS decrease, and thus, to phenotype improvement in vitro. We show that GGHI-mB-3D LV can lead to high and potentially therapeutic HbF levels, reaching a mean 2-fold increase to a mean value of VCN/cell of 1.0 and a mean transduction efficiency of 55%. Furthermore, this increase was accompanied by a significant 1.6-fold HbS decrease, a beneficial therapeutic feature for SCD. In summary, our data demonstrate the efficacy of the optimized γ-globin lentiviral vector to improve the SCD phenotype in vitro, and highlights its potential use in future clinical SCD trials.

Keywords: sickle cell disease; thalassemia; globin gene therapy; γ-globin lentiviral vector; HbF; HPFH enhancers; -117 HPFH type; CD34⁺ hematopoietic stem cells

1. Introduction

Sickle cell disease (SCD) is a monogenic disorder caused by a single amino acid substitution in the β-globin gene (glu(E)6Val(A); GAG → GTG; rs334), which results in a multi-organic disease phenotype [1–3]. It is characterized by the polymerization of deoxygenated sickle hemoglobin (HbS; $a_2\beta^S_2$), leading to sickle-shaped red blood cells, and thus, to vaso-occlusion and hemolytic anemia, which are accompanied by organ damage and painful crisis [4,5]. Patients with SCD have a shortened life span, and suffer from severe clinical manifestations, such as acute painful episodes, leg ulcers [6,7], osteonecrosis, chest pain, priapism, stroke, silent cerebral infarction, systemic high blood pressure, and, less frequently, from sickle vasculopathy [8,9].

The beneficial effect of fetal HbF on SCD outcome becomes apparent as early as 6 months after birth, where high levels of HbF lead to asymptomatic disease. SCD patients who continue to produce increased (>20%) γ-globin levels in adult life exhibit a less severe phenotype [4,10], primarily due to decreased HbS polymerization [8,10]. Therapies utilizing HbF inducers, such as 5-azacytidine [11,12] and hydroxyurea (HU) [13–15], or blood transfusion, combined with iron chelation, offer relief to SCD patients [16], as they alleviate disease symptoms.

Alternative therapeutic approaches, such as gene therapy, can also be curative in SCD, as reported by Ribeil et al. in a clinical trial where a patient with severe SCD underwent gene therapy and exhibited a high proportion of anti-sickling hemoglobin post transplant, which accounted for 45% of the total hemoglobin production [17]. Gene therapy studies utilizing LVs containing either an anti-sickling β^{T87Q}-globin or γ-globin transgenes was shown to ameliorate the SCD clinical manifestation either in vivo in SCD mouse models [18–20] or in vitro using CD34$^+$ cells from SCD patients [21–24]. Additionally, the incorporation of small hairpin RNA (shRNA) in therapeutic globin LVs, either for simultaneous reduction in β^S transcripts [25] or for down-regulation of the BCL11A expression [26], can lead to SCD phenotype correction. Furthermore, the role of many miRNAs that bind to transcription factors such as BCL11A, GATA-1, KLF-1, and MYB, is crucial in reactivating the γ-globin gene expression [27,28]. To this end, Sankaran et al. showed that there is a delay in silencing and persistence of fetal hemoglobin coupled with an elevation in embryonic hemoglobin in newborns, attributed to miR-15a and miR-16-1 directly targeting MYB [29,30]. Finally, following the gene editing approach, Magis et al. recently managed to successfully correct the sickle cell mutation in more than 30% of the long-term engrafting hHSCs, using a high-fidelity Cas9 nucleoprotein (RNP) and a single-stranded oligonucleotide donor [31]. Corrected erythroblasts showed a clear dominance of the corrected allele over null β-thalassemia alleles produced by non-homologous end joining (NHEJ), demonstrating a marked survival advantage in vivo [31].

We have previously demonstrated that the original LCR-free, self-inactivating (SIN), insulated γ-globin lentiviral vector GGHI [32], containing the HPFH-2 enhancer element (shown to lead to elevated HbF levels [33,34]), the -117 activating HPFH mutation of the $^A\gamma$ gene promoter, and the HS-40 enhancer from the α-globin locus [35], led to in vitro correction of the thalassemic phenotype [32]. The use of the HS-40 enhancer element instead of the LCR core elements of the β-globin locus was shown to efficiently enhance the expression of the $^A\gamma$-globin gene, and was also associated with high functional titers and genomic stability [32,34], features that the globin vectors containing several LCR elements usually lack, while they are prone to genomic rearrangements and trans-activation of cancer-related genes [36,37]. Furthermore, in the novel, improved GGHI-mB-3D vector [38], we have incorporated the 3D enhancer element of the naturally occurring HPFH-1 deletion [39,40], along with the β-globin gene 3' UTR [18], and pseudotyped it with the alternative BaEVRless envelope glycoprotein [41,42], which resulted in high and stable HbF expression at low multiplicities of infections (MOIs) in thalassemic CD34$^+$ cells.

Based on the previous successful performance of both vectors in thalassemic CD34$^+$ cells, in this study, we evaluated their ability to correct the SCD phenotype in vitro. To this end, CD34$^+$ cells from non-mobilized peripheral blood of SCD patients were transduced with the GGHI or GGHI-mB-3D vectors, pseudotyped with the VSVG envelope glycoprotein. We show that transduction with the optimized GGHI-mB-3D vector leads to a significant increase in the $^A\gamma/\alpha$ ratio and HbF, along with a concomitant decrease in HbS in our patient cohorts, thus demonstrating an increased potential of improving, as well, the SCD phenotype in vitro compared to GGHI. The above effects were achieved at a transduction efficiency of 55% and a clinically relevant mean vector copy number (VCN)/cell of 1.0.

2. Materials and Methods

2.1. Virus Production and Titration

GGHI and GGHI-mB-3D LVs were produced by transient co-transfection of HEK 293T cells using a third-generation lentiviral system [43,44] comprising the following plasmids: vector (GGHI or GGHI-mB-3D), packaging plasmids (pMDLg/pRRE and pRSV-

Rev), and envelope plasmids encoding the vesicular stomatitis virus glycoprotein VSVG (pMD2.VSVG). All accessory plasmids were generously donated by Dr Luigi Naldini (additional information on these plasmids can be also found at www.addgene.com) (accessed on 1 December 2022). Virus-containing supernatant was collected at 48 h and 72 h following transfection, as previously described by Papanikolaou et al. [45]. Viral titers were determined by transducing 5×10^5 mouse erythroleukemia (MEL-585) cells, using serial dilutions of concentrated virus, followed by their induction to erythroid differentiation using 10 μM hemin (Sigma-Aldrich, St. Louis, MO, USA) and 3 mM of HMBA (N,N′-hexamethylene bisacetamide, Sigma-Aldrich, St. Louis, MO, USA). HbF expression was measured by fluorescence-activated cell sorter scanner (FACS) using an anti-HbF FITC conjugated monoclonal antibody (BD Biosciences, Franklin Lakes, NJ, USA).

2.2. Human CD34⁺ Stem Cell Isolation and Transduction: Sample Collection and Processing

We used CD34$^+$ hematopoietic stem cells isolated from five homozygote patients for SCD ($\beta^S\beta^S$) and five compound heterozygotes ($\beta^S\beta^+$) for SCD and β-thalassemia. All samples were obtained from non-mobilized peripheral blood and harvested from volunteer donors, using protocols approved by the Institutional Review Board of the Laiko General Hospital of Athens, in accordance with the Helsinki declaration of 1975. Informed consent was obtained from all subjects involved in the study. All samples were obtained shortly before the next scheduled blood transfusion. CD34$^+$ cells were isolated from mononuclear cells using an EasySep™ Human CD34 Positive Selection Kit (Stem Cell Technologies, Cambridge, UK), according to the manufacturer's instructions. Samples obtained were >90% enriched with CD34$^+$ cells, and were further cultured for up to 21 days in erythroid liquid cultures, as described previously [32,38]. Analyses of liquid cultures included high-performance liquid chromatography (HPLC) and flow cytometry for the assessment of the $^A\gamma/\alpha$ ratio, and HbF and HbS production, and flow cytometry for the assessment of apoptosis using the Annexin V/7-AAD detection kit (BioLegend, San Diego, CA, USA) on day 20–21. At day 20–21, 10^5 cells were removed from culture and used for RNA isolation, as described below. A schematic representation of the experimental procedure is shown in Figure S1A.

2.3. Reversed-Phase High-Performance Liquid Chromatography (RP-HPLC) Analysis for Globin Chain $^A\gamma/\alpha$ Ratio Quantitation

Reversed-phase high-performance chromatography was performed with slight modification of published methods [46]. In brief, cell material was pelleted at 2000 RCF for 10 min and resuspended in H_2O supplemented with 5 mM dithiothreitol at a concentration of 20,000 cells/μl. Following two freeze thaw cycles and a 10 min centrifugation at 21,000 RCF at 4 °C, the supernatant was transferred to 100-μL HPLC microvials (Altmann Analytik, Munich, Germany), and 10–30 μL were injected per run. Analyses were performed on a Prominence HPLC machine with SPD-M20A diode array detector and LC-20AD pump (Shimadzu, Kyoto, Japan), in combination with a Jupiter 5 μm C18 4.6 mm column and corresponding guard columns (Phenomenex, Torrance, CA, USA) using an increasing gradient of phase B, i.e., 0.1% trifluoroacetic acid in acetonitrile (Sigma-Aldrich, St. Louis, MO, USA), against phase A, i.e., 0.1% trifluoroacetic acid (Sigma-Aldrich, St. Louis, MO, USA) and 6.4 mM sodium hydroxide.

2.4. Hemoglobin Electrophoresis and Cation Exchange HPLC (CE-HPLC) for HbF and HbS Quantitation

HbF and HbS quantification in $\beta^S\beta^+$ samples was performed using the Hydragel-Hemoglobin(e) K20 kit (SEBIA, Lisses, France) according to the manufacturer's instructions. Briefly, 10^6–10^7 cells were harvested from liquid cultures on day 20–21 and hemolyzed using 10–20 μL of hemolyzing solution; 10 μL of the hemolysate was loaded onto a Hydragel K20 application carrier according to the manufacturer's instructions. HbF and HbS quantification was performed using QuantityOne software (Bio-Rad, CA, USA) and densiometric analysis. HbF and HbS quantification in $\beta^S\beta^S$ samples was performed by CE-HPLC as described previously by Papanikolaou et.al [32].

2.5. RNA Analysis and Measurement of γ-Globin Transcript Levels Using Quantitative Real-Time PCR

Quantitation of γ-globin transcripts was carried out employing quantitative real-time PCR (qPCR) using SYBR^{TM} Green mix (Kapa Biosystems, Wilmington, MA, USA), and performed on a CFX Connect^{TM} Real-Time System (Bio-Rad, Hercules, CA, USA). For the quantification of γ-globin production from erythroid cultures, total RNA was isolated from both mock-transduced and transduced CD34$^+$ cells maintained at day 20–21, using the RNeasy kit (Qiagen, Germantown, MD, USA) and according to the manufacturer's instructions. A quantity of 5–500 ng of total RNA was reverse transcribed to cDNA using the Superscript First-Strand Synthesis System for RT-PCR (Invitrogen, Carlsbad, CA, USA), and 20 ng of c-DNA was subjected to qPCR analysis. Production of γ-globin was measured using the following primers: gamma F: 5′-GCCATAAAGCACCTGGATGA-3′, and gamma R: 5′-GATTGCCAAAACGGTCACC-3′. The human α-globin gene was used as a reference gene using the following primers: alpha-globin F: 5′-CACGCTGGCGAGTATGGTG-3′ and alpha-globin R: 5′-TTAACCCTGGGCAGAGCCGT-3′. Fold increase in γ-globin mRNA in transduced and mock-transduced cell populations was calculated using the $\Delta\Delta$Ct [47] method. Measurements of human γ-globin mRNA levels were performed in duplicate for each sample. The amplification was carried out according to the conditions suggested by the manufacturer.

2.6. Flow Cytometry

Transduced MEL-585 cells were induced towards erythroid phenotype and were then processed for flow cytometry analysis as previously described [38], using an anti-HbF FITC-conjugated monoclonal antibody (BD Pharmingen, Franklin Lakes, NJ, USA). Similarly, CD34$^+$ cells from erythroid cultures were stained with both anti-HbF FITC and anti-Glycophorin A-PE conjugated monoclonal antibodies (BD Pharmingen, Franklin Lakes, NJ, USA) and subjected to flow cytometry analysis.

Apoptotic assays were performed on cells derived from erythroid cultures on day 20–21, employing FITC using the Annexin V/7-AAD detection kit with 7-AAD (BioLegend, San Diego, CA, USA) according to the manufacturer's instructions. Briefly, cells were washed twice with cold PBS, containing 1% FBS, and resuspended in Annexin-V Binding buffer at a concentration of $0.25–1 \times 10^7$ cells/mL. Cells were then incubated for 15 min at room temperature in the dark, and analyzed by flow cytometry within an hour.

All samples were analyzed in a Cytomics FC 500 (CXP) Series Flow Cytometry System (Beckman Coulter, Nyon, Switzerland). Flow cytometry analysis was performed using FlowJo 10.8.1 analysis software.

2.7. Determination of Vector Copy Number and Transduction Efficiency

In order to assess the vector transduction efficiency, 10–20 burst-forming units (BFUe) per patient were subjected to semi-quantitative PCR analysis using the primers gamma F/R described above. DNA was isolated with the QIAamp DNA Micro Kit (Qiagen, MD, USA) according to the manufacturer's instructions. Vector transfer efficiency in CD34$^+$ cells was determined by assessing the proportion of BFUe colonies that tested positive for vector sequences. Those tested positive for vector-specific sequences were further subjected to absolute quantitation with qPCR analysis for VCN determination, using the primers gamma F/R and the hRNAseP primers, as previously described [38]. Analysis of each sample was performed in duplicate, using SYBR^{TM} Green mix (Kapa Biosystems, Wilmington, MA, USA) and according to manufacturer's instructions.

2.8. Statistical Analysis

Repeated-measures ANOVA, followed by post hoc Tukey tests, were used to detect statistically significant differences among different treatments, and paired two-tailed Student's *t*-tests for pairwise comparisons, unless stated otherwise. Pearson's *r* correlation coefficient was applied for correlation between values. Statistical analyses were performed using Prism 9.3.1, IBM SPSS 28, and R 4.0.0 software, while graph creation was conducted with Prism 9.3.1 software.

3. Results

3.1. GGHI and GGHI-mB-3D LVs Exhibit High Titers

Both GGHI and GGHI-mB-3D LVs exhibited high mean titers; specifically, 1.63×10^8 TU/mL ($n = 6$) and 1.56×10^8 TU/mL ($n = 5$), respectively (Figure S1B). No statistical difference was observed between the mean values of GGHI and GGHI-mB-3D LVs titers ($p = 0.902$, unpaired two-tailed *t*-test).

3.2. Increased $^A\gamma/\alpha$ Chain Ratio in SCD CD34$^+$ Cells following Transduction with GGHI-mB-3D Lentiviral Vector

Non-mobilized peripheral blood CD34$^+$ cells from ten SCD patients were isolated and processed. Five were homozygotes for the sickle cell disease mutation and five were compound heterozygotes for SCD and β-thalassemia ($\beta^S\beta^+$). All patients had four intact α-globin genes. Typical yields from an initial volume of 20 mL of peripheral blood ranged from 2×10^5 to 10^6 CD34$^+$ cells. All patients were receiving HU, and therefore, initial HbF levels were high, reaching a mean percentage of $68.3 \pm 14.2\%$ in erythroid cultures, as detected by flow cytometry. Baseline levels of HbF and HbS prior to HU treatment are shown in Figure S2A,B (upper panel).

We initially asked whether both GGHI/VSVG and GGHI-mB-3D/VSVG γ-globin vectors (Figure 1A), previously tested on CD34$^+$ cells from thalassemia patients [32,38], could increase the $^A\gamma/\alpha$ ratio in SCD patients as well. To this end, we isolated CD34$^+$ cells from SCD donors and cultured them under serum-free conditions for 18 h, dividing them into three cell aliquots; two subpopulations were transduced with each type of γ-globin vectors for 24 h, while the third was mock-transduced and served as a control. In both cases, and since our vectors were both pseudotyped with the conventional VSVG envelope glycoprotein, the MOI used was 100, as described by Papanikolaou et al. [32]. Next day, the cells were washed twice in PBS and resuspended in erythroid medium, as described in Section 2.

RP-HPLC analysis demonstrated that GGHI-mB-3D-transduced cells led to a highly significant increase in mean $^A\gamma/\alpha$ chain ratio ($p = 0.02$, $n = 9$) (Figure 1B, left panel); specifically, from 0.23 ± 0.08 observed in the control to 0.28 ± 0.08 (Table 1), with a mean difference of 0.056 ± 0.057 (Figure S3A). On the contrary, and under the aforementioned conditions, GGHI LV failed to do so, leading to a mean $^A\gamma/\alpha$ ratio of 0.27 ± 0.11 ($p = 0.30$, $n = 9$), as presented in Table 1. However, no statistical difference was observed between GGHI and GGHI-mB-3D LVs ($p = 0.74$, $n = 9$).

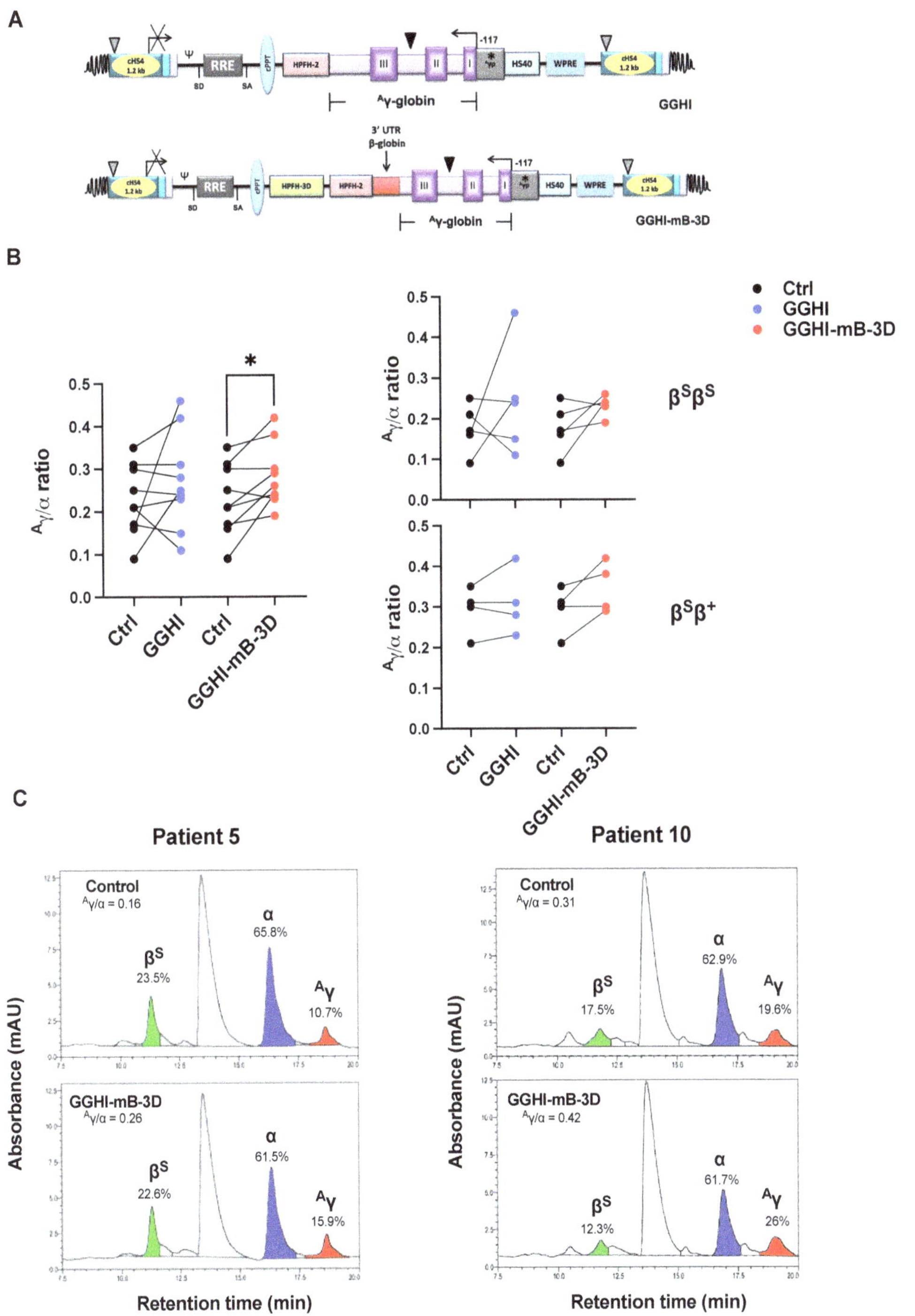

Figure 1. In vitro assessment of γ-globin lentiviral vectors (LVs) in SCD CD34[+] cells: (**A**) The structure and the individual regulatory elements of the GGHI and GGHI-mB-3D LVs shown as proviral elements.

The grey triangle represents the 398 bp deletion in the cHS4, the black triangle shows the 713 bp deletion in the $^A\gamma$ cassette, and the asterisk depicts the -117 point mutation in the $^A\gamma$ promoter. (**B**) Before-and-after plot showing $^A\gamma/\alpha$ globin chain ratios obtained with RP-HPLC across patients, prior and post transduction, with GGHI ($p = 0.297$, $n = 9$, paired two-tailed *t*-test) and GGHI-mB-3D ($p = 0.0189$, $n = 9$, paired two-tailed *t*-test) LVs (left panel). Before-and-after plots showing $^A\gamma/\alpha$ globin chain ratios in the $\beta^S\beta^S$ and $\beta^S\beta^+$ patient cohort, prior and post transduction, with GGHI and GGHI-mB-3D LVs (right panel). For $\beta^S\beta^S$ patient cohort GGHI ($p = 0.41$, $n = 5$, paired two-tailed *t*-test) and GGHI-mB-3D ($p = 0.14$, $n = 5$, paired two-tailed *t*-test), and for $\beta^S\beta^+$ patient cohort GGHI ($p = 0.45$, $n = 4$, paired two-tailed *t*-test) and GGHI-mB-3D ($p = 0.12$, $n = 4$, paired two-tailed *t*-test). (**C**) RP-HPLC chromatograms from Patient 5 (left panel) and Patient 10 (right panel). Percentages shown are relative to the sum of the highlighted peaks. Each dot corresponds to each patient. * $p \leq 0.05$.

When patients were divided according to genotype, transduction of $\beta^S\beta^S$ and $\beta^S\beta^+$ cohorts with both GGHI and GGHI-mB-3D LVs showed a trend, but not a statistically significant increase. Specifically, as shown in Figure 1B (right panel) and Table 2, transduction with the GGHI-mB-3D lentiviral vector led to a $^A\gamma/\alpha$ ratio increase from 0.18 ± 0.06, observed in the mock-transduced cells, to 0.23 ± 0.03 ($p = 0.14$, $n = 5$), while the corresponding value post transduction with GGHI was 0.24 ± 0.14 ($p = 0.41$, $n = 5$). Mean $^A\gamma/\alpha$ ratio achieved by GGHI and GGHI-mB-3D LVs did not test significantly different ($p = 0.86$, $n = 5$). In the case of the compound heterozygote $\beta^S\beta^+$ patient cohort, the majority of patients showed an increase in the $^A\gamma/\alpha$ ratio post transduction, with GGHI-mB-3D reaching a mean 0.35 ± 0.06 compared to 0.29 ± 0.06 seen in mock-transduced cells ($p = 0.12$, $n = 4$) and GGHI 0.31 ± 0.08 (Figure 1B, right panel, Table 3). Figure 1C shows RP-HPLC chromatograms of representative experiments, and specifically, $\beta^S\beta^S$ SCD Patient 5 and $\beta^S\beta^+$ SCD Patient 10, prior and post transduction, with GGHI-mB-3D. Transduction with the aforementioned LV led to a $^A\gamma/\alpha$ ratio increase from 0.16 to 0.26 in Patient 5 (Figure 1C, left panel), and from 0.31 to 0.42 in Patient 10 (Figure 1C, right panel).

3.3. Increased Percentage of F-Cells following Transduction with GGHI-mB-3D Lentiviral Vector

The progression of differentiation in liquid cultures was assessed at day 20–21 by flow cytometry, using anti-HbF and anti-CD235a antibodies. Cells stained positive for both markers will be referred to as F-cells from this point onwards. Summarized results of F-cell percentages are depicted in Tables 1–3. Flow cytometry and repeated-measures ANOVA showed that transduction of only the $\beta^S\beta^S$ cohort with both GGHI and GGHI-mB-3D LVs leads to statistically significant increase in F-cell percentage ($F_{(2,12)} = 5.81$, $p = 0.028$). MFI did not test significantly different between samples. More specifically, and regarding F-cells in the $\beta^S\beta^+$ cohort (Figure 2A), there was no significant increase following transduction with GGHI ($p = 0.746$, $n = 5$) or GGHI-mB-3D ($p = 0.758$, $n = 5$), while in the case of $\beta^S\beta^S$ patients (Figure 2B), both vectors led to significant increase in F-cells. Transduction with either GGHI or GGHI-mB-3D led to $73 \pm 4.5\%$ and $73 \pm 4.8\%$ of F-cells, respectively (mean difference between GGHI and Ctrl of 3.9 ± 2.87, and between GGHI-mB-3D and Ctrl of 3.98 ± 3.12, Figure S3A), compared to $69 \pm 5.7\%$ observed in the control ($p = 0.038$ and $p = 0.046$, $n = 5$, respectively). Mean F-cell increase did not exceed 6% for both vectors (Figure 2B, right panel). In the representative experiment of Patient 5, GGHI led to 74.1% of F-cells and GGHI-mB-3D to 77.1%, compared to 68.4% seen in the control (Figure 2C, Tables 1 and 2). Erythroid differentiation in $\beta^S\beta^S$ liquid cultures (days 20–21), exceeded 90% as measured by CD235a expression, demonstrating no statistical difference prior and post transduction, either with GGHI ($p = 0.27$, $n = 5$) or GGHI-mB-3D ($p = 0.22$, $n = 5$) LVs (Figure 2D (left panel) and Figure S2C). Figure 2D (right panel) shows representative percentages of CD235a expression Patient 5 at the end of differentiation.

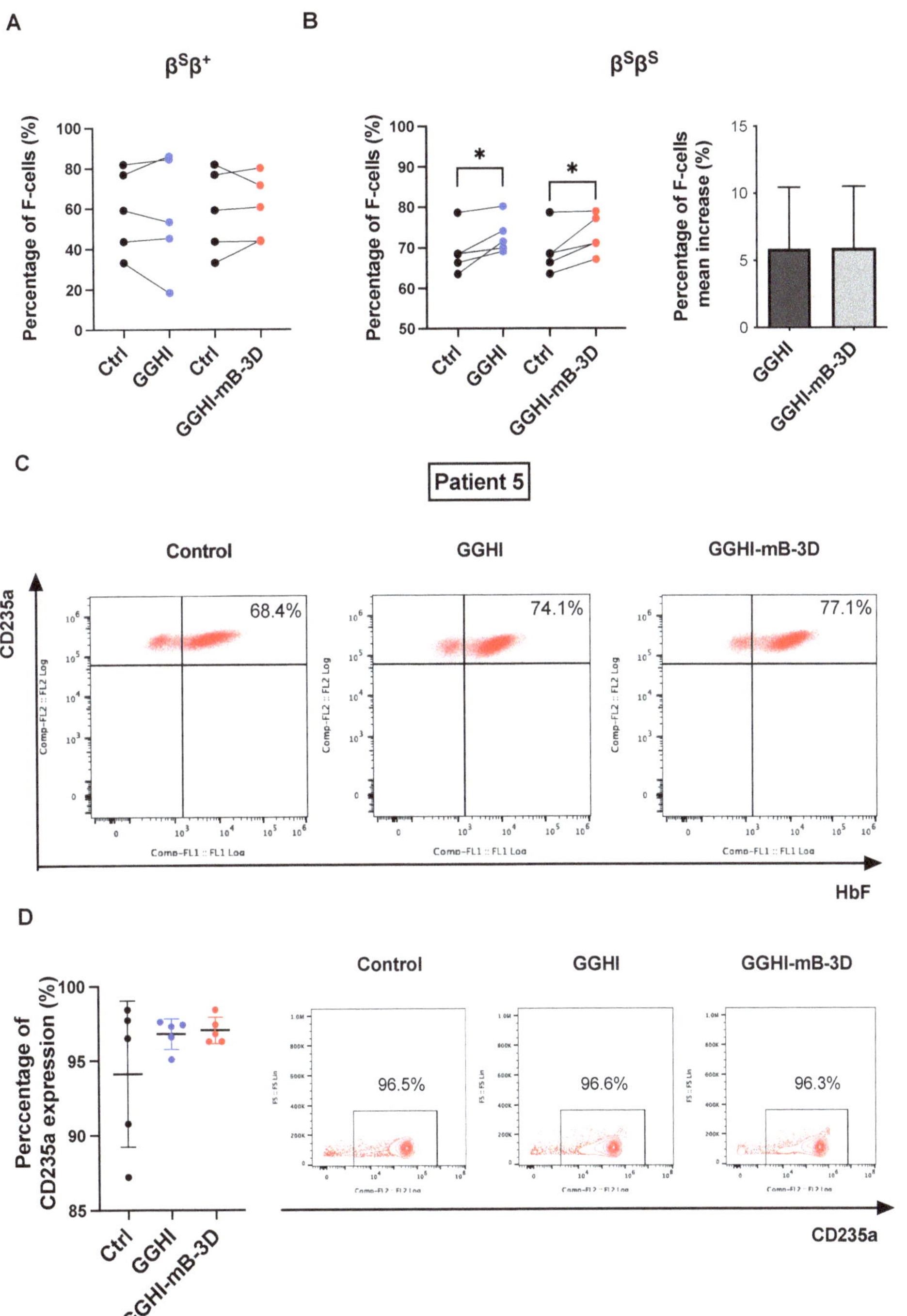

Figure 2. Increased production of F-cells in $\beta^S\beta^S$ SCD CD34$^+$ cells: (**A**) Before-and-after plot showing F-cell percentages before and after transduction of $\beta^S\beta^+$ SCD CD34$^+$ cells with GGHI ($p = 0.746$, $n = 5$,

paired two-tailed *t*-test) and GGHI-mB-3D ($p = 0.758$, $n = 5$, paired two-tailed *t*-test) LVs at MOI 100. (**B**) Before-and-after plot showing F-cell percentages before and after transduction of $\beta^S\beta^S$ SCD CD34$^+$ cells with GGHI ($p = 0.038$, $n = 5$, paired two-tailed *t*-test) and GGHI-mB-3D ($p = 0.046$, $n = 5$, paired two-tailed *t*-test) LVs at MOI 100 (left panel), and bar chart showing the percentage of F-cell mean increase ($p = 0.983$, $n = 5$, unpaired two-tailed *t*-test)(right panel). (**C**) Representative flow cytometry profiles showing F-cell percentages in $\beta^S\beta^S$ Patient 5. (**D**) CD235a expression at the end of in vitro differentiation (days 20–21) in the $\beta^S\beta^S$ patient cohort. Mean percentage of CD235a expression was above 90%; specifically, 94.12 ± 4.89, 96.80 ± 1.02, and 97.04 ± 0.88 in mock-transduced, GGHI-transduced, and GGHI-mB-3D-transduced samples, respectively (left panel). No statistical differences were observed between mock-transduced, and GGHI- and GGHI-mB-3D-transduced samples ($p = 0.27$ and $p = 0.22$, respectively, $n = 5$, unpaired two-tailed *t*-test). Representative flow cytometry profiles of CD235a expression in Patient 5 (right panel). Each dot corresponds to each patient. Error bars represent $\pm$SD, * $p \leq 0.05$.

3.4. Improvement in the SCD CD34$^+$ Cell Phenotype In Vitro following Transduction with GGHI-mB-3D Lentiviral Vector

Due to the HU treatment received by all patients, and hence, to elevated levels in the majority of patients, reliable anti-sickling tests, as previously demonstrated by Urbinati et al. [22], could not be performed. Therefore, in order to demonstrate a potential phenotype improvement following transduction with our lentiviral vectors, we performed Hb electrophoresis/CE-HPLC, aiming to document an HbS decrease, and a concomitant HbF increase, following transduction. Repeated-measures ANOVA showed significant HbS differences between treatments ($F_{(2,24)} = 3.74$, $p = 0.046$) and across all patients. Overall, and as it can be seen in Figure 3A (left panel), transduction with GGHI-mB-3D LV only, led to a marked decrease in HbS. Specifically, the mean HbS percentage reached 41.9 ± 12.3%, compared to 46.6 ± 8.9% observed in mock-transduced cells ($p = 0.022$, $n = 9$), with a mean difference of -4.66 ± 4.94 (Figure S3A), The above decrease was followed by a concomitant HbF increase (Figure 3A, right panel), reaching 37 ± 27% compared to 33 ± 25% in mock-transduced cells ($p = 0.023$, $n = 9$), with a mean difference of 4.47 ± 4.76 (Figure S3A). Transduction with GGHI LV led to a marginal effect, with the mean HbS and HbF percentages post transduction reaching 43.1 ± 11.8% ($p = 0.053$, $n = 9$) and 35.3 ± 27% ($p = 0.091$, $n = 9$), respectively.

Regarding the $\beta^S\beta^S$ cohort, the percentage of HbS did not test significantly different by repeated-measures ANOVA ($F_{(2,3)} = 0.338$, $p = 0.73$). Specifically, transduction with GGHI-mB-3D led to a mean HbS percentage of 51.4 ± 5.2%, compared to 52.7 ± 6.6% observed in the control ($p = 0.23$, $n = 4$), while the corresponding mean HbS percentage achieved by GGHI was 52.95 ± 9.9% ($p = 0.91$, $n = 4$) (Figure 3B, left panel and Table 2). As expected, HbF showed a small increase, without statistical significance, following GGHI and GGHI-mB-3D transduction ($p = 0.79$ and $p = 0.33$, respectively, $n = 4$) (Figure 3B, right panel).

Interestingly, however, in the $\beta^S\beta^+$ patient cohort, both GGHI and GGHI-mB-3D LVs led to a significant reduction in HbS percentage, as demonstrated by repeated-measures ANOVA ($F_{(2,12)} = 5.48$, $p = 0.03$); particularly 35.2 ± 5.5% in the case of GGHI ($p = 0.005$, $n = 5$) (mean difference between GGHI and Ctrl of -6.42 ± 2.54, Figure S3B) and 34.3 ± 11% in the case of GGHI-mB-3D ($p = 0.03$, $n = 5$) (mean difference between GGHI-mB-3D and Ctrl of -7.33 ± 5.15, Figure S3B), compared to 41.6 ± 7.5% observed in the control (Figure 3C, left panel and Table 3). Regarding the HbF levels following transduction with GGHI and GGHI-mB-3D LVs (Figure 3C, right panel), both exhibited a marginally significant increase, reaching a mean of 55 ± 17% and 57 ± 20% (mean difference between GGHI and Ctrl of 4.40 ± 3.82, and GGHI-mB-3D and Ctrl of 5.93 ± 4.84, Figure S3B), compared to 51 ± 18% in mock-transduced cells ($p = 0.062$ and $p = 0.052$, respectively, $n = 5$).

Regarding the effects on apoptosis following vector transduction, no differences were detected between mock-transduced and transduced cells in both patient cohorts.

Table 1. Detailed results from all patients.

Sample	Genotype	Mutation	Source	F-Cells (FACS)			$^{A}\gamma/\alpha$ Ratio (RP-HPLC)			HbS Decrease (%)		TSD Efficiency (%)		Mean VCN/Cell		Relative Fold Difference of γ-mRNA Transcripts	
				Ctrl	GGHI	GGHI-mB-3D	Ctrl	GGHI	GGHI-mB-3D	GGHI	GGHI-mB-3D	GGHI	GGHI-mB-3D	GGHI	GGHI-mB-3D	GGHI	GGHI-mB-3D
#4	$\beta^{S}\beta^{S}$	HbS/HbS	PB	66.3	69.1	71.1	0.21	0.11	0.24	0	3.3	20	20	0.8	0.4	1.04	1
#5	$\beta^{S}\beta^{S}$	HbS/HbS	PB	68.4	74.1	77.1	0.16	0.46	0.26	8	0	-	-	-	-	0.93	0.76
#6	$\beta^{S}\beta^{S}$	HbS/HbS	PB	63.5	71.5	67.1	0.25	0.24	0.23	3.6	0.7	20	30	0.3	0.8	1.15	1.28
#7	$\beta^{S}\beta^{S}$	HbS/HbS	PB	78.6	80.2	78.9	0.17	0.15	0.19	0	6	100	100	1.7	1.9	-	-
#8	$\beta^{S}\beta^{S}$	HbS/HbS	PB	68.5	69.9	71	0.09	0.25	0.24	-	-	56	90	0.4	0.9	1.19	0.79
#9	$\beta^{S}\beta^{+}$	HbS/IVS1-110	PB	81.9	84.5	71.5	0.21	0.23	0.29	20	0	30	10	0.3	0.6	-	-
#10	$\beta^{S}\beta^{+}$	HbS/IVS1-110	PB	59.1	53.4	60.6	0.31	0.31	0.42	15.3	20.5	33.3	35.3	1.8	2.8	1.06	1.09
#11	$\beta^{S}\beta^{+}$	HbS/IVS1-110	PB	43.6	45.3	43.8	0.30	0.28	0.30	17.8	11.7	100	100	0.8	0.5	1.29	1.34
#12	$\beta^{S}\beta^{+}$	HbS/IVS1-1	PB	33.2	18.4	44.3	0.35	0.42	0.38	10.9	29.5	66.7	66.7	0.4	0.5	0.56	0.78
#13	$\beta^{S}\beta^{+}$	HbS/IVS1-110	PB	76.9	86	80.2	-	-	-	11.1	36	37.9	44.4	1.5	1.0	1.06	0.95
	Average			64	65	67	0.23	0.27	0.28	9.6	12	52	55	0.89	1.04	1.03	0.999
	p-value				0.59	0.19		0.30	0.02	0.66		0.82		0.65		0.75	

Note: Dashes indicate that the relevant assays were not conducted due to limited number of cells; PB, peripheral blood; TSD, transduction; VCN, vector copy number; RP-HPLC, reversed-phase HPLC.

Table 2. Detailed results from $\beta^S\beta^S$ patients.

Sample	Genotype	Mutation	F-Cells (FACS)			$^A\gamma/\alpha$ Ratio (RP-HPLC)			HbS (%)			HbF (%)			TSD Efficiency (%)		Mean VCN/cell		Relative Fold Difference of γ-mRNA Transcripts	
			Ctrl	GGHI	GGHI-mB-3D	Ctrl	GGHI	GGHI-mB-3D	Ctrl	GGHI	GGHI-mB-3D	Ctrl	GGHI	GGHI-mB-3D	GGHI	GGHI-mB-3D	GGHI	GGHI-mB-3D	GGHI	GGHI-mB-3D
#4	$\beta^S\beta^S$	HbS/HbS	66.3	69.1	71.1	0.21	0.11	0.24	55.4	57.3	53.6	4.8	3.2	13.3	20	20	0.8	0.4	1.04	1
#5	$\beta^S\beta^S$	HbS/HbS	68.4	74.1	77.1	0.16	0.46	0.26	51.1	47	51.5	10.7	8.1	10.4	-	-	-	-	0.93	0.76
#6	$\beta^S\beta^S$	HbS/HbS	63.5	71.5	67.1	0.25	0.24	0.23	44.5	42.9	44.2	18.7	25.5	22.8	20	30	0.3	0.8	1.15	1.28
#7	$\beta^S\beta^S$	HbS/HbS	78.6	80.2	78.9	0.17	0.15	0.19	59.9	64.6	56.3	6.3	6.2	4.6	100	100	1.7	1.9	-	-
#8	$\beta^S\beta^S$	HbS/HbS	68.5	69.9	71	0.09	0.25	0.24	-	-	-	-	-	-	56	90	0.4	0.9	1.19	0.79
	Average		69	73	73	0.18	0.24	0.23	52.7	53	51.4	10.13	10.75	12.78	49	60	0.8	1	1.08	0.96
	p-value			0.038	0.046		0.41	0.14		0.91	0.23		0.79	0.33	0.71		0.67		0.41	

Note: Dashes indicate that the relevant assays were not conducted due to limited number of cells; PB, peripheral blood; TSD, transduction; VCN, vector copy number; RP-HPLC, reversed-phase HPLC.

Table 3. Detailed results from $\beta^S\beta^+$ patients.

Sample	Genotype	Mutation	F-Cells (FACS)			$^A\gamma/\alpha$ Ratio (RP-HPLC)			HbS (%)			HbF (%)			TSD Efficiency (%)		Mean VCN/Cell		Relative Fold Difference of γ-mRNA Transcripts	
			Ctrl	GGHI	GGHI-mB-3D	Ctrl	GGHI	GGHI-mB-3D	Ctrl	GGHI	GGHI-mB-3D	Ctrl	GGHI	GGHI-mB-3D	GGHI	GGHI-mB-3D	GGHI	GGHI-mB-3D	GGHI	GGHI-mB-3D
#9	$\beta^S\beta^+$	HbS/IVS1-110	81.9	84.5	71.5	0.21	0.23	0.29	44.01	35.19	44.86	55.97	64.81	55.14	30	10	0.3	0.6	-	-
#10	$\beta^S\beta^+$	HbS/IVS1-110	59.1	53.4	60.6	0.31	0.31	0.42	42.45	35.96	33.77	36.77	36.92	42.53	33.3	35.3	1.8	2.8	1.06	1.09
#11	$\beta^S\beta^+$	HbS/IVS1-110	43.6	45.3	43.8	0.30	0.28	0.30	50.50	41.50	44.60	28.90	36.40	33.00	100	100	0.8	0.5	1.29	1.34
#12	$\beta^S\beta^+$	HbS/IVS1-1	33.2	18.4	44.3	0.35	0.42	0.38	41.40	36.90	29.20	58.70	63.10	70.80	66.7	66.7	0.4	0.5	0.56	0.78
#13	$\beta^S\beta^+$	HbS/IVS1-110	76.9	86	80.2	-	-	-	29.70	26.40	19.00	72.50	73.60	81.00	37.9	44.4	1.5	1.0	1.06	0.95
	Average		59	58	60	0.29	0.31	0.35	41.61	35.19	34.29	50.57	54.97	56.49	54	51	0.96	1.1	0.99	1.04
	p-value			0.75	0.76		0.45	0.12		0.005	0.03		0.062	0.052	0.91		0.83		0.82	

Note: Dashes indicate that the relevant assays were not conducted due to limited number of cells; PB, peripheral blood; TSD, transduction; VCN, vector copy number; RP-HPLC, reversed-phase HPLC.

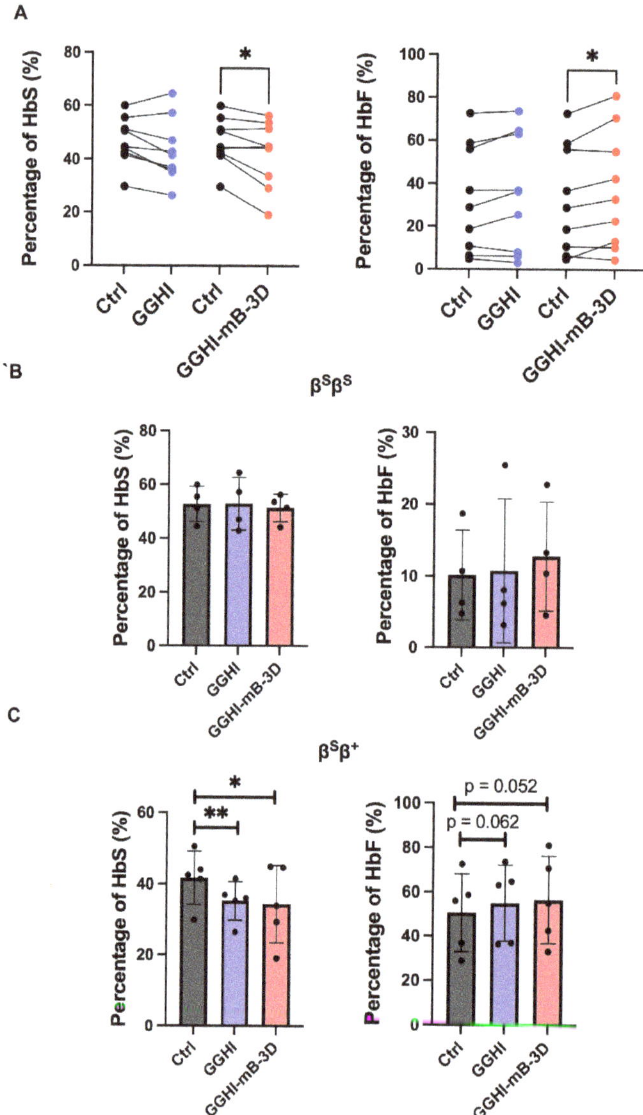

Figure 3. Phenotypic improvement in SCD cells in vitro following transduction with GGHI-mB-3D LV at MOI 100: (**A**) Before-and-after plots showing mean percentage of HbS (left panel) and HbF (right panel) obtained with hemoglobin (Hb) electrophoresis/CE-HPLC across all SCD patients, prior and post transduction, with GGHI ($p = 0.053$ and $p = 0.091$, respectively, $n = 9$, paired two-tailed t-test) and GGHI-mB-3D ($p = 0.022$ and $p = 0.023$, respectively, $n = 9$, paired two-tailed t-test) LVs. (**B**) Scatter plots showing mean percentage of HbS (left panel) and HbF (right panel), prior and post transduction, with GGHI ($p = 0.91$ and $p = 0.79$, respectively, $n = 4$, paired two-tailed t-test) and GGHI-mB-3D ($p = 0.23$ and $p = 0.33$, respectively, $n = 4$, paired two-tailed t-test) LVs in the $\beta^S\beta^S$ patient cohort. (**C**) Scatter plots showing mean percentage of HbS (left panel) and HbF (right panel), prior and post transduction, with GGHI ($p = 0.005$ and $p = 0.062$, respectively, $n = 5$, paired two-tailed t-test) and GGHI-mB-3D ($p = 0.03$ and $p = 0.052$, respectively, $n = 5$, paired two-tailed t-test) LVs in the $\beta^S\beta^+$ patient cohort. Each dot corresponds to each patient. Error bars represent $\pm$SD, * $p \leq 0.05$, ** $p \leq 0.01$.

3.5. Gene Transfer Efficiency and Vector Copy Number in Transduced BFUe from SCD Patients

In order to determine LV transduction efficiency, we plated $CD34^+$ cells derived from both $\beta^S\beta^S$ or $\beta^S\beta^+$ patients that were either transduced with GGHI or GGHI-mB-3D or mock-transduced, in methylcellulose semi-liquid medium for colony-forming assays; 10–20 individual BFUe per patient were analyzed in each experiment by colony PCR scoring and using primers for vector-specific sequences. Gene transfer efficiency of the LVs in the $CD34^+$ progenitors was determined by assessing the proportion of BFUe colonies that tested positive for vector sequences. As shown in Figure 4A (left panel) and Table 1, the mean transduction efficiencies for GGHI and GGHI-mB-3D were similar, i.e., $52 \pm 32\%$ (median 38%) and $55 \pm 35\%$ (median 44%), respectively ($p = 0.82$, $n = 9$, unpaired two-tailed *t*-test). The clonogenicities of both mock- or γ-globin-vector-transduced $CD34^+$ cells were similar. Furthermore, the VCN/cell was calculated for each LV-positive BFUe from each patient, and then a mean value was extracted from all BFUe for each patient. The average VCN for GGHI-transduced cells was 0.89 ± 0.62 (median 0.8), with a range of 0.3 to 1.8, while for GGHI-mB-3D, it was 1.04 ± 0.8 (median 0.8), with a range of 0.4 to 2.8 ($p = 0.65$, $n = 9$, unpaired two-tailed *t*-test) (Figure 5A right panel and Table 1).

Transduction efficiency and VCN in the $\beta^S\beta^S$ and $\beta^S\beta^+$ patient cohorts are shown in Figure 4B,C and Tables 2 and 3. More specifically, transduction efficiency achieved by GGHI and GGHI-mB-3D LVs in $\beta^S\beta^S$ patients was $49 \pm 38\%$ and $60 \pm 41\%$, respectively ($p = 0.71$, $n = 4$, unpaired two-tailed *t*-test) (Figure 4B, left panel and Table 2), while mean VCN/cell was 0.8 ± 0.64 (median 0.6, with a range 0.3–1.7) for GGHI and 1 ± 0.66 (median 0.85, with a range 0.4–1.9) for GGHI-mB-3D ($p = 0.67$, $n = 4$, unpaired two-tailed *t*-test) (Figure 4B, right panel and Table 2). Transduction efficiency achieved by GGHI and GGHI-mB-3D LVs in the $\beta^S\beta^+$ cohort reached $54 \pm 30\%$ and $51 \pm 34\%$, respectively ($p = 0.91$, $n = 5$, unpaired two-tailed *t*-test) (Figure 4C, left panel and Table 3), while mean VCN/cell was 0.96 ± 0.67 (median 0.8, with a range 0.3–1.8) and 1.1 ± 0.98 (median 0.6, with a range 0.5–2.8), respectively ($p = 0.83$, $n = 5$, unpaired two-tailed *t*-test) (Figure 4C, right panel and Table 3).

3.6. γ-Globin mRNA Analysis Using Quantitative Real-Time PCR

Quantification of γ-globin production from SCD erythroid cultures showed that γ-globin mRNA is not increased significantly following transduction either with GGHI or GGHI-mB-3D lentiviral vectors at MOI 100. Specifically, we observed a mean relative fold difference of 1.03 and 0.999 ($p = 0.75$, $n = 8$, unpaired two-tailed *t*-test) in our patient cohorts, following transduction either with GGHI or GGHI-mB-3D, with the greatest increase observed in Patient 11 (Table 1 and Figure S4A). When $\beta^S\beta^S$ and $\beta^S\beta^+$ patients were analyzed separately, we failed again to detect a mean increase in γ-globin at the mRNA level, as a result of transduction with either LV, with mean relative fold difference levels in both cases to be around 1. Specifically, in $\beta^S\beta^S$ patients, the relative fold difference of γ-globin transcript was calculated 1.08 for GGHI and 0.96 for GGHI-mB-3D ($p = 0.41$, $n = 4$, unpaired two-tailed *t*-test), as shown in Table 2 and Figure S4B (left panel), while for the $\beta^S\beta^+$ patient cohort it was 0.99 and 1.04 ($p = 0.82$, $n = 4$, unpaired two-tailed *t*-test), respectively (Table 3 and Figure S4B, right panel).

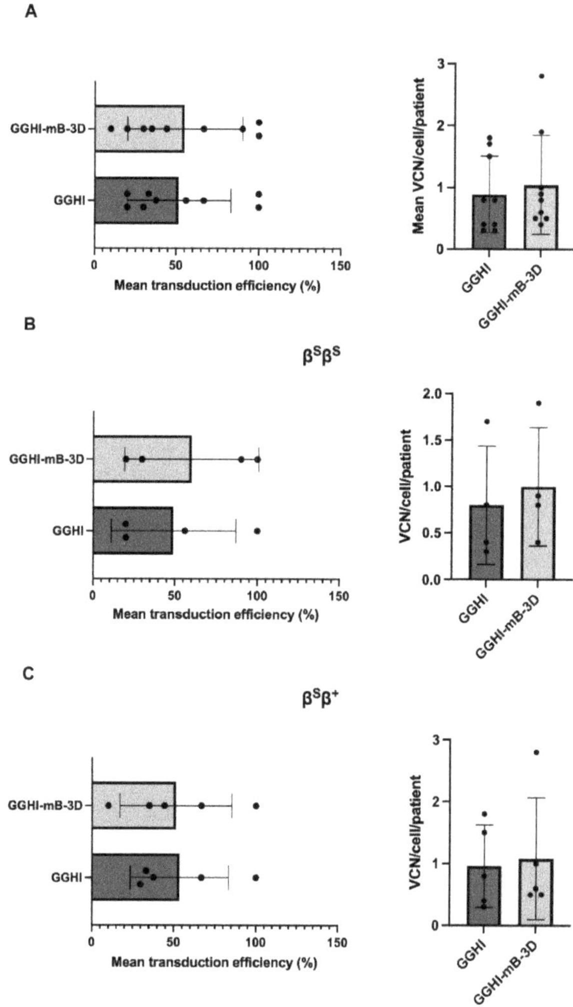

Figure 4. Transduction efficiency and VCN/cell demonstrated by γ-globin LVs in SCD CD34$^+$ cells: (**A**) Scatter plots showing mean percentage of transduction efficiency (left) and mean VCN/cell/patient (right) across all patients and following transduction with GGHI and GGHI-mB-3D LVs at MOI 100. GGHI leads to 52 ± 31.5% and GGHI-mB-3D to 55 ± 35% transduction efficiency (p = 0.82, n = 9, unpaired two-tailed t-test). Mean VCN/cell/patient for GGHI and GGHI-mB-3D is 0.89 ± 0.62 and 1.04 ± 0.8, respectively (p = 0.65, n = 9, unpaired two-tailed t-test). (**B**) Scatter plots showing mean percentage of transduction efficiency (left) and mean VCN/cell/patient (right) in the $\beta^S\beta^S$ patient cohort and following transduction with GGHI and GGHI-mB-3D LVs at MOI 100. GGHI leads to 49 ± 38% and GGHI-mB-3D to 60 ± 40.82% transduction efficiency (p = 0.71, n = 4, unpaired two-tailed t-test). Mean VCN/cell/patient for GGHI and GGHI-mB-3D is 0.8 ± 0.64 and 1 ± 0.66, respectively (p = 0.67, n = 4, unpaired two-tailed t-test). (**C**) Scatter plots showing mean percentage of transduction efficiency (left) and mean VCN/cell/patient (right) in the $\beta^S\beta^+$ patient cohort and following transduction with GGHI and GGHI-mB-3D LVs at MOI 100. GGHI leads to 54 ± 30% and GGHI-mB-3D to 51 ± 34% transduction efficiency (p = 0.91, n = 5, unpaired two-tailed t-test). Mean VCN/cell/patient for GGHI and GGHI-mB-3D is 0.96 ± 0.67 and 1.1 ± 0.98, respectively (p = 0.83, n = 5, unpaired two-tailed t-test). Each dot corresponds to each patient. Error bars represent ±SD.

A

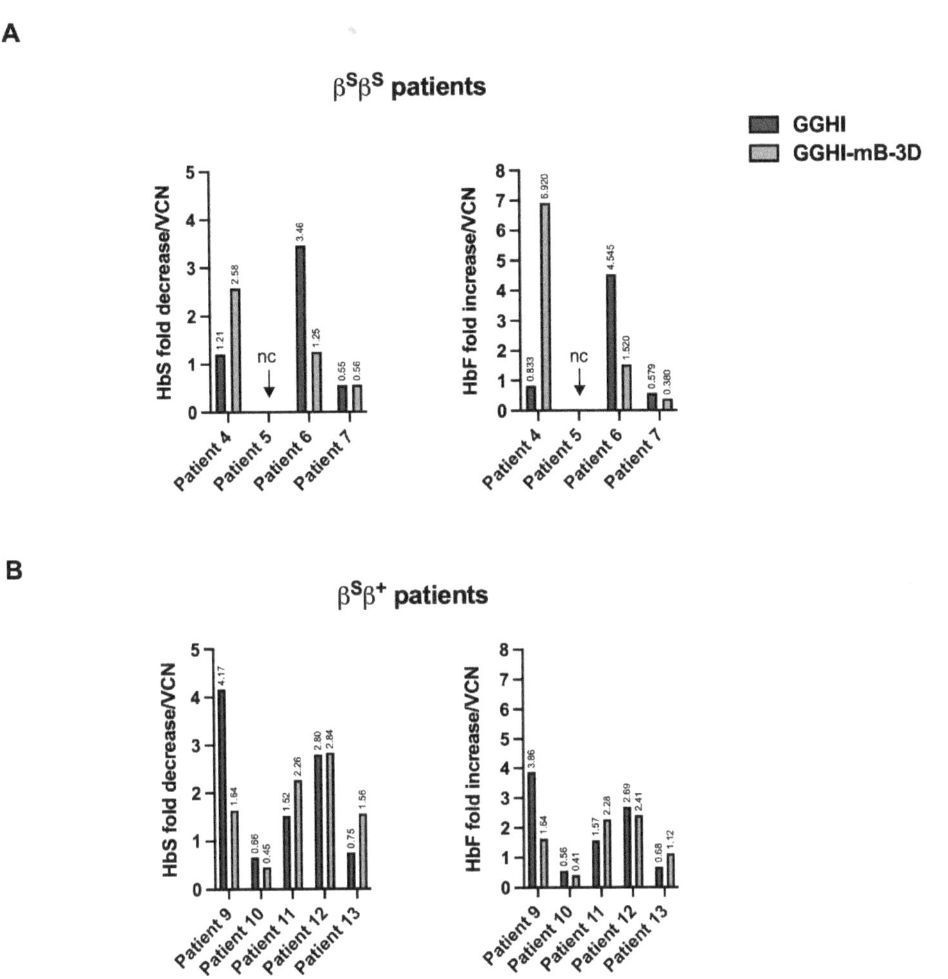

Figure 5. HbF and HbS fold difference in β^S^β^S^ and β^S^β^+^ SCD patients, post transduction: (**A**) Bar chart showing HbS fold decrease (left) and HbF fold increase (right) in β^S^β^S^ patients following normalization to mean VCN/cell/patient; nc, not conducted. (**B**) Bar chart showing HbS fold decrease (left) and HbF fold increase (right) in β^S^β^+^ patients following normalization to mean VCN/cell/patient.

4. Discussion

In this study, we used the previously characterized γ-globin lentiviral vector GGHI [32] and the novel optimized GGHI-mB-3D [38], both successfully assessed using thalassemic CD34^+ cells [32,38], and investigated whether they can also improve or correct the SCD phenotype in vitro. We show that the optimized GGHI-mB-3D vector can significantly increase the $^A\gamma/\alpha$ ratio and HbF percentage in the SCD patient cohort, and lead to significant HbS reduction, at an average VCN of 1.0, calculated per diploid genome. This value represents the ideal target range of VCN per cell for LV-based thalassemia gene therapy [48]. Application of the Pearson's r test across all patients showed very good correlation between normalised HbS fold decrease and normalized $^A\gamma/\alpha$ ratio ($r = 0.9$) or HbF fold increase ($r = 0.7$).

Specifically, results from RP-HPLC regarding $^A\gamma/\alpha$ ratio fold increase, and following normalization to VCN (Figure S5), show that four out of eight patients (Patient 5 was not included) designated as Patients 4, 8, 9, and 12 exhibited a >2-fold increase following transduction with GGHI-mB-3D. Overall, transduction with GGHI-mB-3D led to an average $^A\gamma/\alpha$ ratio fold increase of 1.81 ± 0.97 ($n = 8$) following normalization to VCN. The highest increase was observed in Patient 8, who led to a corrected $^A\gamma/\alpha$ increase of 2.96-fold. Interestingly, despite the achieved maximum $^A\gamma/\alpha$ ratio fold increase in Patient 8, this was not associated with a similar increase in transduction efficiency and VCN/cell (mean transduction efficiency 90% and VCN/cell 0.9). Pearson's r test showed poor positive correlation between $^A\gamma/\alpha$ ratio fold increase and transduction efficiency ($r = 0.16$) and negative correlation between the former and VCN ($r = -0.002$). The above results suggest that expression from GGHI-mB-3D is not entirely dependent on the vector per se, but may also be influenced by the site of integration, an observation also reported by Drakopoulou et al. [38].

Despite the significant increase in the $^A\gamma/\alpha$ ratio observed with RP-HPLC, we did not detect a reciprocally lower β^S/α ratio, following transduction with GGHI-mB-3D lentiviral vector ($p = 0.226$, $n = 9$). This may possibly be due to the restriction of the RP-HPLC analyses to the soluble populations of globin chains. To circumvent the above findings, and in order to demonstrate a potential therapeutic effect of GGHI-mB-3D and/or of GGHI, we performed hemoglobin electrophoresis or CE-HPLC analysis of lysates from mock-transduced and transduced erythroid cultures and assessed HbS and HbF expression. As expected, GGHI-mB-3D LV led to a significantly lower HbS compared with control, demonstrating a mean fold decrease of 1.643 ± 0.88 and a 2.09 ± 2.09 mean HbF fold increase, following normalization to VCN. The respective values for GGHI were 1.890 ± 1.40 and 1.92 ± 1.59.

Following patient classification according to genotype, GGHI-mB-3D failed to demonstrate a marked in vitro improvement in the $\beta^S\beta^S$ patients, suggesting higher HbF requirements for in vitro phenotypic correction in the specific patient cohort, as one would also predict from the high HbS levels before and after HU treatment in the specific patient cohort (Figure S2A,B and Tables 2 and 3). Mean HbS fold decrease following normalization to VCN in the specific cohort was 1.46 ± 1.03 ($n = 3$), with the maximum value observed in Patient 4 who exhibited a 2.58-fold HbS decrease (Figure 5A, left panel), demonstrating also the maximum, i.e., 6.92-fold HbF increase (Figure 5A, right panel). On the contrary, transduction of $\beta^S\beta^+$ cells with GGHI-mB-3D led to a 1.75 ± 0.89 ($n = 5$) mean HbS fold decrease and 1.57 ± 0.83 ($n = 5$) mean HbF fold increase, following normalization to VCN. Out of five $\beta^S\beta^+$ patients, transduction with GGHI-mB-3D led to HbS decrease in four patients, with the highest value observed in Patient 12, who exhibited a HbS decrease of 2.84-fold (Figure 5B, left panel), also showing the highest, i.e., 2.41-fold, HbF increase (Figure 5B, right panel). Pearson's r test showed moderate positive correlation between transduction efficiency and normalized HbS fold decrease ($r = 0.56$) and normalized HbF fold increase ($r = 0.59$).

Regarding the GGHI vector in the $\beta^S\beta^S$ patient cohort, the mean HbS fold decrease following normalization to VCN reached 1.74 ± 1.53, showing no statistical significance. The most profound effect was demonstrated by Patient 6, who demonstrated a 3.46-fold HbS decrease (Figure 5A, left panel), also showing the highest, i.e., 4.55-fold, HbF increase (Figure 5A, right panel). On the contrary, and in the case of $\beta^S\beta^+$ patients, the mean HbS fold decrease and HbF fold increase were 1.98 ± 1.5 and 1.87 ± 1.4, respectively, ($n = 5$), both following normalization to VCN. Pearson's r test showed no positive correlation between transduction efficiency and normalized HbF fold increase ($r = -0.03$) or HbS fold decrease ($r = -0.09$).

With regards to the γ-globin mRNA levels (Figure S4), despite the marked HbF increase observed following transduction with GGHI, and, most importantly, with GGHI-mB-3D γ-globin LVs, we failed to demonstrate a marked mean increase in γ-globin transcripts relative to α-globin. The latter is in contrast with Urbinati et al. [22], in SCD cells, who showed that the γ-globin lentiviral vector V5m3-400 managed to increase γ-globin transcripts by more than 7-fold. However, this finding was based on a rather small number of three patients, while in our series we investigated a more representative cohort of eight informative patients (Patients 7 and 9 were not included due to limited cell number). We also managed to detect a small increase in at least three patients; specifically, Patients

6, 8, and 11, who showed the highest increase following transduction with GGHI, and Patients 6, 10, and 11, who showed increased γ-globin transcripts following transduction with GGHI-mB-3D LV. Most patients with increased γ-globin mRNA levels, displayed a concomitant HbF increase. A possible explanation for not detecting a similar high γ-globin increase at the mRNA level, as Urbinati et al. noted, could be attributed to the late sampling for RNA isolation. Compared to Urbinati et al. [22], we carried out RNA isolation 1 week later during erythroid differentiation, i.e., at day 20–21 instead of day 14, which may have resulted in the underestimation of expression of vector-derived γ-globin mRNA. As documented elsewhere [49,50], in vitro erythroid differentiation partially recapitulates ontogenesis and hemoglobin switching for ES-derived [49], as well as for adult peripheral blood CD34$^+$-derived progenitors [50], with a shift to the contribution of embryonic to fetal and then to adult globin chains on the background of an overall increasing globin expression during erythroid differentiation. Since both the vector-derived $^A\gamma$ gene and the endogenous one share identical cis-acting sequences, and thus undergo the same transcriptional regulation, it is conceivable that an endogenous fetal to adult switch from γ-globin to β-globin and/or β^S at late stages of erythroid differentiation would result in relatively lower γ-globin mRNA levels and a higher overall β^S contribution.

Regarding the differences between the GGHI and GGHI-mB-3D performance, despite the significant $^A\gamma/\alpha$ ratio and HbF increase, as well as the HbS decrease demonstrated only by GGHI-mB-3D LV compared to mock-transduced cells, we did not observe significant differences between the former and GGHI, possibly due to extensive variation among patient samples, an observation also documented in a previous study by Drakopoulou et al. [38].

In line with our previous observations in thalassemia patients, and of others in SCD patients [22], the mean VCN calculated for GGHI and GGHI-mB-3D lentiviral vectors was around 1.0 ($p = 0.65$, $n = 9$, unpaired two-tailed t-test), reflecting a near-ideal value for a clinical setting and a potentially therapeutic outcome, as reported by Kanter et al. [51]. VCN achieved among the $\beta^S\beta^S$ and $\beta^S\beta^+$ patient cohorts did not test significantly different between GGHI ($p = 0.51$, $n = 4$, unpaired two-tailed t-test) and GGHI-mB-3D ($p = 0.66$, $n = 4$, unpaired two-tailed t-test) LVs. With regards to transduction efficiency, GGHI and GGHI-mB-3D both achieved an overall gene transfer of 52% and 55%, respectively ($p = 0.82$, $n = 9$, unpaired two-tailed t-test). Again, no significant differences were observed in transduction efficiency among the $\beta^S\beta^S$ and $\beta^S\beta^+$ cohorts between GGHI ($p = 0.75$, $n = 4$, unpaired two-tailed t-test) and GGHI-mB-3D ($p = 0.74$, $n = 5$, unpaired two-tailed t-test) LVs.

It should be emphasized that our vectors, harboring several erythroid-specific regulatory elements, to the best of our knowledge, are the only LCR-free, SIN, and insulated globin vectors that can be employed to efficiently transduce erythroid progenitor cells and successfully drive the human γ-globin gene to nearly therapeutic levels in both thalassemic [32,38] and SCD CD34$^+$ cells, as shown in this study. In view of the ongoing clinical trials for thalassemia and SCD, one of the main strengths of these vectors is their safety feature, due to their insulation and to the lack of any LCR regulatory elements, which have been recently shown to be active in early hematopoietic progenitor cells [52], and thus are capable of trans-activating cancer-related genes, with all the known associated risks for the patient [36,37,52,53].

In conclusion, despite the significant and potentially therapeutic HbF increase observed in SCD patients with the optimized γ-globin lentiviral vector GGHI-mB-3D, a major limitation of our strategy, primarily regarding the precise assessment and comparison of the HbF increase in each patient and among the $\beta^S\beta^S$ and $\beta^S\beta^+$ cohorts, is the established inherent genetic and epigenetic heterogeneity per se of these patients, coupled with the lack of information regarding important HbF genetic modifiers, such *XmnI, HBS1L-MYB*, and *BCL11A* [54], in the two cohorts. These major SNPs associated with high HbF levels, account for more than 20% of the HbF level variations among SCD patients [55], and affect both the severity and the therapeutic outcome in SCD, including patients under HU treatment [56]. In a recent systematic review by Sales et al. [57], the authors concluded that genetic variations in multiple loci, such as SNPs located at intron 2 of the *BCL11A* gene, can affect both baseline HbF and HbS levels in response to HU treatment in patients with SCD. Therefore, information on these SNPs, together with larger patient cohorts of different ethnic origin, would significantly contribute to a more effective assessment of the GGHI-mB-3D γ-globin vector in the context of sickle cell disease gene therapy [58].

5. Conclusions

In summary, we show that the optimized LCR-free GGHI-mB-3D lentiviral vector, which carries novel regulatory elements, can significantly increase the $^A\gamma/\alpha$ ratio and HbF in CD34$^+$ cells from SCD patients, an increase accompanied by a concomitant HbS decrease, demonstrating a potentially therapeutic outcome in vitro. These data suggest that despite the non-significant differences between the optimized GGHI-mB-3D and GGHI lentiviral vectors, the former demonstrates an increased potential of improving the SCD phenotype in vitro, and thus, can eventually provide a significant therapeutic benefit in the context of future clinical trials for patients with sickle cell disease.

Supplementary Materials: The following supporting information can be downloaded at: https://www.mdpi.com/article/10.3390/v14122716/s1. Figure S1: (**A**) Schematic representation of the experimental procedure. (**B**) Table showing corresponding titers (TU/mL) of different virus batches (LOTs) for GGHI ($n = 6$) and GGHI-mB-3D ($n = 5$) γ-globin LVs (left panel), and bar chart showing mean GGHI and GGHI-mB-3D titers ($p = 0.902$, unpaired two-tailed t-test) (right panel). Figure S2: (**A**) Baseline HbF and HbS expression prior to hydroxyurea (HU) treatment in the $\beta^S\beta^S$ ($n = 5$) (upper panel), and $\beta^S\beta^+$ patient cohort ($n = 5$) (lower panel). (**B**) HbS expression prior and post HU treatment in $\beta^S\beta^S$ and $\beta^S\beta^+$ patients. (**C**) CD235a expression at the end of the in vitro differentiation. Figure S3: Estimation plots for all comparisons with statistical significance. Figure S4: Performance of γ-globin LVs at the RNA level: (**A**) Bar chart showing relative fold difference of γ-mRNA transcripts in different patients ($n = 8$). (**B**) Bar charts showing relative fold difference of γ-mRNA transcripts in the $\beta^S\beta^S$ patient cohort ($n = 4$) (left panel) and in the $\beta^S\beta^+$ patient cohort ($n = 4$) (right panel). Figure S5: $^A\gamma/\alpha$ ratio fold increase per patient post transduction with GGHI and GGHI-mB-3D LVs and following normalization to VCN.

Author Contributions: Conceptualization of the project was conducted by E.D. and N.P.A.; Methodology by E.D., M.G., C.W.L., F.P., E.V., M.K., E.P., D.V. and N.P.A.; Data Curation by E.D. and M.G.; Writing—Original Draft Preparation was carried out by E.D.; Writing—Review and Editing by E.D., C.W.L. and N.P.A.; N.P.A. supervised the whole project and carried out the essential Project Administration and Funding Acquisition. All authors have read and agreed to the published version of the manuscript.

Funding: This research was supported by the European Research Projects on Rare Diseases, E-RARE 2nd Joint Call, *Improvements of vector technology and safety for gene therapy of thalassemia* (GETH-ERTHAL), Grant No. 11E-RARE-09-155 from the European Commission to N.P.A.

Institutional Review Board Statement: The study was conducted according to the guidelines of the Declaration of Helsinki and approved by the Bioethics Committee of School of Medicine, National and Kapodistrian University of Athens, NTUA (protocol code no. 080/2019).

Informed Consent Statement: Informed consent was obtained from all subjects involved in the study.

Data Availability Statement: The data that support the findings of this study are available on request from the first author.

Acknowledgments: We would like to thank Luigi Naldini and Giuliana Ferrari for providing the initial lentiviral plasmids.

Conflicts of Interest: The authors declare no conflict of interest.

References

1. Ingram, V.M. Gene Mutations in Human Hæmoglobin: The Chemical Difference Between Normal and Sickle Cell Hæmoglobin. *Nature* **1957**, *180*, 326–328. [CrossRef] [PubMed]
2. Driss, A.; Asare, K.O.; Hibbert, J.M.; Gee, B.E.; Adamkiewicz, T.V.; Stiles, J.K. Sickle Cell Disease in the Post Genomic Era: A Monogenic Disease with a Polygenic Phenotype. *Genom. Insights* **2009**, *2009*, 23–48. [CrossRef]
3. Milunsky, A. Sickle Cell Disease. *N. Engl. J. Med.* **2017**, *377*, 302–305. [CrossRef]

4. Powars, D.R.; Weiss, J.N.; Chan, L.S.; Schroeder, W.A. Is there a threshold level of fetal hemoglobin that ameliorates morbidity in sickle cell anemia? *Blood* **1984**, *63*, 921–926. [CrossRef] [PubMed]

5. Platt, O.S.; Brambilla, D.J.; Rosse, W.F.; Milner, P.F.; Castro, O.; Steinberg, M.H.; Klug, P.P. Mortality in sickle cell disease—Life expectancy and risk factors for early death. *N. Engl. J. Med.* **1994**, *330*, 1639–1644. [CrossRef] [PubMed]

6. Koshy, M.; Entsuah, R.; Koranda, A.; Kraus, A.P.; Johnson, R.; Bellvue, R.; Flournoy-Gill, Z.; Levy, P. Leg ulcers in patients with sickle cell disease. *Blood* **1989**, *74*, 1403–1408. [CrossRef] [PubMed]

7. Nolan, V.G.; Adewoye, A.; Baldwin, C.; Wang, L.; Ma, Q.; Wyszynski, D.F.; Farrell, J.J.; Sebastiani, P.; Farrer, L.A.; Steinberg, M.H. Sickle cell leg ulcers: Associations with haemolysis and SNPs in Klotho, TEK and genes of the TGF-beta/BMP pathway. *Br. J. Haematol.* **2006**, *133*, 570–578. [CrossRef]

8. Steinberg, M.H.; Forget, B.G.; Higgs, D.R.; Weatherall, D.J. *Disorders of Hemoglobin: Genetics, Pathophysiology and Clinical Management*; Cambridge University Press: Cambridge, UK, 2009.

9. Bunn, H.F.; Forget, B.G. *Hemoglobin-Molecular, Genetic and Clinical Aspects*; W. B. Saunders Co.,Ltd.: Philadelphia, PA, USA, 1986; Volume VII.

10. Steinberg, M.H. Targeting fetal hemoglobin expression to treat β hemoglobinopathies. *Expert Opin. Ther. Targets* **2022**, *26*, 347–359. [CrossRef]

11. Ley, T.J.; DeSimone, J.; Anagnou, N.P.; Keller, G.H.; Humphries, R.K.; Turner, P.H.; Young, N.S.; Heller, P.; Nienhuis, A.W. 5-Azacytidine Selectively Increases γ-Globin Synthesis in a Patient with β⁺Thalassemia. *N. Engl. J. Med.* **1982**, *307*, 1469–1475. [CrossRef]

12. Charache, S.; Dover, G.; Smith, K.; Talbot, C.C., Jr.; Moyer, M.; Boyer, S. Treatment of sickle cell anemia with 5-azacytidine results in increased fetal hemoglobin production and is associated with nonrandom hypomethylation of DNA around the gamma-delta-beta-globin gene complex. *Proc. Natl. Acad. Sci. USA* **1983**, *80*, 4842–4846. [CrossRef]

13. Charache, S.; Terrin, M.L.; Moore, R.D.; Dover, G.J.; Barton, F.B.; Eckert, S.V.; McMahon, R.P.; Bonds, D.R. Effect of Hydroxyurea on the Frequency of Painful Crises in Sickle Cell Anemia. *N. Engl. J. Med.* **1995**, *332*, 1317–1322. [CrossRef] [PubMed]

14. Steinberg, M.H.; Barton, F.; Castro, O.; Pegelow, C.H.; Ballas, S.K.; Kutlar, A.; Orringer, E.; Bellvue, R.; Olivieri, N.; Eckman, J.; et al. Effect of Hydroxyurea on Mortality and Morbidity in Adult Sickle Cell Anemia. *JAMA* **2003**, *289*, 1645–1651. [CrossRef] [PubMed]

15. Steinberg, M.H.; McCarthy, W.F.; Castro, O.; Ballas, S.K.; Armstrong, F.D.; Smith, W.; Ataga, K.; Swerdlow, P.; Kutlar, A.; DeCastro, L.; et al. The risks and benefits of long-term use of hydroxyurea in sickle cell anemia: A 17.5 year follow-up. *Am. J. Hematol.* **2010**, *85*, 403–408. [CrossRef] [PubMed]

16. Olivieri, N.F.; Brittenham, G.M. Management of the Thalassemias. *Cold Spring Harb. Perspect. Med.* **2013**, *3*, a011767. [CrossRef]

17. Ribeil, J.-A.; Hacein-Bey-Abina, S.; Payen, E.; Magnani, A.; Semeraro, M.; Magrin, E.; Caccavelli, L.; Neven, B.; Bourget, P.; El Nemer, W.; et al. Gene Therapy in a Patient with Sickle Cell Disease. *N. Engl. J. Med.* **2017**, *376*, 848–855. [CrossRef] [PubMed]

18. Pestina, T.I.; Hargrove, P.W.; Jay, D.; Gray, J.T.; Boyd, K.M.; Persons, D.A. Correction of murine sickle cell disease using gamma-globin lentiviral vectors to mediate high-level expression of fetal hemoglobin. *Mol. Ther.* **2009**, *17*, 245–252. [CrossRef]

19. Levasseur, D.N.; Ryan, T.M.; Pawlik, K.M.; Townes, T.M. Correction of a mouse model of sickle cell disease: Lentiviral/antisickling beta-globin gene transduction of unmobilized, purified hematopoietic stem cells. *Blood* **2003**, *102*, 4312–4319. [CrossRef]

20. Perumbeti, A.; Higashimoto, T.; Urbinati, F.; Franco, R.; Meiselman, H.J.; Witte, D.; Malik, P. A novel human gamma-globin gene vector for genetic correction of sickle cell anemia in a humanized sickle mouse model: Critical determinants for successful correction. *Blood* **2009**, *114*, 1174–1185. [CrossRef]

21. Tisdale, J.F.; Thein, S.L.; Eaton, W.A. Treating sickle cell anemia. *Science* **2020**, *367*, 1198–1199. [CrossRef]

22. Urbinati, F.; Hargrove, P.W.; Geiger, S.; Romero, Z.; Wherley, J.; Kaufman, M.L.; Hollis, R.P.; Chambers, C.B.; Persons, D.A.; Kohn, D.B.; et al. Potentially therapeutic levels of anti-sickling globin gene expression following lentivirus-mediated gene transfer in sickle cell disease bone marrow CD34+ cells. *Exp. Hematol.* **2015**, *43*, 346–351. [CrossRef]

23. Weber, L.; Poletti, V.; Magrin, E.; Antoniani, C.; Martin, S.; Bayard, C.; Sadek, H.; Felix, T.; Meneghini, V.; Antoniou, M.N.; et al. An Optimized Lentiviral Vector Efficiently Corrects the Human Sickle Cell Disease Phenotype. *Mol. Ther. Methods Clin. Dev.* **2018**, *10*, 268–280. [CrossRef] [PubMed]

24. Breda, L.; Casu, C.; Gardenghi, S.; Bianchi, N.; Cartegni, L.; Narla, M.; Yazdanbakhsh, K.; Musso, M.; Manwani, D.; Little, J.; et al. Therapeutic hemoglobin levels after gene transfer in beta-thalassemia mice and in hematopoietic cells of beta-thalassemia and sickle cells disease patients. *PLoS ONE* **2012**, *7*, e32345. [CrossRef] [PubMed]

25. Samakoglu, S.; Lisowski, L.; Budak-Alpdogan, T.; Usachenko, Y.; Acuto, S.; Di Marzo, R.; Maggio, A.; Zhu, P.; Tisdale, J.F.; Riviere, I.; et al. A genetic strategy to treat sickle cell anemia by coregulating globin transgene expression and RNA interference. *Nat. Biotechnol.* **2006**, *24*, 89–94. [CrossRef] [PubMed]

26. Xu, J.; Peng, C.; Sankaran, V.G.; Shao, Z.; Esrick, E.B.; Chong, B.G.; Ippolito, G.C.; Fujiwara, Y.; Ebert, B.L.; Tucker, P.W.; et al. Correction of sickle cell disease in adult mice by interference with fetal hemoglobin silencing. *Science* **2011**, *334*, 993–996. [CrossRef]

27. Tallack, M.R.; Perkins, A.C. Three fingers on the switch: Kruppel-like factor 1 regulation of gamma-globin to beta-globin gene switching. *Curr. Opin. Hematol.* **2013**, *20*, 193–200. [CrossRef]

28. Sankaran, V.G.; Xu, J.; Ragoczy, T.; Ippolito, G.C.; Walkley, C.R.; Maika, S.D.; Fujiwara, Y.; Ito, M.; Groudine, M.; Bender, M.A.; et al. Developmental and species-divergent globin switching are driven by BCL11A. *Nature* **2009**, *460*, 1093–1097. [CrossRef]

29. Sankaran, V.G.; Menne, T.F.; Scepanovic, D.; Vergilio, J.A.; Ji, P.; Kim, J.; Thiru, P.; Orkin, S.H.; Lander, E.S.; Lodish, H.F. MicroRNA-15a and -16-1 act via MYB to elevate fetal hemoglobin expression in human trisomy 13. *Proc. Natl. Acad. Sci. USA* **2011**, *108*, 1519–1524. [CrossRef]

30. Williams, D.A.; Esrick, E. Investigational curative gene therapy approaches to sickle cell disease. *Blood Adv.* **2021**, *5*, 5452. [CrossRef]

31. Magis, W.; DeWitt, M.A.; Wyman, S.K.; Vu, J.T.; Heo, S.J.; Shao, S.J.; Hennig, F.; Romero, Z.G.; Campo-Fernandez, B.; Said, S.; et al. High-level correction of the sickle mutation is amplified in vivo during erythroid differentiation. *iScience* **2022**, *25*, 104374. [CrossRef]

32. Papanikolaou, E.; Georgomanoli, M.; Stamateris, E.; Panetsos, F.; Karagiorga, M.; Tsaftaridis, P.; Graphakos, S.; Anagnou, N.P. The new self-inactivating lentiviral vector for thalassemia gene therapy combining two HPFH activating elements corrects human thalassemic hematopoietic stem cells. *Hum. Gene Ther.* **2012**, *23*, 15–31. [CrossRef]

33. Katsantoni, E.Z.; Langeveld, A.; Wai, A.W.; Drabek, D.; Grosveld, F.; Anagnou, N.P.; Strouboulis, J. Persistent gamma-globin expression in adult transgenic mice is mediated by HPFH-2, HPFH-3, and HPFH-6 breakpoint sequences. *Blood* **2003**, *102*, 3412–3419. [CrossRef] [PubMed]

34. Fragkos, M.; Anagnou, N.P.; Tubb, J.; Emery, D.W. Use of the hereditary persistence of fetal hemoglobin 2 enhancer to increase the expression of oncoretrovirus vectors for human gamma-globin. *Gene Ther.* **2005**, *12*, 1591–1600. [CrossRef] [PubMed]

35. Li, Q.; Emery, D.W.; Fernandez, M.; Han, H.; Stamatoyannopoulos, G. Development of viral vectors for gene therapy of beta-chain hemoglobinopathies: Optimization of a gamma-globin gene expression cassette. *Blood* **1999**, *93*, 2208–2216. [CrossRef] [PubMed]

36. Cavazzana-Calvo, M.; Payen, E.; Negre, O.; Wang, G.; Hehir, K.; Fusil, F.; Down, J.; Denaro, M.; Brady, T.; Westerman, K.; et al. Transfusion independence and HMGA2 activation after gene therapy of human beta-thalassaemia. *Nature* **2010**, *467*, 318–322. [CrossRef] [PubMed]

37. Boulad, F.; Maggio, A.; Wang, X.; Moi, P.; Acuto, S.; Kogel, F.; Takpradit, C.; Prockop, S.; Mansilla-Soto, J.; Cabriolu, A.; et al. Lentiviral globin gene therapy with reduced-intensity conditioning in adults with beta-thalassemia: A phase 1 trial. *Nat. Med.* **2022**, *28*, 63–70. [CrossRef] [PubMed]

38. Drakopoulou, E.; Georgomanoli, M.; Lederer, C.W.; Kleanthous, M.; Costa, C.; Bernadin, O.; Cosset, F.L.; Voskaridou, E.; Verhoeyen, E.; Papanikolaou, E.; et al. A Novel BaEVRless-Pseudotyped gamma-Globin Lentiviral Vector Drives High and Stable Fetal Hemoglobin Expression and Improves Thalassemic Erythropoiesis In Vitro. *Hum. Gene Ther.* **2019**, *30*, 601–617. [CrossRef] [PubMed]

39. Elder, J.T.; Forrester, W.C.; Thompson, C.; Mager, D.; Henthorn, P.; Peretz, M.; Papayannopoulou, T.; Groudine, M. Translocation of an erythroid-specific hypersensitive site in deletion-type hereditary persistence of fetal hemoglobin. *Mol. Cell. Biol.* **1990**, *10*, 1382–1389. [CrossRef]

40. Tuan, D.; Feingold, E.; Newman, M.; Weissman, S.M.; Forget, B.G. Different 3′ end points of deletions causing delta beta-thalassemia and hereditary persistence of fetal hemoglobin: Implications for the control of gamma-globin gene expression in man. *Proc. Natl. Acad. Sci. USA* **1983**, *80*, 6937–6941. [CrossRef]

41. Frecha, C.; Costa, C.; Negre, D.; Amirache, F.; Trono, D.; Rio, P.; Bueren, J.; Cosset, F.L.; Verhoeyen, E. A novel lentiviral vector targets gene transfer into human hematopoietic stem cells in marrow from patients with bone marrow failure syndrome and in vivo in humanized mice. *Blood* **2012**, *119*, 1139–1150. [CrossRef]

42. Girard-Gagnepain, A.; Amirache, F.; Costa, C.; Levy, C.; Frecha, C.; Fusil, F.; Negre, D.; Lavillette, D.; Cosset, F.L.; Verhoeyen, E. Baboon envelope pseudotyped LVs outperform VSV-G-LVs for gene transfer into early-cytokine-stimulated and resting HSCs. *Blood* **2014**, *124*, 1221–1231. [CrossRef]

43. Dull, T.; Zufferey, R.; Kelly, M.; Mandel, R.J.; Nguyen, M.; Trono, D.; Naldini, L. A third-generation lentivirus vector with a conditional packaging system. *J. Virol.* **1998**, *72*, 8463–8471. [CrossRef] [PubMed]

44. Zufferey, R.; Dull, T.; Mandel, R.J.; Bukovsky, A.; Quiroz, D.; Naldini, L.; Trono, D. Self-inactivating lentivirus vector for safe and efficient in vivo gene delivery. *J. Virol.* **1998**, *72*, 9873–9880. [CrossRef] [PubMed]

45. Papanikolaou, E.; Kontostathi, G.; Drakopoulou, E.; Georgomanoli, M.; Stamateris, E.; Vougas, K.; Vlahou, A.; Maloy, A.; Ware, M.; Anagnou, N.P. Characterization and comparative performance of lentiviral vector preparations concentrated by either one-step ultrafiltration or ultracentrifugation. *Virus Res.* **2013**, *175*, 1–11. [CrossRef] [PubMed]

46. Loucari, C.C.; Patsali, P.; van Dijk, T.B.; Stephanou, C.; Papasavva, P.; Zanti, M.; Kurita, R.; Nakamura, Y.; Christou, S.; Sitarou, M.; et al. Rapid and Sensitive Assessment of Globin Chains for Gene and Cell Therapy of Hemoglobinopathies. *Hum. Gene Ther. Methods* **2018**, *29*, 60–74. [CrossRef] [PubMed]

47. Pfaffl, M.W. A new mathematical model for relative quantification in real-time RT-PCR. *Nucleic Acids Res.* **2001**, *29*, e45. [CrossRef] [PubMed]

48. Lisowski, L.; Sadelain, M. Locus control region elements HS1 and HS4 enhance the therapeutic efficacy of globin gene transfer in beta-thalassemic mice. *Blood* **2007**, *110*, 4175–4178. [CrossRef] [PubMed]

49. Fujita, A.; Uchida, N.; Haro-Mora, J.J.; Winkler, T.; Tisdale, J. beta-Globin-Expressing Definitive Erythroid Progenitor Cells Generated from Embryonic and Induced Pluripotent Stem Cell-Derived Sacs. *Stem Cells* **2016**, *34*, 1541–1552. [CrossRef] [PubMed]

50. Mahajan, M.C.; Karmakar, S.; Newburger, P.E.; Krause, D.S.; Weissman, S.M. Dynamics of alpha-globin locus chromatin structure and gene expression during erythroid differentiation of human CD34(+) cells in culture. *Exp. Hematol.* **2009**, *37*, 1143–1156.e3. [CrossRef]

51. Kanter, J.; Walters, M.C.; Krishnamurti, L.; Mapara, M.Y.; Kwiatkowski, J.L.; Rifkin-Zenenberg, S.; Aygun, B.; Kasow, K.A.; Pierciey, F.J., Jr.; Bonner, M.; et al. Biologic and Clinical Efficacy of LentiGlobin for Sickle Cell Disease. *N. Engl. J. Med.* **2022**, *386*, 617–628. [CrossRef]
52. Cabriolu, A.; Odak, A.; Zamparo, L.; Yuan, H.; Leslie, C.S.; Sadelain, M. Globin vector regulatory elements are active in early hematopoietic progenitor cells. *Mol. Ther.* **2022**, *30*, 2199–2209. [CrossRef]
53. Goyal, S.; Tisdale, J.; Schmidt, M.; Kanter, J.; Jaroscak, J.; Whitney, D.; Bitter, H.; Gregory, P.D.; Parsons, G.; Foos, M.; et al. Acute Myeloid Leukemia Case after Gene Therapy for Sickle Cell Disease. *N. Engl. J. Med.* **2022**, *386*, 138–147. [CrossRef] [PubMed]
54. Lettre, G.; Sankaran, V.G.; Bezerra, M.A.; Araujo, A.S.; Uda, M.; Sanna, S.; Cao, A.; Schlessinger, D.; Costa, F.F.; Hirschhorn, J.N.; et al. DNA polymorphisms at the BCL11A, HBS1L-MYB, and beta-globin loci associate with fetal hemoglobin levels and pain crises in sickle cell disease. *Proc. Natl. Acad. Sci. USA* **2008**, *105*, 11869–11874. [CrossRef] [PubMed]
55. Galarneau, G.; Palmer, C.D.; Sankaran, V.G.; Orkin, S.H.; Hirschhorn, J.N.; Lettre, G. Fine-mapping at three loci known to affect fetal hemoglobin levels explains additional genetic variation. *Nat. Genet.* **2010**, *42*, 1049–1051. [CrossRef] [PubMed]
56. Menzel, S.; Thein, S.L. Genetic Modifiers of Fetal Haemoglobin in Sickle Cell Disease. *Mol. Diagn. Ther.* **2019**, *23*, 235–244. [CrossRef]
57. Sales, R.R.; Nogueira, B.L.; Tosatti, J.A.G.; Gomes, K.B.; Luizon, M.R. Do Genetic Polymorphisms Affect Fetal Hemoglobin (HbF) Levels in Patients With Sickle Cell Anemia Treated With Hydroxyurea? A Systematic Review and Pathway Analysis. *Front. Pharmacol.* **2021**, *12*, 779497. [CrossRef]
58. Iolascon, A.; Rivella, S.; Anagnou, N.P.; Camaschella, C.; Swinkels, D.; Muckenthaler, M.U.; Porto, G.; Barcellini, W.; Andolfo, I.; Risitano, A.M.; et al. The EHA Research Roadmap: Anemias. *Hemasphere* **2021**, *5*, e607. [CrossRef]

Review

Innate Immune Response to Viral Vectors in Gene Therapy

Yixuan Wang and Wenwei Shao *

Academy of Medical Engineering and Translational Medicine, Medical College, Tianjin University, Tianjin 300072, China; 2021235075@tju.edu.cn
* Correspondence: wenwei.shao@tju.edu.cn; Tel.: +86-022-83612122

Abstract: Viral vectors play a pivotal role in the field of gene therapy, with several related drugs having already gained clinical approval from the EMA and FDA. However, numerous viral gene therapy vectors are currently undergoing pre-clinical research or participating in clinical trials. Despite advancements, the innate response remains a significant barrier impeding the clinical development of viral gene therapy. The innate immune response to viral gene therapy vectors and transgenes is still an important reason hindering its clinical development. Extensive studies have demonstrated that different DNA and RNA sensors can detect adenoviruses, adeno-associated viruses, and lentiviruses, thereby activating various innate immune pathways such as Toll-like receptor (TLR), cyclic GMP-AMP synthase–stimulator of interferon genes (cGAS-STING), and retinoic acid-inducible gene I–mitochondrial antiviral signaling protein (RLR-MAVS). This review focuses on elucidating the mechanisms underlying the innate immune response induced by three widely utilized viral vectors: adenovirus, adeno-associated virus, and lentivirus, as well as the strategies employed to circumvent innate immunity.

Keywords: innate immune; adeno-associated virus; lentivirus; adenovirus

Citation: Wang, Y.; Shao, W. Innate Immune Response to Viral Vectors in Gene Therapy. *Viruses* **2023**, *15*, 1801. https://doi.org/10.3390/v15091801

Academic Editors: Ottmar Herchenröder and Brigitte Pützer

Received: 12 July 2023
Revised: 21 August 2023
Accepted: 22 August 2023
Published: 24 August 2023

1. Introduction

Gene therapy exhibits considerable potential for advancement in the 21st century. To achieve therapeutic effects, gene therapy relies on vectors to deliver the required genes to target cells. These vectors can be classified into two main categories: viral and non-viral types. Presently, viral vectors enjoy extensive utilization within the field of gene therapy due to their widespread adoption and effectiveness.

Currently, prevalent viral vectors in gene therapy encompass adenovirus (Ad), adeno-associated virus (AAV), and lentivirus (LV) (Figure 1). Human Ad (HAdV) was first isolated from human adenoid tissue cultured by Wallace Rowe and his colleagues in 1953. It was considered to be a gene therapy vector around the 1980s [1]. In 1992, the first successful utilization of Ad as a vector for gene therapy was reported. AAV, initially discovered in the mid-1960s, served as a vector for ex vivo gene transduction in 1984. It was first employed to treat patients with cystic fibrosis in 1995 and exhibited encouraging therapeutic outcomes in patients with Leber's congenital amaurosis in 2008. Following that, in 2012, Glybera (Table 1) became the pioneering AAV gene therapy drug to be approved by European Medicines Agency (EMA) [2]. LV, derived from the human immunodeficiency virus (HIV), serves as a viral vector. Upon gene transduction by the vector, if the cell can persist over an extended period, the transgene integrates into the host cell's genome, thereby facilitating enduring transgene expression [3]. As a result of the integration capability exhibited by LV vector, it is prioritized as the primary option in gene therapy.

Innate immunity functions as the initial barrier against foreign intrusions and assumes a pivotal role in host protection [4]. Operating through swift and non-specific reactions, it often triggers cytokine responses within an hour. Given the diverse pathways through which viruses can infiltrate cells, the host engages in a sequence of biological processes aimed at countering viral entry. Antigen-presenting cells, including plasmacytoid dendritic cells (DCs), conventional DCs, macrophages, and B cells, emerge as significant generators of antiviral cytokines [5]. When encountering viruses, the innate immune system recognizes these viral entities as foreign by employing pattern-recognition receptors (PRRs). These PRRs have the capacity to identify distinctive structures inherent to non-self agents, commonly denoted as pathogen-associated molecular patterns (PAMPs) [6]. Subsequent to this recognition, PRRs can instigate the activation of an array of intracellular signaling pathways, potentially leading to the initiation of innate immune responses [7]. Among these receptors, Toll-like receptors (TLRs) expressed on cellular membranes, retinoic acid-inducible gene-I (RIG-I)-like receptors (RLRs), nucleotide oligomerization domain (NOD)-like receptors (NLRs), C-type lectin receptors (CLRs), and a spectrum of intracellular DNA sensors such as cyclic GMP-AMP synthase (cGAS) play pivotal roles in transducing these immune-initiating signals [8]. Upon detection of viral PAMPs, PRRs become activated, subsequently engaging NF-κB and interferon regulatory transcription factors IRF3 and IRF7. This orchestrated activation cascade orchestrates the transcription of proinflammatory cytokines and type I interferons within the infected cells. Consequent to their synthesis, these interferons bind to their designated IFN-I receptors, thereby initiating the expression of IFN-stimulated genes (ISGs). These molecular cues serve to attract effector lymphocytes, impede viral transduction, and propel the elimination of transduced cells via the instigation of an adaptive immune response.

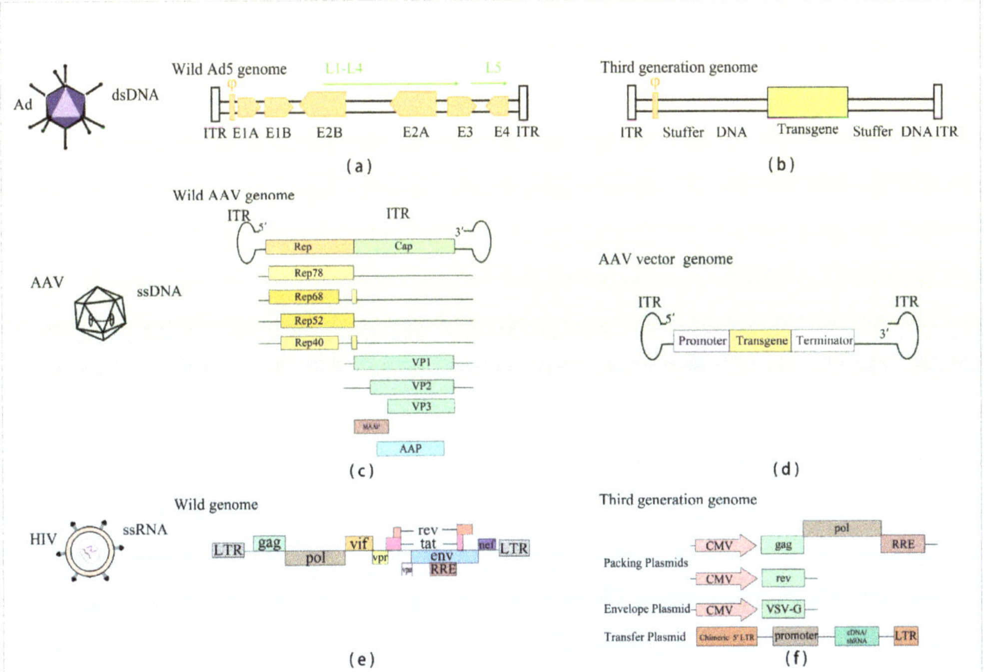

Figure 1. Genome structure description of Ad, AAV, and LV viruses and their viral vectors. (**a**) Adenovirus genome composition. (**b**) Genome composition of third-generation adenovirus vectors currently used for gene therapy. (**c**) Adeno-associated virus genome composition. (**d**) Genome composition of recombinant AAV vectors currently used in gene therapy. (**e**) Genome composition of HIV. (**f**) Genome composition of third-generation lentiviral vectors presently utilized in gene therapy.

Adenovirus disassembly in the endosomes, accompanied by the exposure of viral DNA (Table 2) in plasmacytoid dendritic cells, precipitates the initiation of the TLR9/MyD88 axis. This process subsequently leads to IRF7-dependent transcription of IFN-I genes.

Alternatively, the perception of viral DNA in the cytoplasm provokes TLR-independent activation of type I interferon through the cGAS/STING/TBK1/IRF3 signaling cascade [9]. The manner of HAdV recognition is contingent upon cell type, involving both cytoplasmic and nuclear sensing mechanisms, thereby inciting an innate immune response to the infection. In human cell lines, the Ad genome triggers activation of the cytoplasmic cGAS-STING pathway, which in turn activates TBK1 and IRF3, ultimately culminating in the expression of IFN-β genes. The release of IL-1β induced by Ad hinges on the detection of the Ad genome by the DNA sensor TLR9. In murine models, Ad infection traverses plasmacytoid dendritic cells (pDCs). Notably, conventional dendritic cells and macrophages elicit substantial levels of type I interferon. Recognition of Ad in pDCs is mediated by TLR9, whereas non-pDC recognition of Ad is TLR9-independent and instead contingent upon the cytosolic sensing of viral DNA. In murine antigen-presenting cells (APCs), Ad infection triggers IRF6-mediated IFN responses via TLR-independent DNA sensing, partly reliant on the cGAS-STING pathway [10]. Several studies have attested to the activation of the RLR-MAVS signaling pathway during Ad virus infection. In Ad-transduced host cells, Ad-associated RNA (VA-RNA) engages RIG-I, thereby instigating the RLR-MAVS pathway and subsequently fostering the production of IFN-I [11].

DNA viruses, including AAV (Table 2), have the capability to undergo uncoating and discharge their genomes within the endosome. This action facilitates the engagement of TLR9, thereby instigating the initiation of the innate immune response. As AAV undergoes endosomal trafficking, there arises the potential for virion uncoating or partial genome exposure, which could consequently enable TLR9 detection of CpG motifs. This process transpires rapidly and consequently activates NF-κB, ultimately leading to the synthesis of proinflammatory cytokines [12]. Within the cellular cytosol, the DNA sensor cGAS can effectively bind double-stranded DNA, subsequently eliciting a signaling cascade that involves STING and TBK1. Notably, AAV ITRs exhibit promoter activity, thereby implying that the presence of both 5′- and 3′-ITRs within the AAV genome could potentially result in the generation of sense and antisense RNAs. These RNA molecules possess the capability to form double-stranded RNA intermediates within the cytoplasm. It is pertinent to mention that double-stranded RNA molecules are duly recognized by MDA5 and RIG-I, subsequently instigating the production of IFN-β [13].

HIV-1 (Table 2) infection triggers immune system responses through the recognition of various HIV-associated PAMPs by PRRs. Both cell membrane-expressed TLRs and endosomal TLRs recognize specific PAMPs in HIV, thus initiating the innate immune response. TLR2 and TLR4 present on the cell surface play roles in detecting HIV glycoprotein gp120, leading to NF-κB activation and subsequent production of inflammatory cytokines. Within phagocytic immune cells, TLR7/8 can identify single-stranded DNA (ssDNA), resulting in the secretion of IFN-α and inflammatory cytokines. The HIV virus generates intermediates that replicate within the cytoplasm, where the cytosolic HIV genomic RNA is recognized by RIG-I, prompting the expression of ISGs. Furthermore, beyond the viral genome, RIG-I in virus-infected macrophages can also detect newly synthesized viral mRNA post-viral replication, subsequently promoting ISG expression. As a retrovirus, HIV-1 produces a nucleic acid intermediate, double-stranded DNA (dsDNA), during reverse transcription. This dsDNA within the cytoplasm can be detected by cGAS, leading to the induction of IFN-β production [14].

This review summarizes the mechanisms of innate immune response and immune escape measures after the transduction of therapeutic genes by three viral vectors, Ad, AAV, and LV, which are widely used in gene therapy. Viruses inherently trigger innate immune responses; nevertheless, within this review, our concentration is solely on the innate responses provoked by genetically modified viral vectors utilized in gene therapy, along with strategies to counteract these innate immune reactions. Endeavors to circumvent immune responses in gene therapy can be classified into two primary categories: strategies aimed at shielding the vector and its transgene product from immune surveillance, and those designed to obscure the vector/transgene product from immune detection [5]. While

RNA modification or vector alteration can curtail the likelihood of an innate immune response, they cannot entirely avert its occurrence. In practical application, the emergence of an innate immune response is an unavoidable outcome.

Table 1. Currently authorized viral vector drugs.

Date	Drugs	Regulatory Approval	Application	Vector
2003	Gendicine (recombinant human p53 adenovirus) [15]	China Food and Drug Administration (CFDA)	Head and neck squamous cell carcinoma (HNSCC)	Ad-p53
2006	Oncorine [16]	Chinese SFDA	• Head and neck cancers, liver cancers, pancreatic cancers, cervical cancers, and other cancers	H101
2012	Alipogene tiparvovec (Glybera) (it was withdrawn from the market in 2017) [17]	EMA	• Lipoprotein lipase (LPL) deficiency	AAV1-LPL
2017	Luxturna [18]	FDA	Leber congenital amaurosis caused by RPE65 mutations	AAV2-RPE65
2017	Kymriah (tisagenlecleucel) [19]	FDA	• Acute lymphoblastic leukemia and diffuse large B-cell lymphoma	LV-CD19
2019	Zolgensma [20]	FDA	• Spinal muscular atrophy (SMA)	scAAV9-SMN1
2020	Libmeldy [21]	EU	Metachromatic leukodystrophy (MLD)	SIN LV vector
2021	Elivaldogene autotemcel (Skysona, eli-cel) [22]	EU	Cerebral adrenoleukodystrophy (CALD)	LV. ABCD1
2021	Breyanzi (lisocabtagene maraleucel) [23]	FDA	• Patients with relapsed or refractory large B-cell lymphomas	LV-CD19
2021	Abecma (idecabtagene vicleucel, ide-cel) [23]	FDA	Relapsed or refractory multiple myeloma (R/R MM)	LV-CD19
2022	Eladocagene exuparvovec (Upstaza) [24]	EMA	• Human aromatic L-amino acid decarboxylase (AADC) deficiency	rAAV2-hAADC
2022	Roctavian (valoctocogene roxaparvovec) [25]	EMA	• Severe hemophilia A [congenital factor VIII (FVIII) deficiency] in adults	BMN 270: AAV5-hFVIII-SQ
2022	Adstiladrin [26]	FDA	Patients with NMIBC who do not respond to BCG	rAd-IFNα/Syn3
2022	Zynteglo (betibeglogene autotemcel, beti-cel) [27]	FDA	β-thalassemia patients	BB305 LV vector

Table 2. The differences between wild-type viruses and viral vectors.

	Virus Type	Size	Genome Structure	Genome Type	Immunogenicity	Integration
Ad	Wild-type Virus	26–45 kb	ITR, φ, E1A, E1B, E2, E3, E4, L1–L5	dsDNA	High	Rarely
	Viral Vector	OAd: 3 kb; HDAd: 34 kb	ITR, φ, transgene			
AAV	Wild-type Virus	4.7 kb	ITR, Rep, Cap, AAP, MAAP	ssDNA	Low	Rarely
	Viral Vector	4.7 kb	ITR, transgene			
LV	Wild-type Virus	8–9 kb	LTR, gag, pol, env, rev, tat, vpr, vpu, vif, nef	ssRNA	Moderate	Random
	Viral Vector	<5 kb	1: gag, pol, rre, transgene; 2: rev, transgene			

2. Adenovirus Vector Therapy

2.1. Introduction to Adenovirus Vectors

2.1.1. Principle of Gene Therapy with Adenoviral Vectors

Ad is an icosahedral virus with a diameter of 70–100 mm and no envelope capsids. Its genetic material is double-stranded DNA (dsDNA), with a length of 25–46 kb [28] and a 103 bp inverted terminal repeat (ITR) at each end of the genome, crucial for viral DNA replication (Figure 1). Ad can infect both dividing and non-dividing cells, is able to replicate in the infected nucleus with high transfection efficiencies and is not integrated into the host genome [29]. Nevertheless, Ad is not without limitations. It demonstrates heightened immunogenicity, leading to notable inflammatory responses. Additionally, it exhibits suboptimal infection efficiencies in certain cancer contexts. Ad is classified into seven types, A–G. To date, a cumulative count of 111 HAdV genotypes has been documented and categorized into seven species (A–G), all of which have the potential to infect individuals across various age groups, albeit only a limited subset leading to severe infections [30]. These vectors are mainly derived from Ad serotype 2 (Ad2) and Ad serotype 5 (Ad5) of species C, with Ad5 being the most widespread [31]. Different HAdV serotypes exhibit distinct tropism, capabilities, and clinical manifestations [32]. Among them, the HAdV-5 vector is currently the most used vector for cancer treatment. During adenovirus (Ad) infection, transcription occurs in distinct regions known as early (E), intermediate (I), and late (L) regions at different stages. The E region includes E1A, E1B, E2, E3, and E4 genes, primarily contributing to Ad replication. The L region comprises genes L1–L5, which are mainly responsible for coding structural and non-structural proteins. These proteins play essential roles in capsid formation, DNA packaging, and maturation of offspring Ad [33].

Ad5 can be categorized into two types based on their replication abilities: replication-deficient viruses and replication-competent oncolytic viruses (Table 2).

Over the last three decades, remarkable progress has been achieved in the manipulation of Ad genomes. The most recent iteration of Ad vector technology is represented by the helper-dependent Ad (HDAd), wherein all viral coding sequences have been excised from the genome, retaining solely the cis-acting ITRs and packaging sequences. This configuration offers a transgene capacity of up to 34 kb [34]. The E1-deficient adenovirus, being replication-defective, serves as an ideal shuttle vector for applications in gene therapy and vaccination [1].

Adenoviruses capable of conditionally activating the E1 gene for replication within tumor cells are termed conditionally replicating adenoviruses (CRAd), which are clinically referred to as oncolytic adenoviruses (OAds). The current adenovirus vectors utilized in clinical applications encompass four strategies to attain conditional replication. One approach involves governing E1A and adenovirus replication via cancer cell-specific promoters. These exogenous promoters are generally inserted into the E1A regulatory region, exploiting the scarcity or complete absence of tumor-specific promoters. This characteristic hinges on their capacity to induce robust expression of specific genes crucial for the malignant phenotype. The insertion of exogenous promoters predominantly occurs in the E1A regulatory region without substantial deletions. The remaining three methodologies predominantly entail modifications to the transcription units E1A and E1B [35]. Currently, the new generation of OAds is created through the deletion of 24 amino acids in the CR2 domain of the E1A protein, leading to the generation of the AdΔ24 vector. The CR2 domain binds to the retinoblastoma protein (pRb) and facilitates the release of S-phase activating transcription factors (E2F) essential for viral replication in normal cells. These vectors have demonstrated a remarkable combination of high replication efficiency and selectivity across diverse tumor cells. OAds necessitate the preservation of the majority of Ad genes to ensure efficient replication and cleavage function. Consequently, their transgene capacity is confined to approximately 3 kb.

2.1.2. Application of Adenoviral Vectors in Gene Therapy

Currently, several vector gene therapy drugs for Ad have been certified (Table 1). In 2003, Gendicine was approved by the CFDA as a rAd-p53 drug for the treatment of HNSCC [15]. Most studies have combined rAd-p53 with other traditional therapies, and the results show that combination therapy is more effective than traditional treatment [36]. In 2006, Oncorine received approval from the State Food and Drug Administration for marketing [16]. It is primarily employed in the treatment of head and neck tumors, specifically nasopharyngeal carcinoma. The treatment regimen involves local injections combined with chemotherapy [37]. In 2022, Adstiladrin received FDA approval as a rAd-IFNα/Syn3 vector gene therapy for the treatment of adults with high-risk BCG non-muscle-invasive bladder cancer (NMIBC) [26]. In clinical studies, about 53% of patients achieved complete remission within 3 months of treatment [38].

For certain types of cancers, clinical trials have demonstrated that gene therapy using Ad vectors can effectively enhance therapeutic outcomes. Ad vectors have shown promise in the treatment of metastatic diseases, such as prostate cancer, where limited treatment. At present, a number of clinical trials utilizing Ad5 as a vector for localized prostate cancer treatment have exhibited favorable safety profiles and notable efficacy [37].

Tahir Muhammad et al. [39] conducted a study utilizing OAd packaged with Ad E1A/B gene-modified human mesenchymal stem cells, which could effectively suppress the growth of prostate cancer cells in mouse models. Furthermore, Tien V Nguyen et al. [40] observed minimal mouse injury upon administration of Ad657 compared to the widely used Ad5. This resulted in significant inhibition of tumor growth, prolonged survival time in mice, and notably improved therapeutic effects [40].

OAd vectors are also widely used in the treatment of liver cancers. Jian Meng et al. [41] conducted a study involving surgery combined with Oad administration in patients with small lung cancer. The gene therapy group received gene therapy prior to surgery, resulting in a notable increase in the overall survival rate of the gene therapy group and a significant reduction in postoperative recurrence probability. In vitro cell experiments have demonstrated the effectiveness of recombinant OAds, such as silica-coated OAds encoding anti-cancer gene [42], KGHV500 carrying anti-p21ras scFv [43], and chemically synthesized EpDT3-PEG-Ad5-PTEN (EPAP) [44], in inhibiting the growth of HCC cells. Ad vectors play an important role in the treatment of various diseases including ovarian cancer, glioblastoma, and uterine sarcoma lung cancer when combined with other therapeutic modalities. The use of Enadenotucirev OAd [45] or of ranergene obadenovec (VB-111) [46] combined with paclitaxel for platinum-resistant ovarian cancer treatment has shown a significant increase in median progression-free survival (PFS) and tumor immune cell infiltration. In a mouse model of lung cancer, the delivery of rAd-p53 and IL-2 using an OAd vector, along with concurrent treatment of the chemotherapeutic drug paclitaxel, resulting in significantly inhibited tumor growth [47]. Furthermore, the combination of the rAd-p53 with chemotherapy in the treatment of patients with advanced breast cancer has demonstrated favorable safety and efficacy outcomes [48].

Ad vectors have been successfully utilized in the treatment of patients with breast cancer [49]. In the context of breast cancer mouse models, the utilization of OA AdLyp.sT and mHAdLyp.sT resulted in the safe administration of Ad vector-based gene therapy.

2.2. Innate Immune Responses against Adenovirus Vectors

2.2.1. Occurrence of Innate Immune Response

Activation of innate immune responses after Ad vectors infusion has been well documented in clinical trials. For instance, in a clinical trial involving the combined utilization of Ad vectors and chemotherapy for the management of malignant pleural mesothelioma, a substantial 97.5% of patients encountered adverse reactions characterized by cytokine release and interferon syndrome [50]. In a separate clinical trial targeting colon cancer patients with Ad vectors, a surge in the multifunctional cytokine IL-6, signifying the early innate immune response to Ad vectors, was observed within 6 h post administration [51].

Within an hour following intravenous administration of the HAdV vector, the plasma concentration of the complement component C3a, which exhibits heightened efficacy in initiating innate immune signaling, reached its zenith, subsequently provoking systemic thrombocytopenia. Furthermore, upon intravenous delivery of the HAdV vector, the possibility of inducing shock emerged, encompassing symptoms like hypotension, hemoconcentration, tissue edema, and vascular congestion. In a rat model, the virus-triggered lipid mediator platelet-activating factor (PAF) became activated merely 10 min post injection of the HAdV vector, escalating by more than fivefold, eventually culminating in shock within the animals [52].

Ad infection triggers a robust innate immune response, especially during the early stage of viral entry (Figure 2). The innate responses may occur within minutes to hours, resulting in alterations in blood pressure, thrombocytopenia, inflammation, fever, and other related symptoms [3]. Ad entry represents a critical stage in initiating the innate response, as it triggers viral sensing and signaling primarily in the cells of the innate immune system such as macrophages and dendritic cells (DC). Ad vectors can activate numerous innate immune pathways. Toll-like receptors (TLRs) have been identified as key components in the innate sensing of Ad. In plasmacytoid dendritic cells (pDCs), TLR9 is involved in the process of Ad-induced production of IFN, while TLR2, TLR4, and TLR9 induce IL-2 responses when Ad infects mouse or mononuclear phagocyte in vitro [53]. Upon release, type I interferons activate a comparable set of interferon-stimulated genes by binding to IFN-α receptors. The majority of proteins encoded by these genes can regulate signaling pathways or transcription factors, thereby substantially amplifying the synthesis of IFNs, and augmenting the antiviral signal as well as the antiviral state. In essence, the innate immune response provoked by Ad vectors has the potential to diminish the efficacy of gene therapy.

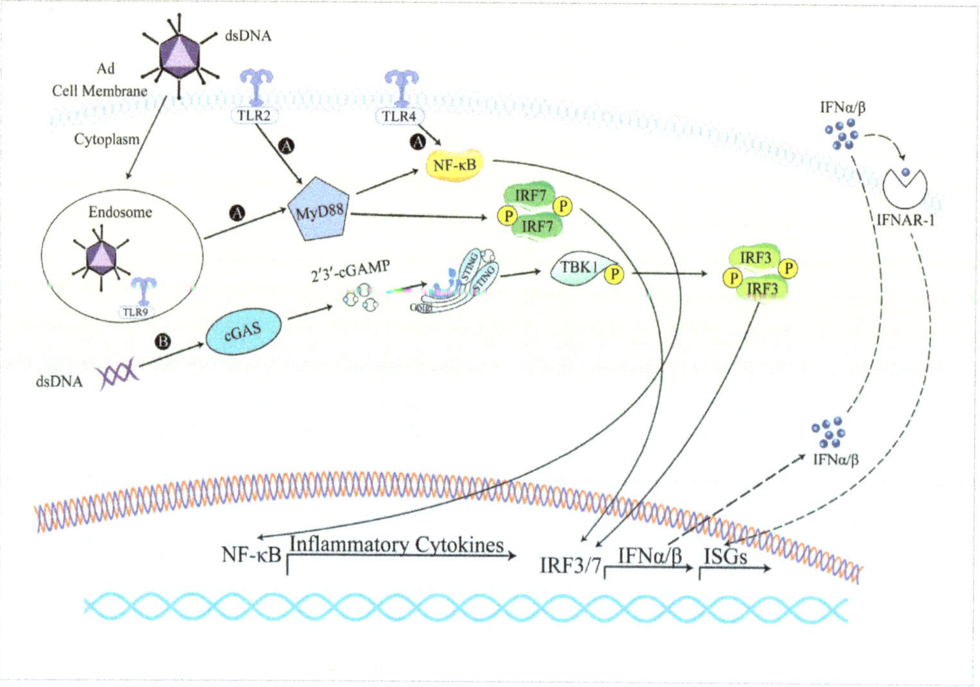

Figure 2. An overview of the innate response to Ad vectors. (A) Upon Ad infection, the genomic DNA of Ad can serve as a PAMP, leading to the activation of TLR2 and TLR9. TLRs subsequently

trigger the activation of NF-κB and IRF7. This cascade of activation culminates in the regulation of the production of IFN-I and ISGs [53]. (B) DsDNA within the Ad genome propels the engagement of the cGAS-STING signaling pathway [10].

Preexisting antibodies against Ad are prevalent within the population, and individuals who have been infected with the virus develop lifelong immunity, resulting in decreased expression of the transgene and potentially exacerbating virulence of vector transduction. In a study conducted by Maria Bottermann et al. [54], it was discovered that anti-Ad5 antibodies obstructed transgene expression by engaging intracellular Fc receptor tripartite motif-containing protein 21 (TRIM21). Furthermore, the immune response was enhanced by intravenous administration of the vector in individuals with preexisting Ad immunity, and transcriptional analysis revealed that TRIM21 specifically upregulated numerous immune genes, thereby inducing an innate response.

Activation of the innate immune response is independent of the transduction process, and Ad vectors activate the innate immune response in a manner similar to that of biologically active particles. The activation of innate immune response is associated with the capsid proteins on the surface of Ad vectors, and the activation of complement, rather than the genes expressed by the virus. Since most of the viral genome of HAdV is eliminated, transduced cells stop encoding wild-type Ad gene products, including VA-RNA. Thus, the immunotoxicity of HAdV is greatly attenuated, enabling sustained initiation of transgene expression [11,55,56].

After infection with normal non-cancer cells, the OAd exhibits an inability to replicate. However, upon infecting tumor cells, these viruses undergo successful replication, leading to amplified viral production, subsequent release into the tumor microenvironment, and consequential infection of previously uninfected tumor cells [35]. A preclinical investigation has demonstrated the efficacy of Dlta-24-RGD, a cancer-selective oncolytic adenovirus, in the dissolution of malignant gliomas. Within the context of normal immune responses, the viral infection itself and the subsequent viral lysis of cancer cells prompt the release of damage-associated molecular patterns (DAMPs). These DAMPs are identifiable by PRRs expressed on innate immune system cells, culminating in the activation of type I interferon production [57]. Consequently, within an immunocompetent murine glioma model, the oncolytic adenovirus Delta-24-RGD was employed to assess its impact on the immune microenvironment at the tumor site. Findings revealed an augmented population of NK cells at the tumor site post virus injection, indicating the induction of innate immunity due to virus infection [57]. In the treatment of ovarian cancer, oncolytic adenovirus dl922–947 has demonstrated utility. Notably, carvedilol, a drug with the potential to enhance oncolytic adenovirus activity, has been investigated. In a murine ovarian cancer model, the combination of virus treatment with carvedilol exhibited more substantial adenovirus expression compared to virus treatment alone. This combination treatment also led to enhanced infiltration of macrophages and NK cells into the tumor. Consequently, it was inferred that the augmented anticancer effect following combination therapy was attributed to the induction of innate immunity [58].

2.2.2. Evasion of Innate Immune Response

Ad vectors employ strategies to evade innate immune responses, specifically IFN responses and DNA damage responses.

Innate immune evasion strategies of Ad parental viruses include the selection of low immunogenicity serotypes [5,59] and the up-regulation of MYSM1 during viral infection, inhibiting the activation of innate immune signaling pathways [60]. Considering the double-stranded DNA nature of the Ad genome, infection by DNA virus can induce the upregulation of deubiquitinase MYSM1. MYSM1 has the ability to interact with STING and inhibit the ubiquitination of STING K63, thereby leading to the inhibition of the cGAS-STING signaling pathway [60]. Improving the efficiency of Ad delivery can enhance the therapeutic effect by producing more therapeutic proteins and reducing the amount of virus used, thereby inhibiting the activation of the innate immune response. For example, the use of blocking catheters to block hepatic blood flow before Ad vector infusion has shown the potential to enhance the efficiency of vector delivery [5].

Modification of the viral vector to evade the innate immune response is a commonly employed strategy.

Studies have revealed that genome modification can efficiently suppress the activation of an innate immune response [61]. The development of a helper-dependent Ad vector, incorporating numerous gene modifications, has demonstrated increased transgene expression levels and reduced vector toxicity. However, it may trigger an acute inflammatory reaction. To mitigate this adverse reaction, chemical modification of the viral capsid has been explored. For instance, Yasmine Gabal et al. [62] utilized polyethylene glycol (PEG) conjugation to Ad, along with cell-penetrating peptides on the cell surface, to minimize the interaction between the carrier and the host. Consequently, this strategy reduced the likelihood of recognition by immune cells, thereby attenuating the occurrence of the innate immune response. In the case of liposome-encapsulated Ad vectors, PEGylation can offer additional advantages, including the reduction of cytotoxicity, hemolytic activity, anti-vector immunity, and innate immune response [59].

Given the role of complement system activation in promoting the development of innate immunity, inhibiting the activation of the complement system could potentially suppress innate immunity. In pursuit of this objective, Christopher M Gentile et al. [63] conducted genetic modifications to the hypervariable region of the Ad capsid protein. As a result, the capsid surface of the Ad vector displayed the rH17d sequence, thereby inhibiting the classical complement pathway.

The reduction of the likelihood of innate immune response activation can also be achieved through capsid modification of the Ad vector. Since Ad entry into cells primarily hinges on its binding to the coxsackie–adenovirus receptor (CAR), the presence of CARs on various tissue surfaces is constrained. Capsid modification strategies, such as the genetic incorporation of specific motifs or the chemical conjugation of polymers linked to ligands, could potentially enhance gene transduction efficiency into tissues with lower CAR density, consequently diminishing the probability of innate immune response activation [5].

Notably, Svetlana Atasheva et al. [9] employed capsid mutations in a HAdV-C5 vector, affecting the altered HVR1 region. This modified vector displayed no binding affinity to IgM in human and mouse serum and evaded recognition by Kupffer cells following administration [9]. Another approach involved the substitution of the RGD amino acid in the adenovirus penton protein with a laminin-derived peptide incapable of interacting with macrophage β3 integrin. The resultant mutant virus exhibited attenuated cytokine activation in the spleen subsequent to intravenous administration [64]. In summary, structural modifications to the adenovirus capsid protein can effectively circumvent various stages of innate immune recognition, leading to the development of safer adenovirus vectors.

The sequence in the hypervariable region (HVR) of the HAdV5 hexahedron was reciprocally substituted with the corresponding region derived from HAdV48, yielding novel HAdV vectors. These resultant vectors demonstrated considerable immunogenicity even when confronted with elevated levels of preexisting neutralizing antibodies targeting HAdV5. A novel HAdV vector was further engineered by substituting the relevant structure of HAdV5 with the knob structure derived from HAdV3. This modification substantially enhanced the immunogenicity of both the newly created and synthesized HAdV5 vectors [59].

The deletion of E1A has demonstrated significant significance in the suppression of the immune response. Through the incorporation of the E3 region into the recombinant Ad vector, the capacity to mitigate the maximal innate immune response has been achieved, consequently enabling the sustained expression of Ad genes over an extended period [65].

In light of the presence of preexisting neutralizing antibodies, Peng Lv et al. [66] devised a strategy involving bioengineered membrane nanovesicles (BCMNs). They employed in vitro genetic membrane engineering and CRISPR engineering methods on erythrocyte membranes to maintain the infectivity and replication ability of OAd. Notably, encapsulation of OAd with BCMN resulted in a significant reduction in serum levels of IL-6 and TNF-α, indicating that encapsulation of OAd with BCMN significantly inhibited the OAd vector-induced innate immune response by protecting the OAd surface from the immune system.

3. AAV Vector Therapy

3.1. Introduction to AAV Vector

3.1.1. Principle of Action of AAV Vector

AAV is composed of a 26 nm diameter icosahedral protein capsid and a single-stranded DNA (ssDNA) genome measuring approximately 4.7 kb in length (Figure 1). At each end of the genome, there are inverted terminal repeat (ITR) sequences spanning 145 nucleotides. These ITR sequences adopt a hairpin structure, which is a cis-acting element essential for initiating DNA replication and packaging the recombinant AAV genome into infectious virions. Positioned between the ITR sequence is the viral coding region, which encompasses two open reading frames (ORFs). These ORFs give rise to the replication protein Rep, capsid protein Cap, and the packaging activation protein AAP [67]. The Rep gene encodes four non-structural proteins: Rep78, Rep68, Rep52, and Rep40, which are mainly involved in viral genome replication and integration into the host genome. The Cap gene, on the other hand, encodes three structural capsid proteins, VP1, VP2, and VP3. Assembly activation protein AAP primarily facilitates viral genome packaging and the subsequent release from host cells [68]. Among the structural capsid proteins, VP1, VP2, and VP3, there exist 60 copies in a ratio of VP1: VP2: VP3 = 1:1:10. AAV is incapable of self-replication and relies on co-infection with other helper viruses for replication, such as Ad and herpes simplex virus (HSV), for its replication process.

Based on variations of AAV capsid protein, AAV can be categorized into 13 wild types and over 100 mutants [69]. Each serotype exhibits distinct infection efficiencies across various tissues and cell types. Notably, AAV demonstrates exceptional stability, rendering it more resistant to temperature and pH fluctuations compared to other viral vectors. Furthermore, AAV exhibits lower immunogenicity than other viruses [70], and most AAV vectors do not integrate their DNA into the genome of patients. Consequently, the risk of insertion mutations associated with AAV vectors is minimal. These inherent properties render AAV highly suitable for specific gene therapy applications.

The primary AAV utilized in gene therapy is recombinant AAV (rAAV), which has the same capsid sequence and structural organization as wild-type (WT) AAV. However, the genome packaged within rAAV has undergone modifications, wherein the AAV protein coding sequence is deleted and a therapeutic gene expression cassette is inserted. Notably, ITR represents the only viral-derived sequence present in rAAV. Furthermore, rAAV exhibits optimal genome loading capacity, accommodating sequences of up to 5.0 kb or less [2].

Although AAV is less immunogenic than Ad, the capsid proteins, as well as the delivered nucleic acid sequences, can trigger various components of our immune systems. Most people have already been exposed to AAV and have already developed an immune response to the specific variant to which they were previously exposed, resulting in a pre-existing immune response in the sense that neutralizing antibodies (NAbs) against the AAV are already present. NAbs against AAV2 are the most prevalent in the human body, and the prevalence of NAbs against AAV8 and AAV5 is the lowest, so these serotypes are more conducive to use as gene therapy vectors [71].

3.1.2. Application of AAV Vectors in Gene Therapy

A range of approved AAV vector-based drugs is presently available (Table 1). In 2012, Glybera marked the first recommendation of an AAV vector gene therapy for treating lipoprotein lipase deficiency [17]. Nonetheless, due to economic considerations, Glybera was withdrawn from the market in 2017 [72]. In 2017, Luxturna, an AAV2 vector gene therapy, received FDA approval for addressing Leber congenital amaurosis attributed to

RPE65 mutations [18,73]. In 2019, Zolgensma, a scAAV-SMN1 gene therapy, gained FDA approval to treat spinal muscular atrophy (SMA) [20]. Eladocagene exuparvovec (Upstaza), a gene therapy drug vector founded on rAAV2-hAADC, obtained EMA approval in 2022 to address severe ADCC deficiency in patients [24]. Likewise, in 2022, valoctocogene roxaparvovec (Roctavian), a gene therapy agent vector-built upon AAV5-hFVIII-SQ, secured EMA approval for treating severe hemophilia A (congenital factor VIII deficiency) in adults without a history of FVIII inhibitors and without detectable antibodies to AAV5 [25].

Most of the current studies involve the direct administration of AAV vectors to patients [74]. Gene therapy has emerged as a promising therapeutic approach for hemophilia, and AAV2 stands as the pioneering vector employed in its treatment [75]. Furthermore, AAV3 and AAV5 exhibit potential for hemophilia therapy, demonstrating improvements in disease severity [76–78]. AAV8 is considered the optimal vector for the treatment of hemophilia [79]. Compared with AAV2 and AAV5, AAV8 boasts a lower prevalence in human seroprevalence, elicits a reduced immune response towards the vector capsid, and exhibits a pronounced affinity for the liver. Consequently, AAV8 facilitates efficient transduction of hepatocytes when administered to animal models through the peripheral circulation [80,81].

AAV holds promise in the treatment of ocular disorders. Given the highly compartmentalized nature of the eye and the absence of lymphatic vessels, it is generally believed that the risk of activating the innate immune response is low when AAV vectors are administered into the eyes [82]. It has been reported that subretinal vector delivery is associated with lower immunogenicity compared to intravitreal injection [83]. Studies employing subretinal injection of AAV2 and AAV8 in patients with Leber's congenital amaurosis, retinal degeneration, and CNGA3-linked achromatopsia have demonstrated improvements in visual acuity and favorable safety profiles [84–86].

AAV vectors have emerged as a potential therapeutic option for neurological disorders. Commonly utilized vectors include AAV1, AAV2, AAV5, and AAV8. AAV2, as a first-generation vector, has demonstrated its safety and efficacy in the treatment of various neurological conditions, including Parkinson's disease [87], Alzheimer's disease [88,89], AADC deficiency [90], and several genetic diseases. As the second-generation vector, AAV5 has shown promise in ameliorating Huntington's disease (HD) [91] and spinocerebellar ataxia type 3 (SCA3), respectively [92]. AAV9, belonging to the third generation of vectors, exhibits the capability to efficiently target the brain and spinal cord. Compared with other serotypes, AAV9 can demonstrate enhanced blood–brain barrier penetration following intravenous administration, offering the potential for minimally invasive therapeutic interventions [93].

AAV vectors have emerged as a promising approach for the treatment of muscular diseases. For instance, AAV9 has shown efficacy in reducing the severity of spinal muscular atrophy (SMA) in patients [94]. In the case of Duchenne muscular dystrophy (DMD), patients treated with rAAVrh74 experienced significant improvements in their motor abilities after one year of treatment [95]. Furthermore, the administration of AAV8 vectors to patients with X-linked myotubular myopathy (XLMTM) resulted in the recovery of exercise capacity [96].

AAV vectors have also shown potential in the treatment of hearing impairment. Studies utilizing AAV1-VEGFA165, exo-AAV1-GFP, AAV9-PHP.B, and other vectors have demonstrated improved blood supply and alleviation of hearing loss in animal models, highlighting their safety and therapeutic potential in this field [97–99].

3.2. Innate Immune Responses against AAV Vectors

3.2.1. Occurrence of Innate Immune Response

Despite the diminished innate immune response observed in AAV compared to Ad vectors, the host immune response remains a significant hurdle in achieving sustained and effective therapeutic gene expression [100]. These immune responses encompass cytotoxic T lymphocyte (CTL) responses against AAV capsids and therapeutic proteins, the presence of NAbs against AAV viral particles, the production of antibodies against therapeutic proteins, and innate immune responses triggered by AAV transduction [101].

Observation of activated innate immune responses subsequent to AAV vector infusion has also been documented in clinical trials. In a clinical trial involving AAV5-hFIX gene therapy among adults with hemophilia B, the onset of fever and elevated ALT levels within 24 h post-vector delivery indicated the activation of the innate immune response [77].

Previous investigations demonstrated that administration of a high dosage of AAV vector intravenously to mice for 1 h elicited escalated transcription of genes associated with inflammatory cytokines and chemokines, including TNF-α, RANTES, MIP-1β, MIP-2, MCP-1, and IP-10 [102].

TLR9-mediated sensing of CpG motifs in the AAV genome is likely to occur during endosomal trafficking, where partially exposed virions can be recognized (Figure 3). Alternatively, viral capsid degradation within lysosomes can expose the genome to TLR9, leading to the activation of nuclear factor kappa-B (NF-κB) and interferon regulatory factor 7 (IRF7) via myeloid differentiation primary response protein88 (MyD88) signaling pathway. This activation ultimately regulates the production of IFN-I and ISGs [103]. In addition to TLR9 activation, an upregulation in the transcription of TLR2, which recognizes microbial proteins and glycolipid structures, has also been observed following rAAV infection, suggesting the potential involvement of TLR2 in innate immunity against rAAV vectors [13,104].

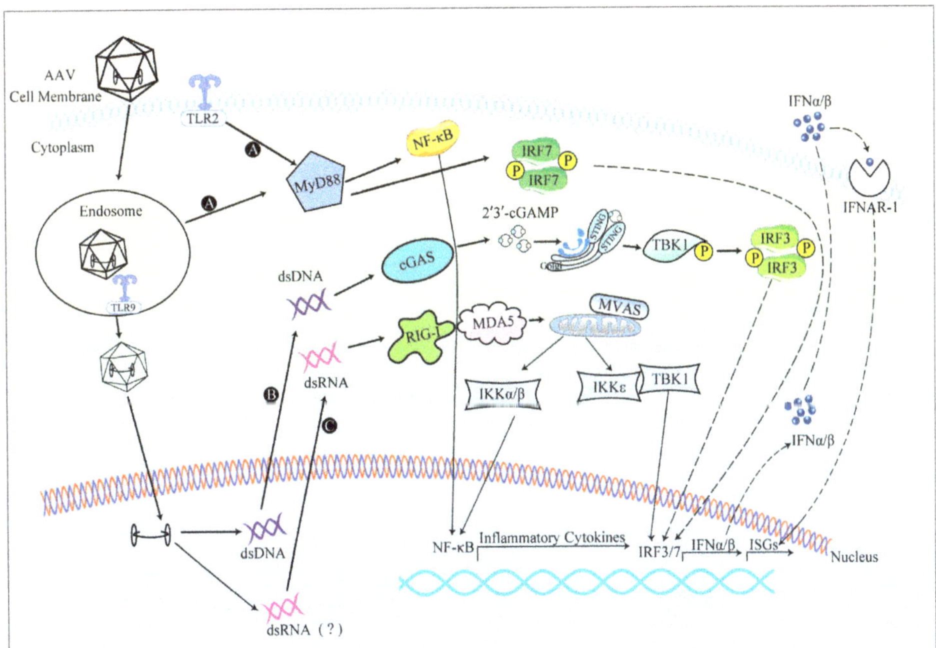

Figure 3. An overview of the innate response to AAV vectors. (A) Upon AAV infection, AAV's genome DNA can act as PAMP, leading to the activation of TLR2 and TLR9. Then, TLRs activate NF-κB and IRF7 via MyD88 signaling pathway, this activation ultimately regulates the production of IFN-I and ISGs [13,103,104]. (B) The ITR structure of the AAV genome may activate cytoplasmic DNA to induce cGAS and antiviral genes and activate the cGAS-STING signaling pathway [105]. (C) Due to the promoter activity of the ITR of AAV, AAV may form dsRNA in target cells and trigger the RIG-I/MDA5-mediated RLR-MAVS innate immune signaling pathway [83,106]. The "(?)" in the figure implies that dsRNA could be theoretically formed during this process, but has not been demonstrated so far.

The presence of NAbs not only reduces the efficiency of vector transduction and compromises the therapeutic effect but also triggers complement activation in the presence of NAbs against the capsid. The complement system, consisting of a variety of proteins, constitutes a significant component of the host's innate immune system, encompassing the classical pathway and the lectin pathway. While it is commonly believed that the complement system plays a limited role in the innate immune response against AAV, Manish Muhuri [13] argues that the evidence from pathological observations suggests the activation of complement in the human body in response to AAV.

The utilization of self-complementary AAV (scAAV) vectors enhances the efficacy of the vector by facilitating faster and more robust transgene expression. Thus, in turn, allows for the administration of lower and safer vector doses. However, when compared to the single-stranded AAV (ssAAV) vector, the use of the scAAV genome leads to an augmented innate immune response towards the transgene, primarily due to activation of the TLR9/MyD88 pathway. The DNA component of the AAV vector, as well as the potential production of dsRNA [106], resulting from the ITR promoter activity, can function as an adjuvant, activating innate immunity alongside other host-specific factors. The buildup of dsRNA would subsequently trigger the MDA5 sensor within human hepatocytes transduced with AAV, resulting in the induction of type I interferons (IFNs) expression [83,107]. Following their release, type I IFNs activate a similar array of ISGs through their interaction with IFN-α receptors. A significant portion of the proteins encoded by these genes function in the regulation of signaling pathways or transcription factors, thereby substantially amplifying IFN synthesis and augmenting both the antiviral signal and antiviral state. In summary, the innate immune response elicited by AAV vectors may lead to decreased efficacy of gene therapy.

The ITR structure of the AAV genome may activate cytoplasmic DNA receptors. Studies have demonstrated that rAAV can induce cGAS and antiviral genes such as TNF-α and IFN-γ [105]. The ITR, located at either end of the AAV or within the bidirectional promoter, can generate reverse complementary positive-strand RNA and negative-strand RNA. These RNA molecules form dsRNA in target cells, subsequently triggering the innate immune response mediated by the dsRNA recognition receptor. As cytoplasmic RNA sensors, RLRs can activate downstream innate immune signaling pathways through adaptor protein MAVS. RLRs include three RNA-binding proteins: MDA5, RIG-I, and LGP2. MDA5 and RIG-I can recognize the dsRNA formed by AAV, thereby activating the RLR-MAVS signaling pathway to induce the production of IFN-β [13].

3.2.2. Evasion of Innate Immune Response

Various strategies exist that do not entail genetic alteration of AAV vectors but aid in circumventing the innate immune response. For instance, within AAV gene therapy, distinct serotypes exhibit varying affinities for different tissues, thereby necessitating the choice of serotypes compatible with specific disease contexts [5]. Additionally, the adoption of less immunogenic vector delivery approaches can be considered [83].

Given that AAV serotypes exhibit differential tissue tropism, selecting an appropriate serotype in line with the target tissue becomes imperative in gene therapy. For instance, employing AAV8 vectors for treating liver disorders can enhance the efficacy of gene therapy [5]. The immunogenicity of AAV vectors is partly dependent on the dosage administered. Low doses of vectors are more likely to induce mild inflammation, which can be controlled and dose not lead to complete loss of transgene expression [83]. The transgene immune response, which plays a crucial role in the production of anti-capsid antibodies, is influenced by various factors, including the AAV capsid, the method of vector delivery, and the tissue specificity of the promoter driving gene expression. Notably, systemic, and intramuscular vector delivery methods have shown greater immunogenicity compared to systemic delivery with gene transfer to immune organs or the use of liver-specific promoters [83].

Utilizing immunosuppressive agents in gene therapy can impede DNA synthesis and cellular signaling essential for lymphocyte activation and proliferation. For instance, employing glucocorticoid dexamethasone can mitigate the innate immune response, thereby curbing cytokine storms [108].

Furthermore, altering the viral structure to evade the innate immune response stands as a commonly employed approach.

In clinical trials involving AAV vectors, the used of corticosteroids has been a common approach to suppress the immune response [109]. Alternatively, more targeted approaches have been explored, such as modifying the capsids for immune evasion through techniques like site-directed mutagenesis, directed evolution, and utilizing exosomes. To reduce the presentation of AAV2 capsid-derived peptide epitopes on MHC class I, the proteasome inhibitor bortezomib has been employed in hepatocytes [109].

Inhibiting TLR9 activation can diminish innate immune activation probability. In a study, a short single-stranded DNA oligonucleotide (TLR9i) antagonizing TLR9 activation was employed. TLR9i sequence was inserted into scAAV vector plasmid encoding human Factor IX (FIX) to form scAAV8.FIX.io1. In mice, equal amounts of scAAV8.FIX and scAAV8.FIX.io1 showed minimal innate immune response activation in the liver. Compared to the PBS injection group, scAAV8.FIX.io1 injection did not elevate IFN gene expression after 4 h. The TLR9i-modified single-stranded AAV2 vector (AAV2.GFP.wpre.io2) was used with primary human peripheral blood mononuclear cells from 13 healthy individuals. AAV2.GFP.wpre induced a stronger cytokine response than TLR9i modification [100]. Recent accounts indicate that the integration of TLR9 inhibitory sequences has led to reduced immune reactions related to rAAV in mice and pigs [13].

Narendra Maheshri et al. conducted AAV capsid engineering to generate AAV2 variants with enhanced attributes. Their findings highlighted that these AAV mutants exhibited altered affinities for heparin and demonstrated heightened efficacy in evading neutralizing antibodies to AAV. This manipulation also bolstered gene delivery, even in the presence of preexisting neutralizing antibodies to AAV within the population [110].

Research has evidenced that restraining the recognition signaling of TLR9 through ODN antagonists and inhibiting the NF-κB pathway can decrease innate immune responses, although full elimination remains elusive. Predominantly, NF-κB inhibitors impede the initial TLR9 activation while not fundamentally curtailing IFN production. In efforts to mitigate TLR9-mediated innate immune responses, AAV transgene cassettes are tailored to diminish TLR9 recognition of CpG sites. This strategic approach has been applied in clinical trials involving patients with hemophilia [111].

4. Lentiviral Vector Therapy

4.1. Introduction to Lentiviral Vectors

4.1.1. Principle of Action of Lentiviral Vector

LV vectors are spherical enveloped viruses, with a diameter ranging from 80 to 120 nm. They carry a single-stranded RNA genome with two copies. LV belongs to the retrovirus genus, capable of infecting both dividing and non-dividing cells, and exhibits a relatively prolonged incubation period and low pathogenicity [112]. Reverse transcriptase encoded by the LV genome transcribes viral RNA into dsDNA, which is subsequently integrated into the host cell genome by viral integrase [112]. The HIV genome consists of two positive-sense RNA strands, approximately 9 kb in length, encoding nine regulatory and auxiliary genes gag, pol, env, rev, tat, vpr, vpu, vif, and nef. Within the HIV genome, trans elements encode functional, structural, and accessory proteins, while non-coding cis elements such as the long terminal repeat (LTR) are also present.

LV vectors can be categorized into primate LV vectors and non-primate LV vectors based on the source of the virus. Primate LV vectors consist of human immunodeficiency virus type 1 (HIV-1), type 2 (HIV-2), and simian immunodeficiency virus (SIV). Among these, HIV-1 is one of the most widely studied viruses. LV vectors share common characteristics in terms of viral structure, phagocytosis, and the elicited immune response [113].

In the current third generation of LV vectors, a conditional packaging platform is created by removing the tat and a portion of the 3′-LTR, while providing the rev gene required for replication in trans. Furthermore, the U3 region in the 5′-LTR of the transfer plasmid is replaced with a constitutive promoter, enabling transcription in the absence of tat, thereby improving safety and reducing immunogenicity [114–116].

LV generally primarily targets immune cells, particularly macrophages and T lymphocytes [113]. HIV-1 mainly infects and replicates within $CD4^+$ T cells, with a smaller fraction infecting $CD4^+$ DCs and macrophages.

4.1.2. Application of Lentiviral Vectors in Gene Therapy

A range of approved lentiviral (LV) vector drugs currently exist. In 2017, the FDA granted approval to Kymriah (tisagenlecleucel), a gene therapy drug rooted in LV-CD19, intended for managing acute lymphoblastic leukemia and diffuse large B-cell lymphoma [19]. The year 2020 saw the EU's endorsement of Libmeldy, a gene therapy drug grounded in SIN LV, for addressing metachromatic leukodystrophy (MLD) [21]. Similarly, in 2021, the European Union approved Elivaldogene autotemcel (Skysona, eli-cel), a gene therapy drug built on LV. ABCD1 is designed for managing cerebral adrenoleukodystrophy (CALD) [22]. Breyanzi (Lisocabtagene Maraleucel), a gene therapy drug founded on LV-CD19, was also sanctioned by the FDA in 2021, targeting patients with relapsed or refractory large B-cell lymphomas [23]. Additionally, in 2021, the FDA endorsed Abecma (Idecabtagene Vicleucel, ide-cel), a gene therapy drug employing LV-CD19 as a vector, for treating relapsed or refractory multiple myeloma (R/R MM) [23]. Finally, in 2022, Zynteglo (Betibeglogene Autotemcel, beti-cel), a gene therapy drug hinging on the BB305 LV vector, received FDA approval for managing β-Thalassemia patients [27].

LV vectors have been utilized as gene therapy vectors in the treatment of genetic disorders. It has been applied to the treatment of chronic granulomatous disease (CGD). In a study by Donald B Kohn et al. [117], LV vectors were first utilized to treat X-linked CGD (X-CGD), resulting in the majority of patients being able to discontinue antibiotic prophylaxis without associated infections. For patients with transfusion-dependent β-thalassemia (TDT), LV vector gene therapy has proven to be a safe and effective treatment. Autologous $CD34^+$ cells transduced with LV vector BB305 have been administrated to patients with TDT [118,119]. The findings demonstrate that a substantial number of patients achieve remission of clinical response, experienced a reduction in transfusion frequency, or even discontinued transfusions altogether. The treatment of infants with server combined immune-deficiency (SCID) through LV vector transduction has demonstrated remarkable clinical success. The administration of LV-transduced $CD34^+$ cells to infants with SCID, followed by targeted low-exposure busulfan infusion, effectively restored immune function and facilitated normal growth with minimal toxicity and high safety level [120,121]. Mutations in ATP binding cassette subfamily D member 1 (ABCD1) lead to the loss of adrenoleukodystrophy (ALD) protein function, resulting in X-linked adrenoleukodystrophy, Florian Eichler et al. [122] transduced autologous $CD34^+$ cells from patients with LV vectors and observed the expression of ALD protein in all patients, with majority patients surviving and experiencing few adverse reactions. LV vector gene therapy may serve as a viable alternative to allogeneic hematopoietic stem cell transplantation for this patient population.

Allogeneic hematopoietic stem cell transplantation is also considered the standard treatment for Hurler syndrome, known as mucopolysaccharidosis type I Hurler variant (MPSIH). However, this treatment only provides a partial cure and is associated with a higher incidence of adverse reactions. In a study conducted by Bernhard Gentner et al. [123], autologous hematopoietic stem and progenitor cells were transduced with LV vectors encoding α-L-iduronidase (IDUA) in eight children with MPSIH. Interim findings indicated that the safety profile of this approach was comparable to that of allogeneic hematopoietic stem cell transplantation. Moreover, gene therapy resulted in previously undetectable levels of IDUA activity and demonstrated improvements in the patients' motor abilities and cognitive performance. In another application case, the use of LV-transduced bone marrow-derived CD34$^+$ hematopoietic stem cells (HSCs) in the treatment of children with leukodystrophy showed promising results. This therapeutic approach effectively decelerated disease progression, ameliorated neurological abnormalities, and proved to be safe, delivering substantial treatment benefits [124].

LVs have been investigated in clinical trials as vectors for cancer immunotherapy. In these trials, LVs are used to modify patients' immune cells, such as T cells, to express chimeric antigen receptors (CARs) or T-cell receptors (TCRs) targeting specific tumor antigens. Acute lymphoblastic leukemia (ALL) carries a bleak prognosis upon relapse following allogeneic transplantation. In an effort to address this, LV vectors were employed to transduce a CAR encoding CD3ζ and 4–1BB. Subsequently, gene therapy using LV vectors yielded a marked enhancement in the long-term survival of patients [125]. LV vector-transduced fully human belantamab mafodotin (BCMA)-specific CART cells were utilized to treat multiple myeloma (MM) clinically. The results demonstrated that all patients successfully expanded CART-BCMA cells [126].

Preclinical investigations have explored the application of LV vector gene therapy for several diseases prior to clinical implementation. Promising outcomes have been observed in murine models of breast cancer [127], liver cancer [128], bladder cancer [129], hemophilia [130–133], and AD [134–136], following treatment with LV vectors.

4.2. Innate Immune Responses against Lentiviral Vectors

4.2.1. Occurrence of Innate Immune Response

LV vectors are mainly derived from HIV-1, necessitating a comprehensive examination of the innate immune response following HIV-1 infection (Figure 4).

The initiation of innate immune responses following the administration of lentiviral (LV) vectors has been documented in clinical trials. For instance, in a clinical trial involving a patient with relapsed and refractory acute myeloid leukemia who underwent LV vector treatment, pronounced fever episodes manifested within 0.5 to 1 h following daily vector infusion, indicating the activation of innate immune responses triggered by LV vector gene therapy [137]. In a preclinical experiment involving mice, intravenous injection of LV vector induced a swift and transient interferon (IFN) response, marked by the production of IFN-α and IFN-β [138]. Within four hours of intravenous administration of LV vector to mice, a rapid and transient upsurge of type I interferon was detected in their serum [102].

PRRs in host cells play a crucial role in recognizing PAMP to drive IFN-mediated immune responses against the virus. Multiple PAMPs associated with HIV-1 have been identified, including capsids, genomic RNA, reverse transcription products such as ssDNA, dsRNA, and RNA-DNA hybrids [139]. When the LV vectors reach the target nucleus and integrate into the host genome, innate sensors recognize LV nucleic acids and proteins. Notably, the transduction of LV vectors into immune cells and stem cells activates innate immune responses through TLR recognition [140]. Certain sensors such as TLR7, TLR8 and unidentified DNA sensors generally recognize nucleic acids associated with retroviral infection.

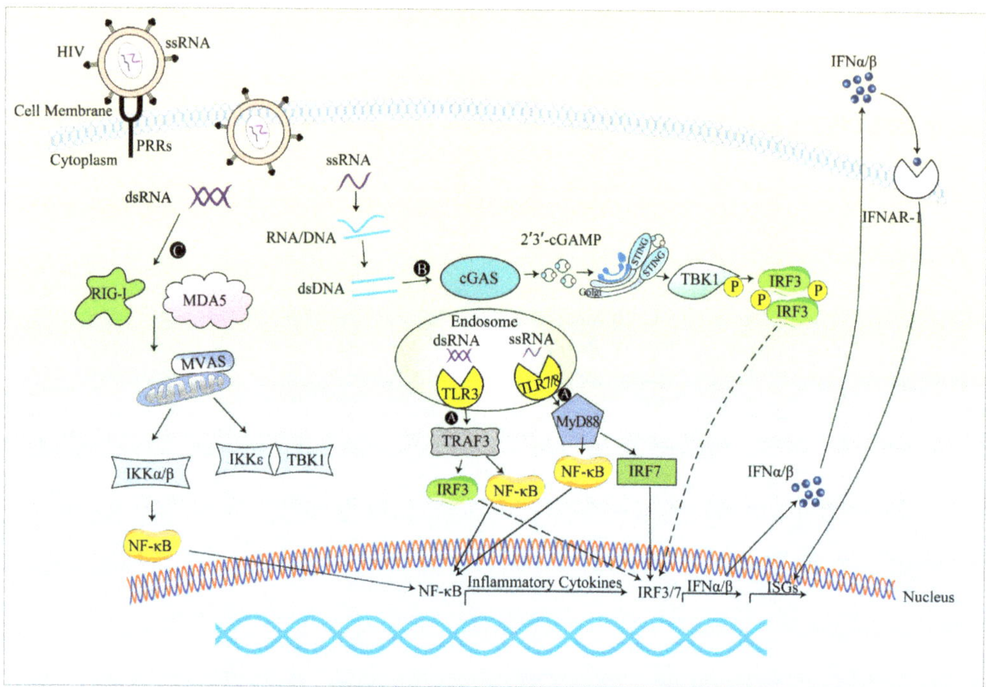

Figure 4. An overview of the innate response to LV vectors. (A) LV RNA genome can activate TLR3, TLR7 and TLR8, resulting in production of IFN-I and activating innate immunity [140]. (B) In certain cell types, reverse transcription gives rise to cytoplasmic HIV DNA, thereby leading to the activation of the cGAS-STING innate immune signaling pathway [139]. (C) The secondary structure of HIV-1 genome RNA can elicit innate immune responses similar to those triggered by full-length genome RNA, activating the RNA sensor RIG-I in primary human peripheral blood mononuclear cells and macrophages. The secondary structure of HIV-1 genome RNA can activate the RNA sensor RIG-I in primary human peripheral blood mononuclear cells and macrophages. Thus, RIG-I-MAVS innate immune signaling pathway is activated [139,141].

In humans, preexisting immunity to lentiviral (LV) vectors is generally limited. However, upon the transduction of LV vectors, the nucleic acids and proteins within the vector can be identified by restriction factors (RFs) which serve as innate immune sensors [142].

HIV-1 RNA, encompassing both genome RNA and newly synthesized RNA, has the capacity to engage RNA sensors. The secondary structure of HIV-1 genome RNA can elicit innate immune responses similar to those triggered by full-length genome RNA, activating the RNA sensor RIG-I in primary human peripheral blood mononuclear cells and macrophages. Thus, the RIG-I-MAVS innate immune signaling pathway is activated [139,141]. Notably, Andrea Annoni et al. [143] demonstrated that the LV RNA genome can activate TLR3 and TLR7, resulting in the production of IFN-I and activating innate immunity.

In certain cell types, reverse transcription gives rise to cytoplasmic HIV DNA. Acting as a PRR, cGAS selectively binds to the stem-loop structure of HIV-1 ssDNA in a sequence-specific manner and becomes activated by the reverse transcription product, thereby leading to the activation of the cGAS-STING innate immune signaling pathway [139].

LATS1/2, the core kinase of the Hippo pathway, has been implicated in the regulation of anti-tumor immunity. Tiansheng He et al. [144] discovered that LATS2 has the ability to interact with PQBP1, a cofactor of cGAS. This interaction leads to the augmentation of the antiviral response mediated by the cGAS-STING signaling pathway, subsequently resulting in enhanced IFN-I production. Consequently, LATS2 plays a pivotal role in promoting the development of innate immune responses.

The magnitude of the innate immune response can be further augmented by additional factors. These factors encompass an elevated vector dose exceeding the initial dosage administered during HIV-1 infection, loss of viral accessory proteins that aid in immune evasion during HIV-1 infection, and the stimulation of the cGAS pathway through the presence of DNA within the vector [140].

4.2.2. Evasion of the Innate Immune Response

The evasion of innate immunity in parental viruses is exemplified by the m6A modification of the HIV virus, which inhibits the activation of IRF3 and IRF7 [145]. N6-methyladenosine (m6A) is a type of RNA modification that plays a significant role in various biological processes. HIV-1 RNA, which contains m6A modifications, has been observed to regulate the viral infection of CD4$^+$ T cells. A study conducted by Shuliang Chen et al. [145] investigated the impact of m6A modification of HIV-1 RNA. It was found that these modifications suppressed the activation levels of the upstream transcription factors IRF3 and IRF7, which are involved in the production of IFN-I and are crucial for evading RIG-I-mediated RNA sensing. Consequently, this evasion strategy facilitated the escape from innate immune responses.

Similarly, akin to AAV and Ad, the evasion of innate immune responses can be attained through the modification of LV vector or drug regulation.

LVs produced by MHC-free 293T cells or CD47hi LVs were administered intravenously to rhesus monkeys. The results revealed that CD47hi LV-treated non-human primates exhibited reduced elevation of macrophage-associated cytokines, indicating a safer and well-tolerated response. This result suggests that CD47hi LV has the potential to mitigate the activation of the innate immune system [146].

Furthermore, studies have demonstrated that LV vectors modified by the CD47ED fusion gene, incorporating extracellular domain-core of CD47 and streptavidin, can trigger phagocytosis of LV by antiphagocytic cells [147]. Notably, when BX795, an inhibitor of RIG-I, MDA5, and TBK1/IKKε complex downstream of TLR3, was employed to inhibit innate immune signals in cell lines, it resulted in significantly improved LV gene modification efficiency and enhanced LV transduction efficiency. This practical and safe approach shows promise for future applications [148].

At present, the most commonly used method for transduction of human hematopoietic stem and progenitor cells (HSPCs) involves the use of LVs. However, it has been observed that the expression of LVs in stem cells can trigger the innate immune response. Carolina Petrillo et al. [149] found that cyclosporine H (CSH), a new drug, could enhance the efficiency of LV transduction and gene editing in HSPCs. Notably, CSH effectively blocked the IFN response induced by LV vectors, demonstrating a favorable anti-innate immune response effect.

In a preclinical investigation, Judith Agudo et al. found that mice receiving treatment with GC dexamethasone (Dex) concurrent with LV vector administration exhibited significant upregulation of IFN-induced genes in LV-treated mice. Notably, in the liver of LV + Dex-treated mice, the expression of these genes was normalized. Furthermore, the total number of neutrophils and macrophages in the liver was decreased, indicating that the administration of GC could effectively impede the innate immune response triggered by the LV vector [150].

In order to reduce the innate immune response caused by LV vectors, specific antibodies can be used to block type I interferon receptor, IL-6, IL-6 receptor, and other specific antibodies, and applied to LV vectors gene therapy can reduce the occurrence probability of innate immune response to LV vectors [143].

5. Conclusions and Prospects

After decades of research and development, the application of viral vectors, including Ad, AAV, and LV vectors, in gene therapy has made significant strides and is now extensively employed in clinical settings. However, the innate immune response to those vectors and transgenes poses challenges to improving therapeutic efficacies. Investigating the underlying mechanisms of these innate immune responses can pave the way for vector evasion, thereby expanding the scope of gene therapy to encompass a wider range of diseases. At present, ongoing research has demonstrated that the utilization of inhibitors or modification of viral vectors can facilitate vector evasion of the innate immune response, ultimately enhancing the effectiveness of gene therapy. Nonetheless, several factors hinder the optimal efficacy of gene therapy. Notably, preexisting vector immunity and the propensity for augmented innate immunity at elevated vector dosages emerge as substantial limitations within clinical applications, exerting potential impact on the gene therapy vectors' effectiveness. In the realm of gene therapy, the comprehension of host- and vector-related determinants that intricately influence the genesis of cytotoxic reactions, thereby contributing to the instability of transgene expression duration, remains limited [107]. Endeavors aimed at refining viral vector transduction strategies, particularly towards the utilization of diminished dosages, persist as requisite measures to enhance therapeutic efficacy, a notion especially relevant in clinical settings [142]. Hence, further investigations are imperative, encompassing not solely the scrutiny of the innate immune response, but also a comprehensive exploration of the adaptive immune response. Such in-depth inquiry holds the potential to ameliorate the therapeutic impact of gene therapy.

Author Contributions: Conceptualization, W.S.; writing, Y.W.; review editing, W.S. All authors have read and agreed to the published version of the manuscript.

Funding: This research was funded by the Ministry of Science and Technology of the People's Republic of China, National Key Research and Development Program of China (Grant numbers: 2021YFF1200800, 2022YFF1202903) and the National Natural Science Foundation of China, the Youth Fund of the National Natural Science Foundation of China (Grant number: 82101853).

Institutional Review Board Statement: Not applicable.

Informed Consent Statement: Not applicable.

Data Availability Statement: Not applicable.

References

1. Sato-Dahlman, M.; LaRocca, C.J.; Yanagiba, C.; Yamamoto, M. Adenovirus and Immunotherapy: Advancing Cancer Treatment by Combination. *Cancers* **2020**, *12*, 1295. [CrossRef] [PubMed]
2. Wang, D.; Tai, P.W.L.; Gao, G.P. Adeno-associated virus vector as a platform for gene therapy delivery. *Nat. Rev. Drug Discov.* **2019**, *18*, 358–378. [CrossRef] [PubMed]
3. Shirley, J.L.; de Jong, Y.P.; Terhorst, C.; Herzog, R.W. Immune Responses to Viral Gene Therapy Vectors. *Mol. Ther.* **2020**, *28*, 709–722. [CrossRef] [PubMed]
4. Hensley, S.E.; Cun, A.S.; Giles-Davis, W.; Li, Y.; Xiang, Z.Q.; Lasaro, M.O.; Williams, B.R.G.; Silverman, R.H.; Ertl, H.C.J. Type I interferon inhibits antibody responses induced by a chimpanzee adenovirus vector. *Mol. Ther.* **2007**, *15*, 393–403. [CrossRef] [PubMed]
5. Sack, B.K.; Herzog, R.W. Evading the immune response upon in vivo gene therapy with viral vectors. *Curr. Opin. Mol. Ther.* **2009**, *11*, 493–503. [PubMed]
6. Rogers, G.L.; Martino, A.T.; Aslanidi, G.V.; Jayandharan, G.R.; Srivastava, A.; Herzog, R.W. Innate Immune Responses to AAV Vectors. *Front. Microbiol.* **2011**, *2*, 194. [CrossRef] [PubMed]
7. Faure, M.; Rabourdin-Combe, C. Innate immunity modulation in virus entry. *Curr. Opin. Virol.* **2011**, *1*, 6–12. [CrossRef]
8. Decout, A.; Katz, J.D.; Venkatraman, S.; Ablasser, A. The cGAS-STING pathway as a therapeutic target in inflammatory diseases. *Nat. Rev. Immunol.* **2021**, *21*, 548–569. [CrossRef]
9. Atasheva, S.; Shayakhmetov, D.M. Cytokine Responses to Adenovirus and Adenovirus Vectors. *Viruses* **2022**, *14*, 888. [CrossRef]
10. Sohn, S.Y.; Hearing, P. Adenoviral strategies to overcome innate cellular responses to infection. *FEBS Lett.* **2019**, *593*, 3484–3495. [CrossRef]

11. Vachon, V.K.; Conn, G.L. Adenovirus VA RNA: An essential pro-viral non-coding RNA. *Virus Res.* **2016**, *212*, 39–52. [CrossRef] [PubMed]

12. Rabinowitz, J.; Chan, Y.K.; Samulski, R.J. Adeno-associated Virus (AAV) versus Immune Response. *Viruses* **2019**, *11*, 102. [CrossRef] [PubMed]

13. Muhuri, M.; Maeda, Y.; Ma, H.; Ram, S.; Fitzgerald, K.A.; Tai, P.W.; Gao, G. Overcoming innate immune barriers that impede AAV gene therapy vectors. *J. Clin. Investig.* **2021**, *131*, e143780. [CrossRef] [PubMed]

14. Bergantz, L.; Subra, F.; Deprez, E.; Delelis, O.; Richetta, C. Interplay between Intrinsic and Innate Immunity during HIV Infection. *Cells* **2019**, *8*, 922. [CrossRef] [PubMed]

15. Wilson, J.M. Gendicine: The first commercial gene therapy product. *Hum. Gene Ther.* **2005**, *16*, 1014–1015. [CrossRef] [PubMed]

16. Liang, M. Oncorine, the World First Oncolytic Virus Medicine and its Update in China. *Curr. Cancer Drug Targets* **2018**, *18*, 171–176. [CrossRef] [PubMed]

17. Ylä-Herttuala, S. Endgame: Glybera finally recommended for approval as the first gene therapy drug in the European union. *Mol. Ther.* **2012**, *20*, 1831–1832. [CrossRef] [PubMed]

18. Rodrigues, G.A.; Shalaev, E.; Karami, T.K.; Cunningham, J.; Slater, N.K.H.; Rivers, H.M. Pharmaceutical Development of AAV-Based Gene Therapy Products for the Eye. *Pharm. Res.* **2018**, *36*, 29. [CrossRef]

19. Ali, S.; Kjeken, R.; Niederlaender, C.; Markey, G.; Saunders, T.S.; Opsata, M.; Moltu, K.; Bremnes, B.; Grønevik, E.; Muusse, M.; et al. The European Medicines Agency Review of Kymriah (Tisagenlecleucel) for the Treatment of Acute Lymphoblastic Leukemia and Diffuse Large B-Cell Lymphoma. *Oncologist* **2020**, *25*, e321–e327. [CrossRef]

20. Day, J.W.; Mendell, J.R.; Mercuri, E.; Finkel, R.S.; Strauss, K.A.; Kleyn, A.; Tauscher-Wisniewski, S.; Tukov, F.F.; Reyna, S.P.; Chand, D.H. Clinical Trial and Postmarketing Safety of Onasemnogene Abeparvovec Therapy. *Drug Saf.* **2021**, *44*, 1109–1119. [CrossRef]

21. Jensen, T.L.; Gøtzsche, C.R.; Woldbye, D.P.D. Current and Future Prospects for Gene Therapy for Rare Genetic Diseases Affecting the Brain and Spinal Cord. *Front. Mol. Neurosci.* **2021**, *14*, 695937. [CrossRef] [PubMed]

22. Keam, S.J. Elivaldogene Autotemcel: First Approval. *Mol. Diagn. Ther.* **2021**, *25*, 803–809. [CrossRef] [PubMed]

23. Watanabe, N.; Mo, F.; McKenna, M.K. Impact of Manufacturing Procedures on CAR T Cell Functionality. *Front. Immunol.* **2022**, *13*, 876339. [CrossRef] [PubMed]

24. Keam, S.J. Eladocagene Exuparvovec: First Approval. *Drugs* **2022**, *82*, 1427–1432. [CrossRef]

25. Blair, H.A. Valoctocogene Roxaparvovec: First Approval. *Drugs* **2022**, *82*, 1505–1510. [CrossRef] [PubMed]

26. Lee, A. Correction to: Nadofaragene Firadenovec: First Approval. *Drugs* **2023**, *83*, 951. [CrossRef] [PubMed]

27. Schuessler-Lenz, M.; Enzmann, H.; Vamvakas, S. Regulators' Advice Can Make a Difference: European Medicines Agency Approval of Zynteglo for Beta Thalassemia. *Clin. Pharmacol. Ther.* **2020**, *107*, 492–494. [CrossRef]

28. Watanabe, M.; Nishikawaji, Y.; Kawakami, H.; Kosai, K.I. Adenovirus Biology, Recombinant Adenovirus, and Adenovirus Usage in Gene Therapy. *Viruses* **2021**, *13*, 2502. [CrossRef]

29. Athanasopoulos, T.; Munye, M.M.; Yanez-Munoz, R.J. Nonintegrating Gene Therapy Vectors. *Hematol. Oncol. Clin. N. Am.* **2017**, *31*, 753–770. [CrossRef]

30. Fang, B.; Lai, J.; Liu, Y.; Yu, T.T.; Yu, X.; Li, X.; Dong, L.; Zhang, X.; Yang, W.; Yan, Q.; et al. Genetic characterization of human adenoviruses in patients using metagenomic next-generation sequencing in Hubei, China, from 2018 to 2019. *Front. Microbiol.* **2023**, *14*, 1153728. [CrossRef]

31. Wu, C.; Wei, F.K.; Xu, Z.Y.; Wen, R.M.; Chen, J.C.; Wang, J.Q.; Mao, L.J. Tropism and transduction of oncolytic adenovirus vectors in prostate cancer therapy. *Front. Biosci.* **2021**, *26*, 866–872. [CrossRef]

32. Crenshaw, B.J.; Jones, L.B.; Bell, C.R.; Kumar, S.; Matthews, Q.L. Perspective on Adenoviruses: Epidemiology, Pathogenicity, and Gene Therapy. *Biomedicines* **2019**, *7*, 61. [CrossRef] [PubMed]

33. Kulanayake, S.; Tikoo, S.K. Adenovirus Core Proteins: Structure and Function. *Viruses* **2021**, *13*, 388. [CrossRef] [PubMed]

34. Rosewell Shaw, A.; Porter, C.; Biegert, G.; Jatta, L.; Suzuki, M. HydrAd: A Helper-Dependent Adenovirus Targeting Multiple Immune Pathways for Cancer Immunotherapy. *Cancers* **2022**, *14*, 2769. [CrossRef] [PubMed]

35. Mantwill, K.; Klein, F.G.; Wang, D.; Hindupur, S.V.; Ehrenfeld, M.; Holm, P.S.; Nawroth, R. Concepts in Oncolytic Adenovirus Therapy. *Int. J. Mol. Sci.* **2021**, *22*, 10522. [CrossRef]

36. Xia, Y.; Li, X.; Sun, W. Applications of Recombinant Adenovirus-p53 Gene Therapy for Cancers in the Clinic in China. *Curr. Gene Ther.* **2020**, *20*, 127–141. [CrossRef] [PubMed]

37. Yang, K.; Feng, S.; Luo, Z. Oncolytic Adenovirus, a New Treatment Strategy for Prostate Cancer. *Biomedicines* **2022**, *10*, 3262. [CrossRef]

38. The Medical Letter, Inc. In brief: Adstiladrin—A gene therapy for bladder cancer. *Med. Lett. Drugs Ther.* **2023**, *65*, e40–e41. [CrossRef]

39. Muhammad, T.; Sakhawat, A.; Khan, A.A.; Ma, L.; Gjerset, R.A.; Huang, Y. Mesenchymal stem cell-mediated delivery of therapeutic adenoviral vectors to prostate cancer. *Stem Cell Res. Ther.* **2019**, *10*, 190. [CrossRef]

40. Nguyen, T.V.; Crosby, C.M.; Heller, G.J.; Mendel, Z.I.; Barry, M.E.; Barry, M.A. Oncolytic adenovirus Ad657 for systemic virotherapy against prostate cancer. *Oncolytic Virother.* **2018**, *7*, 43–51. [CrossRef]

41. Meng, J.; Zhang, J.G.; Du, S.T.; Li, N. The effect of gene therapy on postoperative recurrence of small hepatocellular carcinoma (less than 5cm). *Cancer Gene Ther.* **2019**, *26*, 114–117. [CrossRef] [PubMed]

42. Kong, H.; Zhao, R.; Zhang, Q.; Iqbal, M.Z.; Lu, J.; Zhao, Q.; Luo, D.; Feng, C.; Zhang, K.; Liu, X.; et al. Biosilicified oncolytic adenovirus for cancer viral gene therapy. *Biomater. Sci.* **2020**, *8*, 5317–5328. [CrossRef] [PubMed]

43. Dai, F.; Zhang, P.B.; Feng, Q.; Pan, X.Y.; Song, S.L.; Cui, J.; Yang, J.L. Cytokine-induced killer cells carrying recombinant oncolytic adenovirus expressing p21Ras scFv inhibited liver cancer. *J. Cancer* **2021**, *12*, 2768–2776. [CrossRef] [PubMed]

44. Xiao, S.; Liu, Z.; Deng, R.; Li, C.; Fu, S.; Chen, G.; Zhang, X.; Ke, F.; Ke, S.; Yu, X.; et al. Aptamer-mediated gene therapy enhanced antitumor activity against human hepatocellular carcinoma in vitro and in vivo. *J. Control Release* **2017**, *258*, 130–145. [CrossRef] [PubMed]

45. Moreno, V.; Barretina-Ginesta, M.P.; García-Donas, J.; Jayson, G.C.; Roxburgh, P.; Vázquez, R.M.; Michael, A.; Antón-Torres, A.; Brown, R.; Krige, D.; et al. Safety and efficacy of the tumor-selective adenovirus enadenotucirev with or without paclitaxel in platinum-resistant ovarian cancer: A phase 1 clinical trial. *J. Immunother. Cancer* **2021**, *9*, e003645. [CrossRef] [PubMed]

46. Arend, R.C.; Monk, B.J.; Herzog, T.J.; Moore, K.N.; Shapira-Frommer, R.; Ledermann, J.A.; Tewari, K.S.; Secord, A.A.; Minei, T.R.; Freedman, L.S.; et al. Utilizing an interim futility analysis of the OVAL study (VB-111-701/GOG 3018) for potential reduction of risk: A phase III, double blind, randomized controlled trial of ofranergene obadenovec (VB-111) and weekly paclitaxel in patients with platinum resistant ovarian cancer. *Gynecol. Oncol.* **2021**, *161*, 496–501. [CrossRef] [PubMed]

47. Qiao, H.B.; Li, J.; Lv, L.J.; Nie, B.J.; Lu, P.; Xue, F.; Zhang, Z.M. The effects of interleukin 2 and rAd-p53 as a treatment for glioblastoma. *Mol. Med. Rep.* **2018**, *17*, 4853–4859. [CrossRef]

48. Xia, Y.; Du, Z.H.; Wang, X.Y.; Li, X.Q. Treatment of Uterine Sarcoma with rAd-p53 (Gendicine) Followed by Chemotherapy: Clinical Study of TP53 Gene Therapy. *Hum. Gene Ther.* **2018**, *29*, 242–250. [CrossRef]

49. Tan, T.J.; Ang, W.X.G.; Wang, W.W.; Chong, H.S.; Tan, S.H.; Cheong, R.; Chia, J.W.; Syn, N.L.; Shuen, W.H.; Ba, R.; et al. A phase I study of an adenoviral vector delivering a MUC1/CD40-ligand fusion protein in patients with advanced adenocarcinoma. *Nat. Commun.* **2022**, *13*, 6453. [CrossRef]

50. Sterman, D.H.; Alley, E.; Stevenson, J.P.; Friedberg, J.; Metzger, S.; Recio, A.; Moon, E.K.; Haas, A.R.; Vachani, A.; Katz, S.I.; et al. Pilot and Feasibility Trial Evaluating Immuno-Gene Therapy of Malignant Mesothelioma Using Intrapleural Delivery of Adenovirus-IFNα Combined with Chemotherapy. *Clin. Cancer Res.* **2016**, *22*, 3791–3800. [CrossRef]

51. Ben-Gary, H.; McKinney, R.L.; Rosengart, T.; Lesser, M.L.; Crystal, R.G. Systemic interleukin-6 responses following administration of adenovirus gene transfer vectors to humans by different routes. *Mol. Ther.* **2002**, *6*, 287–297. [CrossRef] [PubMed]

52. Atasheva, S.; Yao, J.; Shayakhmetov, D.M. Innate immunity to adenovirus: Lessons from mice. *FEBS Lett.* **2019**, *593*, 3461–3483. [CrossRef] [PubMed]

53. Fejer, G.; Freudenberg, M.; Greber, U.F.; Gyory, I. Adenovirus-triggered innate signalling pathways. *Eur. J. Microbiol. Immunol.* **2011**, *1*, 279–288. [CrossRef] [PubMed]

54. Bottermann, M.; Foss, S.; van Tienen, L.M.; Vaysburd, M.; Cruickshank, J.; O'Connell, K.; Clark, J.; Mayes, K.; Higginson, K.; Hirst, J.C.; et al. TRIM21 mediates antibody inhibition of adenovirus-based gene delivery and vaccination. *Proc. Natl. Acad. Sci. USA* **2018**, *115*, 10440–10445. [CrossRef] [PubMed]

55. Minamitani, T.; Iwakiri, D.; Takada, K. Adenovirus virus-associated RNAs induce type I interferon expression through a RIG-I-mediated pathway. *J. Virol.* **2011**, *85*, 4035–4040. [CrossRef] [PubMed]

56. Yamaguchi, T.; Kawabata, K.; Kouyama, E.; Ishii, K.J.; Katayama, K.; Suzuki, T.; Kurachi, S.; Sakurai, F.; Akira, S.; Mizuguchi, H. Induction of type I interferon by adenovirus-encoded small RNAs. *Proc. Natl. Acad. Sci. USA* **2010**, *107*, 17286–17291. [CrossRef] [PubMed]

57. Jiang, H.; Clise-Dwyer, K.; Ruisaard, K.E.; Fan, X.; Tian, W.; Gumin, J.; Lamfers, M.L.; Kleijn, A.; Lang, F.F.; Yung, W.K.; et al. Delta-24-RGD oncolytic adenovirus elicits anti-glioma immunity in an immunocompetent mouse model. *PLoS ONE* **2014**, *9*, e97407. [CrossRef]

58. Hoare, J.I.; Osmani, B.; O'Sullivan, E.A.; Browne, A.; Campbell, N.; Metcalf, S.; Nicolini, F.; Saxena, J.; Martin, S.A.; Lockley, M. Carvedilol targets β-arrestins to rewire innate immunity and improve oncolytic adenoviral therapy. *Commun. Biol.* **2022**, *5*, 106. [CrossRef]

59. Wang, W.C.; Sayedahmed, E.E.; Mittal, S.K. Significance of Preexisting Vector Immunity and Activation of Innate Responses for Adenoviral Vector-Based Therapy. *Viruses* **2022**, *14*, 2727. [CrossRef]

60. Tian, M.; Liu, W.; Zhang, Q.; Huang, Y.; Li, W.; Wang, W.; Zhao, P.; Huang, S.; Song, Y.; Shereen, M.A.; et al. MYSM1 Represses Innate Immunity and Autoimmunity through Suppressing the cGAS-STING Pathway. *Cell Rep.* **2020**, *33*, 108297. [CrossRef]

61. Kreppel, F.; Hagedorn, C. Capsid and Genome Modification Strategies to Reduce the Immunogenicity of Adenoviral Vectors. *Int. J. Mol. Sci.* **2021**, *22*, 2417. [CrossRef] [PubMed]

62. Gabal, Y.; Ramsey, J.D. Surface Modification of Adenovirus Vector to Improve Immunogenicity and Tropism. *Methods Mol. Biol.* **2021**, *2183*, 357–366. [CrossRef] [PubMed]

63. Gentile, C.M.; Borovjagin, A.V.; Richter, J.R.; Jani, A.H.; Wu, H.; Zinn, K.R.; Warram, J.M. Genetic strategy to decrease complement activation with adenoviral therapies. *PLoS ONE* **2019**, *14*, e0215226. [CrossRef] [PubMed]

64. Atasheva, S.; Emerson, C.C.; Yao, J.; Young, C.; Stewart, P.L.; Shayakhmetov, D.M. Systemic cancer therapy with engineered adenovirus that evades innate immunity. *Sci. Transl. Med.* **2020**, *12*, eabc6659. [CrossRef] [PubMed]

65. Thaci, B.; Ulasov, I.V.; Wainwright, D.A.; Lesniak, M.S. The challenge for gene therapy: Innate immune response to adenoviruses. *Oncotarget* **2011**, *2*, 113–121. [CrossRef] [PubMed]

66. Lv, P.; Liu, X.; Chen, X.; Liu, C.; Zhang, Y.; Chu, C.; Wang, J.; Wang, X.; Chen, X.; Liu, G. Genetically Engineered Cell Membrane Nanovesicles for Oncolytic Adenovirus Delivery: A Versatile Platform for Cancer Virotherapy. *Nano Lett.* **2019**, *19*, 2993–3001. [CrossRef] [PubMed]

67. Chen, Y.H.; Keiser, M.S.; Davidson, B.L. Viral Vectors for Gene Transfer. *Curr. Protoc. Mouse Biol.* **2018**, *8*, e58. [CrossRef]

68. Laredj, L.N.; Beard, P. Adeno-associated virus activates an innate immune response in normal human cells but not in osteosarcoma cells. *J. Virol.* **2011**, *85*, 13133–13143. [CrossRef]

69. Issa, S.S.; Shaimardanova, A.A.; Solovyeva, V.V.; Rizvanov, A.A. Various AAV Serotypes and Their Applications in Gene Therapy: An Overview. *Cells* **2023**, *12*, 785. [CrossRef]

70. Hareendran, S.; Balakrishnan, B.; Sen, D.; Kumar, S.; Srivastava, A.; Jayandharan, G.R. Adeno-associated virus (AAV) vectors in gene therapy: Immune challenges and strategies to circumvent them. *Rev. Med. Virol.* **2013**, *23*, 399–413. [CrossRef]

71. Calcedo, R.; Vandenberghe, L.H.; Gao, G.; Lin, J.; Wilson, J.M. Worldwide epidemiology of neutralizing antibodies to adeno-associated viruses. *J. Infect. Dis.* **2009**, *199*, 381–390. [CrossRef] [PubMed]

72. Keeler, A.M.; Flotte, T.R. Recombinant Adeno-Associated Virus Gene Therapy in Light of Luxturna (and Zolgensma and Glybera): Where Are We, and How Did We Get Here? *Annu. Rev. Virol.* **2019**, *6*, 601–621. [CrossRef]

73. Prado, D.A.; Acosta-Acero, M.; Maldonado, R.S. Gene therapy beyond luxturna: A new horizon of the treatment for inherited retinal disease. *Curr. Opin. Ophthalmol.* **2020**, *31*, 147–154. [CrossRef] [PubMed]

74. High, K.A.; Roncarolo, M.G. Gene Therapy. *N. Engl. J. Med.* **2019**, *381*, 455–464. [CrossRef] [PubMed]

75. George, L.A.; Ragni, M.V.; Rasko, J.E.J.; Raffini, L.J.; Samelson-Jones, B.J.; Ozelo, M.; Hazbon, M.; Runowski, A.R.; Wellman, J.A.; Wachtel, K.; et al. Long-Term Follow-Up of the First in Human Intravascular Delivery of AAV for Gene Transfer: AAV2-hFIX16 for Severe Hemophilia B. *Mol. Ther.* **2020**, *28*, 2073–2082. [CrossRef] [PubMed]

76. George, L.A.; Monahan, P.E.; Eyster, M.E.; Sullivan, S.K.; Ragni, M.V.; Croteau, S.E.; Rasko, J.E.J.; Recht, M.; Samelson-Jones, B.J.; MacDougall, A.; et al. Multiyear Factor VIII Expression after AAV Gene Transfer for Hemophilia A. *N. Engl. J. Med.* **2021**, *385*, 1961–1973. [CrossRef]

77. Miesbach, W.; Meijer, K.; Coppens, M.; Kampmann, P.; Klamroth, R.; Schutgens, R.; Tangelder, M.; Castaman, G.; Schwäble, J.; Bonig, H.; et al. Gene therapy with adeno-associated virus vector 5-human factor IX in adults with hemophilia B. *Blood* **2018**, *131*, 1022–1031. [CrossRef] [PubMed]

78. Pasi, K.J.; Rangarajan, S.; Mitchell, N.; Lester, W.; Symington, E.; Madan, B.; Laffan, M.; Russell, C.B.; Li, M.; Pierce, G.F.; et al. Multiyear Follow-up of AAV5-hFVIII-SQ Gene Therapy for Hemophilia A. *N. Engl. J. Med.* **2020**, *382*, 29–40. [CrossRef]

79. Nathwani, A.C.; Reiss, U.M.; Tuddenham, E.G.; Rosales, C.; Chowdary, P.; McIntosh, J.; Della Peruta, M.; Lheriteau, E.; Patel, N.; Raj, D.; et al. Long-term safety and efficacy of factor IX gene therapy in hemophilia B. *N. Engl. J. Med.* **2014**, *371*, 1994–2004. [CrossRef]

80. Sands, M.S. AAV-mediated liver-directed gene therapy. *Methods Mol. Biol.* **2011**, *807*, 141–157. [CrossRef]

81. Nathwani, A.C.; Tuddenham, E.G.; Rangarajan, S.; Rosales, C.; McIntosh, J.; Linch, D.C.; Chowdary, P.; Riddell, A.; Pie, A.J.; Harrington, C.; et al. Adenovirus-associated virus vector-mediated gene transfer in hemophilia B. *N. Engl. J. Med.* **2011**, *365*, 2357–2365. [CrossRef] [PubMed]

82. Russell, S.; Bennett, J.; Wellman, J.A.; Chung, D.C.; Yu, Z.-F.; Tillman, A.; Wittes, J.; Pappas, J.; Elci, O.; McCague, S.; et al. Efficacy and safety of voretigene neparvovec (AAV2-hRPE65v2) in patients with RPE65-mediated inherited retinal dystrophy: A randomised, controlled, open-label, phase 3 trial. *Lancet* **2017**, *390*, 849–860. [CrossRef] [PubMed]

83. Verdera, H.C.; Kuranda, K.; Mingozzi, F. AAV Vector Immunogenicity in Humans: A Long Journey to Successful Gene Transfer. *Mol. Ther.* **2020**, *28*, 723–746. [CrossRef] [PubMed]

84. Maguire, A.M.; Simonelli, F.; Pierce, E.A.; Pugh, E.N., Jr.; Mingozzi, F.; Bennicelli, J.; Banfi, S.; Marshall, K.A.; Testa, F.; Surace, E.M.; et al. Safety and efficacy of gene transfer for Leber's congenital amaurosis. *N. Engl. J. Med.* **2008**, *358*, 2240–2248. [CrossRef] [PubMed]

85. Fischer, M.D.; Michalakis, S.; Wilhelm, B.; Zobor, D.; Muehlfriedel, R.; Kohl, S.; Weisschuh, N.; Ochakovski, G.A.; Klein, R.; Schoen, C.; et al. Safety and Vision Outcomes of Subretinal Gene Therapy Targeting Cone Photoreceptors in Achromatopsia: A Nonrandomized Controlled Trial. *JAMA Ophthalmol.* **2020**, *138*, 643–651. [CrossRef] [PubMed]

86. Pennesi, M.E.; Weleber, R.G.; Yang, P.; Whitebirch, C.; Thean, B.; Flotte, T.R.; Humphries, M.; Chegarnov, E.; Beasley, K.N.; Stout, J.T.; et al. Results at 5 Years After Gene Therapy for RPE65-Deficient Retinal Dystrophy. *Hum. Gene Ther.* **2018**, *29*, 1428–1437. [CrossRef] [PubMed]

87. Kaplitt, M.G.; Feigin, A.; Tang, C.; Fitzsimons, H.L.; Mattis, P.; Lawlor, P.A.; Bland, R.J.; Young, D.; Strybing, K.; Eidelberg, D.; et al. Safety and tolerability of gene therapy with an adeno-associated virus (AAV) borne GAD gene for Parkinson's disease: An open label, phase I trial. *Lancet* **2007**, *369*, 2097–2105. [CrossRef] [PubMed]

88. Tenenbaum, L.; Chtarto, A.; Lehtonen, E.; Velu, T.; Brotchi, J.; Levivier, M. Recombinant AAV-mediated gene delivery to the central nervous system. *J. Gene Med.* **2004**, *6* (Suppl. 1), S212–S222. [CrossRef]

89. Castle, M.J.; Baltanas, F.C.; Kovacs, I.; Nagahara, A.H.; Barba, D.; Tuszynski, M.H. Postmortem Analysis in a Clinical Trial of AAV2-NGF Gene Therapy for Alzheimer's Disease Identifies a Need for Improved Vector Delivery. *Hum. Gene Ther.* **2020**, *31*, 415–422. [CrossRef]

90. Pearson, T.S.; Gupta, N.; San Sebastian, W.; Imamura-Ching, J.; Viehoever, A.; Grijalvo-Perez, A.; Fay, A.J.; Seth, N.; Lundy, S.M.; Seo, Y.; et al. Gene therapy for aromatic L-amino acid decarboxylase deficiency by MR-guided direct delivery of AAV2-AADC to midbrain dopaminergic neurons. *Nat. Commun.* **2021**, *12*, 4251. [CrossRef]

91. Spronck, E.A.; Brouwers, C.C.; Valles, A.; de Haan, M.; Petry, H.; van Deventer, S.J.; Konstantinova, P.; Evers, M.M. AAV5-miHTT Gene Therapy Demonstrates Sustained Huntingtin Lowering and Functional Improvement in Huntington Disease Mouse Models. *Mol. Ther. Methods Clin. Dev.* **2019**, *13*, 334–343. [CrossRef] [PubMed]

92. Martier, R.; Sogorb-Gonzalez, M.; Stricker-Shaver, J.; Hubener-Schmid, J.; Keskin, S.; Klima, J.; Toonen, L.J.; Juhas, S.; Juhasova, J.; Ellederova, Z.; et al. Development of an AAV-Based MicroRNA Gene Therapy to Treat Machado-Joseph Disease. *Mol. Ther. Methods Clin. Dev.* **2019**, *15*, 343–358. [CrossRef] [PubMed]

93. Saraiva, J.; Nobre, R.J.; Pereira de Almeida, L. Gene therapy for the CNS using AAVs: The impact of systemic delivery by AAV9. *J. Control Release* **2016**, *241*, 94–109. [CrossRef] [PubMed]

94. Kaifer, K.A.; Villalon, E.; Smith, C.E.; Simon, M.E.; Marquez, J.; Hopkins, A.E.; Morcos, T.I.; Lorson, C.L. AAV9-DOK7 gene therapy reduces disease severity in Smn(2B/-) SMA model mice. *Biochem. Biophys. Res. Commun.* **2020**, *530*, 107–114. [CrossRef] [PubMed]

95. Mendell, J.R.; Sahenk, Z.; Lehman, K.; Nease, C.; Lowes, L.P.; Miller, N.F.; Iammarino, M.A.; Alfano, L.N.; Nicholl, A.; Al-Zaidy, S.; et al. Assessment of Systemic Delivery of rAAVrh74.MHCK7.micro-dystrophin in Children with Duchenne Muscular Dystrophy: A Nonrandomized Controlled Trial. *JAMA Neurol.* **2020**, *77*, 1122–1131. [CrossRef] [PubMed]

96. Annoussamy, M.; Lilien, C.; Gidaro, T.; Gargaun, E.; Che, V.; Schara, U.; Gangfuss, A.; D'Amico, A.; Dowling, J.J.; Darras, B.T.; et al. X-linked myotubular myopathy: A prospective international natural history study. *Neurology* **2019**, *92*, e1852–e1867. [CrossRef] [PubMed]

97. Zhang, J.; Hou, Z.; Wang, X.; Jiang, H.; Neng, L.; Zhang, Y.; Yu, Q.; Burwood, G.; Song, J.; Auer, M.; et al. VEGFA165 gene therapy ameliorates blood-labyrinth barrier breakdown and hearing loss. *JCI Insight* **2021**, *6*, e143285. [CrossRef]

98. Gyorgy, B.; Sage, C.; Indzhykulian, A.A.; Scheffer, D.I.; Brisson, A.R.; Tan, S.; Wu, X.D.; Volak, A.; Mu, D.K.; Tamvakologos, P.I.; et al. Rescue of Hearing by Gene Delivery to Inner-Ear Hair Cells Using Exosome-Associated AAV. *Mol. Ther.* **2017**, *25*, 379–391. [CrossRef]

99. György, B.; Meijer, E.J.; Ivanchenko, M.V.; Tenneson, K.; Emond, F.; Hanlon, K.S.; Indzhykulian, A.A.; Volak, A.; Karavitaki, K.D.; Tamvakologos, P.I.; et al. Gene Transfer with AAV9-PHP.B Rescues Hearing in a Mouse Model of Usher Syndrome 3A and Transduces Hair Cells in a Non-human Primate. *Mol. Ther. Methods Clin. Dev.* **2019**, *13*, 1–13. [CrossRef]

100. Chan, Y.K.; Wang, S.K.; Chu, C.J.; Copland, D.A.; Letizia, A.J.; Costa Verdera, H.; Chiang, J.J.; Sethi, M.; Wang, M.K.; Neidermyer, W.J., Jr.; et al. Engineering adeno-associated viral vectors to evade innate immune and inflammatory responses. *Sci. Transl. Med.* **2021**, *13*, eabd3438. [CrossRef]

101. Li, C.; Samulski, R.J. Engineering adeno-associated virus vectors for gene therapy. *Nat. Rev. Genet.* **2020**, *21*, 255–272. [CrossRef] [PubMed]

102. Shayakhmetov, D.M.; Di Paolo, N.C.; Mossman, K.L. Recognition of virus infection and innate host responses to viral gene therapy vectors. *Mol. Ther.* **2010**, *18*, 1422–1429. [CrossRef] [PubMed]

103. Zhu, J.; Huang, X.; Yang, Y. The TLR9-MyD88 pathway is critical for adaptive immune responses to adeno-associated virus gene therapy vectors in mice. *J. Clin. Investig.* **2009**, *119*, 2388–2398. [CrossRef] [PubMed]

104. Martino, A.T.; Suzuki, M.; Markusic, D.M.; Zolotukhin, I.; Ryals, R.C.; Moghimi, B.; Ertl, H.C.; Muruve, D.A.; Lee, B.; Herzog, R.W. The genome of self-complementary adeno-associated viral vectors increases Toll-like receptor 9-dependent innate immune responses in the liver. *Blood* **2011**, *117*, 6459–6460. [CrossRef] [PubMed]

105. Kessler, N.; Viehmann, S.F.; Krollmann, C.; Mai, K.; Kirschner, K.M.; Luksch, H.; Kotagiri, P.; Böhner, A.M.C.; Huugen, D.; de Oliveira Mann, C.C.; et al. Monocyte-derived macrophages aggravate pulmonary vasculitis via cGAS/STING/IFN-mediated nucleic acid sensing. *J. Exp. Med.* **2022**, *219*, e20220759. [CrossRef] [PubMed]

106. Shao, W.; Earley, L.F.; Chai, Z.; Chen, X.; Sun, J.; He, T.; Deng, M.; Hirsch, M.L.; Ting, J.; Samulski, R.J.; et al. Double-stranded RNA innate immune response activation from long-term adeno-associated virus vector transduction. *JCI Insight* **2018**, *3*, e120474. [CrossRef] [PubMed]

107. Ronzitti, G.; Gross, D.A.; Mingozzi, F. Human Immune Responses to Adeno-Associated Virus (AAV) Vectors. *Front. Immunol.* **2020**, *11*, 670. [CrossRef] [PubMed]

108. Seregin, S.S.; Appledorn, D.M.; McBride, A.J.; Schuldt, N.J.; Aldhamen, Y.A.; Voss, T.; Wei, J.; Bujold, M.; Nance, W.; Godbehere, S.; et al. Transient pretreatment with glucocorticoid ablates innate toxicity of systemically delivered adenoviral vectors without reducing efficacy. *Mol. Ther.* **2009**, *17*, 685–696. [CrossRef]

109. Dauletbekov, D.L.; Pfromm, J.K.; Fritz, A.K.; Fischer, M.D. Innate Immune Response Following AAV Administration. *Adv. Exp. Med. Biol.* **2019**, *1185*, 165–168. [CrossRef]

110. Maheshri, N.; Koerber, J.T.; Kaspar, B.K.; Schaffer, D.V. Directed evolution of adeno-associated virus yields enhanced gene delivery vectors. *Nat. Biotechnol.* **2006**, *24*, 198–204. [CrossRef]

111. Bertolini, T.B.; Shirley, J.L.; Zolotukhin, I.; Li, X.; Kaisho, T.; Xiao, W.; Kumar, S.R.P.; Herzog, R.W. Effect of CpG Depletion of Vector Genome on CD8(+) T Cell Responses in AAV Gene Therapy. *Front. Immunol.* **2021**, *12*, 672449. [CrossRef] [PubMed]

112. Munis, A.M. Gene Therapy Applications of Non-Human Lentiviral Vectors. *Viruses* **2020**, *12*, 1106. [CrossRef] [PubMed]

113. de Pablo-Maiso, L.; Domenech, A.; Echeverria, I.; Gomez-Arrebola, C.; de Andres, D.; Rosati, S.; Gomez-Lucia, E.; Reina, R. Prospects in Innate Immune Responses as Potential Control Strategies against Non-Primate Lentiviruses. *Viruses* **2018**, *10*, 435. [CrossRef] [PubMed]

114. Bulcha, J.T.; Wang, Y.; Ma, H.; Tai, P.W.L.; Gao, G. Viral vector platforms within the gene therapy landscape. *Signal Transduct. Target. Ther.* **2021**, *6*, 53. [CrossRef] [PubMed]

115. Perry, C.; Rayat, A. Lentiviral Vector Bioprocessing. *Viruses* **2021**, *13*, 268. [CrossRef] [PubMed]

116. Ferreira, C.B.; Sumner, R.P.; Rodriguez-Plata, M.T.; Rasaiyaah, J.; Milne, R.S.; Thrasher, A.J.; Qasim, W.; Towers, G.J. Lentiviral Vector Production Titer Is Not Limited in HEK293T by Induced Intracellular Innate Immunity. *Mol. Ther. Methods Clin. Dev.* **2020**, *17*, 209–219. [CrossRef] [PubMed]

117. Kohn, D.B.; Booth, C.; Kang, E.M.; Pai, S.Y.; Shaw, K.L.; Santilli, G.; Armant, M.; Buckland, K.F.; Choi, U.; De Ravin, S.S.; et al. Lentiviral gene therapy for X-linked chronic granulomatous disease. *Nat. Med.* **2020**, *26*, 200–206. [CrossRef]

118. Thompson, A.A.; Walters, M.C.; Kwiatkowski, J.; Rasko, J.E.J.; Ribeil, J.A.; Hongeng, S.; Magrin, E.; Schiller, G.J.; Payen, E.; Semeraro, M.; et al. Gene Therapy in Patients with Transfusion-Dependent beta-Thalassemia. *N. Engl. J. Med.* **2018**, *378*, 1479–1493. [CrossRef]

119. Magrin, E.; Semeraro, M.; Hebert, N.; Joseph, L.; Magnani, A.; Chalumeau, A.; Gabrion, A.; Roudaut, C.; Marouene, J.; Lefrere, F.; et al. Long-term outcomes of lentiviral gene therapy for the β-hemoglobinopathies: The HGB-205 trial. *Nat. Med.* **2022**, *28*, 81–88. [CrossRef]

120. Mamcarz, E.; Zhou, S.; Lockey, T.; Abdelsamed, H.; Cross, S.J.; Kang, G.; Ma, Z.J.; Condori, J.; Dowdy, J.; Triplett, B.; et al. Lentiviral Gene Therapy Combined with Low-Dose Busulfan in Infants with SCID-X1. *N. Engl. J. Med.* **2019**, *380*, 1525–1534. [CrossRef]

121. Cowan, M.J.; Yu, J.S.; Facchino, J.; Fraser-Browne, C.; Sanford, U.; Kawahara, M.; Dara, J.; Long-Boyle, J.; Oh, J.; Chan, W.Y.; et al. Lentiviral Gene Therapy for Artemis-Deficient SCID. *N. Engl. J. Med.* **2022**, *387*, 2344–2355. [CrossRef]

122. Eichler, F.; Duncan, C.; Musolino, P.L.; Orchard, P.J.; De Oliveira, S.; Thrasher, A.J.; Armant, M.; Dansereau, C.; Lund, T.C.; Miller, W.P.; et al. Hematopoietic Stem-Cell Gene Therapy for Cerebral Adrenoleukodystrophy. *N. Engl. J. Med.* **2017**, *377*, 1630–1638. [CrossRef] [PubMed]

123. Gentner, B.; Tucci, F.; Galimberti, S.; Fumagalli, F.; De Pellegrin, M.; Silvani, P.; Camesasca, C.; Pontesilli, S.; Darin, S.; Ciotti, F.; et al. Hematopoietic Stem- and Progenitor-Cell Gene Therapy for Hurler Syndrome. *N. Engl. J. Med.* **2021**, *385*, 1929–1940. [CrossRef] [PubMed]

124. Sessa, M.; Lorioli, L.; Fumagalli, F.; Acquati, S.; Redaelli, D.; Baldoli, C.; Canale, S.; Lopez, I.D.; Morena, F.; Calabria, A.; et al. Lentiviral haemopoietic stem-cell gene therapy in early-onset metachromatic leukodystrophy: An ad-hoc analysis of a non-randomised, open-label, phase 1/2 trial. *Lancet* **2016**, *388*, 476–487. [CrossRef]

125. Liu, S.; Deng, B.; Yin, Z.; Lin, Y.; An, L.; Liu, D.; Pan, J.; Yu, X.; Chen, B.; Wu, T.; et al. Combination of CD19 and CD22 CAR-T cell therapy in relapsed B-cell acute lymphoblastic leukemia after allogeneic transplantation. *Am. J. Hematol.* **2021**, *96*, 671–679. [CrossRef] [PubMed]

126. Cohen, A.D.; Garfall, A.L.; Stadtmauer, E.A.; Melenhorst, J.J.; Lacey, S.F.; Lancaster, E.; Vogl, D.T.; Weiss, B.M.; Dengel, K.; Nelson, A.; et al. B cell maturation antigen-specific CAR T cells are clinically active in multiple myeloma. *J. Clin. Investig.* **2019**, *129*, 2210–2221. [CrossRef]

127. Bryson, P.D.; Han, X.; Truong, N.; Wang, P. Breast cancer vaccines delivered by dendritic cell-targeted lentivectors induce potent antitumor immune responses and protect mice from mammary tumor growth. *Vaccine* **2017**, *35*, 5842–5849. [CrossRef] [PubMed]

128. Lee, S.; Kim, Y.Y.; Ahn, H.J. Systemic delivery of CRISPR/Cas9 to hepatic tumors for cancer treatment using altered tropism of lentiviral vector. *Biomaterials* **2021**, *272*, 120793. [CrossRef]

129. Matsunaga, W.; Ichikawa, M.; Nakamura, A.; Ishikawa, T.; Gotoh, A. Lentiviral Vector-mediated Gene Transfer in Human Bladder Cancer Cell Lines. *Anticancer Res.* **2018**, *38*, 2015–2020. [CrossRef]

130. Russell, A.L.; Prince, C.; Lundgren, T.S.; Knight, K.A.; Denning, G.; Alexander, J.S.; Zoine, J.T.; Spencer, H.T.; Chandrakasan, S.; Doering, C.B. Non-genotoxic conditioning facilitates hematopoietic stem cell gene therapy for hemophilia A using bioengineered factor VIII. *Mol. Ther. Methods Clin. Dev.* **2021**, *21*, 710–727. [CrossRef]

131. Olgasi, C.; Borsotti, C.; Merlin, S.; Bergmann, T.; Bittorf, P.; Adewoye, A.B.; Wragg, N.; Patterson, K.; Calabria, A.; Benedicenti, F.; et al. Efficient and safe correction of hemophilia A by lentiviral vector-transduced BOECs in an implantable device. *Mol. Ther. Methods Clin. Dev.* **2021**, *23*, 551–566. [CrossRef] [PubMed]

132. Milani, M.; Canepari, C.; Liu, T.; Biffi, M.; Russo, F.; Plati, T.; Curto, R.; Patarroyo-White, S.; Drager, D.; Visigalli, I.; et al. Liver-directed lentiviral gene therapy corrects hemophilia A mice and achieves normal-range factor VIII activity in non-human primates. *Nat. Commun.* **2022**, *13*, 2454. [CrossRef]

133. Chen, Y.; Schroeder, J.A.; Gao, C.; Li, J.; Hu, J.; Shi, Q. In vivo enrichment of genetically manipulated platelets for murine hemophilia B gene therapy. *J. Cell Physiol.* **2021**, *236*, 354–365. [CrossRef] [PubMed]

134. Ryan, M.; Tan, V.T.Y.; Thompson, N.; Guevremont, D.; Mockett, B.G.; Tate, W.P.; Abraham, W.C.; Hughes, S.M.; Williams, J. Lentivirus-Mediated Expression of Human Secreted Amyloid Precursor Protein-Alpha Promotes Long-Term Induction of Neuroprotective Genes and Pathways in a Mouse Model of Alzheimer's Disease. *J. Alzheimers Dis.* **2021**, *79*, 1075–1090. [CrossRef]

135. Zeng, C.Y.; Yang, T.T.; Zhou, H.J.; Zhao, Y.; Kuang, X.; Duan, W.; Du, J.R. Lentiviral vector-mediated overexpression of Klotho in the brain improves Alzheimer's disease-like pathology and cognitive deficits in mice. *Neurobiol. Aging* **2019**, *78*, 18–28. [CrossRef] [PubMed]
136. Van Kampen, J.M.; Kay, D.G. Progranulin gene delivery reduces plaque burden and synaptic atrophy in a mouse model of Alzheimer's disease. *PLoS ONE* **2017**, *12*, e0182896. [CrossRef] [PubMed]
137. Wang, Q.S.; Wang, Y.; Lv, H.Y.; Han, Q.W.; Fan, H.; Guo, B.; Wang, L.L.; Han, W.D. Treatment of CD33-directed chimeric antigen receptor-modified T cells in one patient with relapsed and refractory acute myeloid leukemia. *Mol. Ther.* **2015**, *23*, 184–191. [CrossRef] [PubMed]
138. Brown, B.D.; Sitia, G.; Annoni, A.; Hauben, E.; Sergi, L.S.; Zingale, A.; Roncarolo, M.G.; Guidotti, L.G.; Naldini, L. In vivo administration of lentiviral vectors triggers a type I interferon response that restricts hepatocyte gene transfer and promotes vector clearance. *Blood* **2007**, *109*, 2797–2805. [CrossRef]
139. Chintala, K.; Mohareer, K.; Banerjee, S. Dodging the Host Interferon-Stimulated Gene Mediated Innate Immunity by HIV-1: A Brief Update on Intrinsic Mechanisms and Counter-Mechanisms. *Front. Immunol.* **2021**, *12*, 716927. [CrossRef]
140. Piras, F.; Riba, M.; Petrillo, C.; Lazarevic, D.; Cuccovillo, I.; Bartolaccini, S.; Stupka, E.; Gentner, B.; Cittaro, D.; Naldini, L.; et al. Lentiviral vectors escape innate sensing but trigger p53 in human hematopoietic stem and progenitor cells. *EMBO Mol. Med.* **2017**, *9*, 1198–1211. [CrossRef]
141. Kajaste-Rudnitski, A.; Naldini, L. Cellular innate immunity and restriction of viral infection: Implications for lentiviral gene therapy in human hematopoietic cells. *Hum. Gene Ther.* **2015**, *26*, 201–209. [CrossRef] [PubMed]
142. Coroadinha, A.S. Host Cell Restriction Factors Blocking Efficient Vector Transduction: Challenges in Lentiviral and Adeno-Associated Vector Based Gene Therapies. *Cells* **2023**, *12*, 732. [CrossRef] [PubMed]
143. Annoni, A.; Gregori, S.; Naldini, L.; Cantore, A. Modulation of immune responses in lentiviral vector-mediated gene transfer. *Cell Immunol.* **2019**, *342*, 103802. [CrossRef] [PubMed]
144. He, T.S.; Dang, L.L.; Zhang, J.H.; Zhang, J.Q.; Wang, G.P.; Wang, E.N.; Xia, H.; Zhou, W.H.; Wu, S.; Liu, X.Q. The Hippo signaling component LATS2 enhances innate immunity to inhibit HIV-1 infection through PQBP1-cGAS pathway. *Cell Death Differ.* **2022**, *29*, 192–205. [CrossRef] [PubMed]
145. Xiao, Q.; Guo, D.; Chen, S. Application of CRISPR/Cas9-Based Gene Editing in HIV-1/AIDS Therapy. *Front. Cell Infect. Microbiol.* **2019**, *9*, 69. [CrossRef] [PubMed]
146. Milani, M.; Annoni, A.; Moalli, F.; Liu, T.; Cesana, D.; Calabria, A.; Bartolaccini, S.; Biffi, M.; Russo, F.; Visigalli, I.; et al. Phagocytosis-shielded lentiviral vectors improve liver gene therapy in nonhuman primates. *Sci. Transl. Med.* **2019**, *11*, eaav7325. [CrossRef] [PubMed]
147. Alyami, E.M.; Tarar, A.; Peng, C.A. Less phagocytosis of viral vectors by tethering with CD47 ectodomain. *J. Mater. Chem. B* **2021**, *10*, 64–77. [CrossRef] [PubMed]
148. Sutlu, T.; Nystrom, S.; Gilljam, M.; Stellan, B.; Applequist, S.E.; Alici, E. Inhibition of intracellular antiviral defense mechanisms augments lentiviral transduction of human natural killer cells: Implications for gene therapy. *Hum. Gene Ther.* **2012**, *23*, 1090–1100. [CrossRef]
149. Petrillo, C.; Thorne, L.G.; Unali, G.; Schiroli, G.; Giordano, A.M.S.; Piras, F.; Cuccovillo, I.; Petit, S.J.; Ahsan, F.; Noursadeghi, M.; et al. Cyclosporine H Overcomes Innate Immune Restrictions to Improve Lentiviral Transduction and Gene Editing in Human Hematopoietic Stem Cells. *Cell Stem Cell* **2018**, *23*, 820–832.e9. [CrossRef]
150. Agudo, J.; Ruzo, A.; Kitur, K.; Sachidanandam, R.; Blander, J.M.; Brown, B.D. A TLR and non-TLR mediated innate response to lentiviruses restricts hepatocyte entry and can be ameliorated by pharmacological blockade. *Mol. Ther.* **2012**, *20*, 2257–2267. [CrossRef]

Review

Suitable Disinfectants with Proven Efficacy for Genetically Modified Viruses and Viral Vectors

Maren Eggers [1,2,*], Ingeborg Schwebke [2], Johannes Blümel [3], Franziska Brandt [4], Helmut Fickenscher [5], Jürgen Gebel [6], Nils Hübner [7], Janis A. Müller [8], Holger F. Rabenau [9], Ingrid Rapp [10], Sven Reiche [11], Eike Steinmann [12], Jochen Steinmann [13], Paula Zwicker [7] and Miranda Suchomel [14,15]

[1] Laboratory Prof. Dr. G. Enders MVZ GbR, Rosenbergstr. 85, 70193 Stuttgart, Germany
[2] Expert Committee on Virus Disinfection of the German Association for the Control of Viral Diseases (DVV) e.V. and the Society for Virology (GfV) e.V., 69126 Heidelberg, Germany; inge.schwebke@gmx.de
[3] Paul-Ehrlich-Institute, Department of Virology, Paul-Ehrlich-Straße 51-56, 63225 Langen, Germany; johannes.bluemel@pei.de
[4] Federal Institute for Drugs and Medical Devices, Kurt-Georg-Kiesinger-Allee 3, 53175 Bonn, Germany; franziska.brandt@bfarm.de
[5] Institute for Infection Medicine, Christian-Albrechts-University Kiel, University Clinic Schleswig-Holstein, Bruinswiker Straße 4, 24105 Kiel, Germany; fickenscher@infmed.uni-kiel.de
[6] VAH c/o Institute for Hygiene and Public Health, Venusberg-Campus 1, 53127 Bonn, Germany; juergen.gebel@ukbonn.de
[7] Institute of Hygiene and Environmental Medicine, University Medicine Greifswald, W. Rathenaustr. 49, 17475 Greifswald, Germany; nils.huebner@med.uni-greifswald.de (N.H.); paula.zwicker@med.uni-greifswald.de (P.Z.)
[8] Institute of Virology, Hans-Meerwein Straße 2, 35043 Marburg, Germany; janismueller@uni-marburg.de
[9] Institute for Medical Virology, University Hospital, Goethe University Frankfurt am Main, 60596 Frankfurt, Germany; rabenau@em.uni-frankfurt.de
[10] Boehringer Ingelheim Therapeutics GmbH, Beim Braunland 1, 88416 Ochsenhausen, Germany; ingrid.rapp@boehringer-ingelheim.com
[11] Friedrich-Loeffler-Institute, Federal Research Institute for Animal Health, Department of Experimental Animal Facilities and Biorisk Management, Suedufer 10, 17493 Greifswald-Insel Riems, Germany; sven.reiche@fli.de
[12] Department for Molecular & Medical Virology, Ruhr University Bochum, 44801 Bochum, Germany; eike.steinmann@ruhr-uni-bochum.de
[13] Dr. Bill + Partner GmbH Institute for Hygiene and Microbiology, Norderoog 2, 28259 Bremen, Germany; jochen-steinmann@web.de
[14] Institute of Hygiene and Applied Immunology, Medical University of Vienna, Kinderspitalgasse 15, 1090 Vienna, Austria; miranda.suchomel@meduniwien.ac.at
[15] Austrian Society for Hygiene, Microbiology and Preventive Medicine (ÖGHMP) c/o MAW, Freyung 6/3, 1010 Vienna, Austria
* Correspondence: eggers@labor-enders.de; Tel.: +49-711-6357-130

Citation: Eggers, M.; Schwebke, I.; Blümel, J.; Brandt, F.; Fickenscher, H.; Gebel, J.; Hübner, N.; Müller, J.A.; Rabenau, H.F.; Rapp, I.; et al. Suitable Disinfectants with Proven Efficacy for Genetically Modified Viruses and Viral Vectors. *Viruses* **2023**, *15*, 2179. https://doi.org/10.3390/v15112179

Academic Editors: Brigitte Pützer and Ottmar Herchenröder

Received: 18 September 2023
Revised: 20 October 2023
Accepted: 23 October 2023
Published: 30 October 2023

Abstract: Viral disinfection is important for medical facilities, the food industry, and the veterinary field, especially in terms of controlling virus outbreaks. Therefore, standardized methods and activity levels are available for these areas. Usually, disinfectants used in these areas are characterized by their activity against test organisms (i.e., viruses, bacteria, and/or yeasts). This activity is usually determined using a suspension test in which the test organism is incubated with the respective disinfectant in solution to assess its bactericidal, yeasticidal, or virucidal activity. In addition, carrier methods that more closely reflect real-world applications have been developed, in which microorganisms are applied to the surface of a carrier (e.g., stainless steel frosted glass, or polyvinyl chloride (PVC)) and then dried. However, to date, no standardized methods have become available for addressing genetically modified vectors or disinfection-resistant oncolytic viruses such as the H1-parvovirus. Particularly, such non-enveloped viruses, which are highly resistant to disinfectants, are not taken into account in European standards. This article proposes a new activity claim known as "virucidal activity PLUS", summarizes the available methods for evaluating the virucidal activity of chemical disinfectants against genetically modified organisms (GMOs) using current European standards, including the activity against highly resistant parvoviridae such as the adeno-associated virus (AAV), and provides guidance on the selection of disinfectants for pharmaceutical manufacturers, laboratories, and clinical users.

Keywords: genetically modified organisms (GMOs); viral vectors; virucidal activity; suspension tests; virus inactivation; disinfection; AAV; vector

1. Introduction

In Europe, the spectrum of the efficacy of disinfection processes against viruses currently includes the following three activity levels [1,2], which are "active against enveloped viruses", "limited spectrum of virucidal activity", and "virucidal activity" (which includes both activity against enveloped viruses and a limited spectrum of virucidal activity). Depending on the risk assessment and area of application of the disinfectant (hands, surfaces, instruments, and laundry), different activity levels are required and can be declared on the basis of existing standardized test methods. For hygienic hand disinfection and surface disinfection, the required activity levels are "active against enveloped viruses", "limited spectrum of virucidal activity", and "virucidal activity". For instrument disinfection they are "active against enveloped viruses" and "virucidal activity", and for laundry disinfection it is "virucidal activity".

The above-mentioned activity levels were originally developed for the medical area. For other areas, additional terms with different content are under discussion in the technical committee (TC) 216 of the European Committee for Standardisation (CEN, Comité Européen de Normalisation in French).

In addition to the efficacy of disinfection processes against viruses, this overview draws attention to genetically modified organisms (GMOs) of viral origin. They are used in numerous laboratories for very different purposes, e.g., for basic research (viral transduction), for research on and the production of vaccines (e.g., recombinant vaccine viruses), for gene transfer (viral vectors), or for direct tumour therapy (e.g., native oncolytic viruses (OV) such as H1-Parvovirus or recombinant oncolytic viruses such as the approved 2015 Talimogene Laherparepvec based on HSV-1) [3,4]. OVs include numerous virus families that exhibit natural tropisms for specific extracellular oncogenic receptors or have been genetically engineered to display targeted tropisms for tumour cells while being restricted in their effects on normal cells [5]. Furthermore, engineered OVs often additionally express immunotherapeutic factors. In contrast to gene therapy vectors, which are used to transfer therapeutic genetic material to target tissues and cannot actively infect the host cells, OVs are less attenuated and can replicate in infected tissues [6].

The continuous progress of research in this field and the constantly evolving technical possibilities have also led to the development of new perspectives and applications for cell- and gene-based medicines. For example, viral vectors are successfully used in the production of functionally modified T cells with chimeric antigen receptors (CAR-T cells) [7] or offer promising tools for the treatment of genetic diseases. Other gene therapy treatments have been designed to treat haemophilia A, haemophilia B, or patients with inherited rare diseases [8,9]. However, in the wake of the COVID-19 pandemic, there is a public perception and a growing trend of scepticism towards research institutions regarding poor lab safety standards or even lab leaks due to inadequate biosafety levels. Viral vectors or genetically modified viruses are assigned to different genetic safety levels. Initially, this occurs independently of the type and function of the nucleic acid insertions. In Germany, the classification of viral vectors or GMOs is carried out by the Central Commission for Biological Safety (ZKBS) on the basis of a risk assessment of the GMOs or viral vectors to determine their safety for humans, animals and the environment. Viral vectors are classified as GMOs. As [10] the ZKBS points out, organisms are not only biological entities that are capable of reproducing but also transmitting genetic material. In addition to the typical GMOs of retroviral, adenoviral, or parvoviral origin, oncolytic viruses that are not genetically modified, such as H-1 parvovirus, are also the subject of this overview.

In the production and use of GMOs or native oncolytic viruses, the question arises as to which disinfectants should be used when handling them (e.g., in research, development, production, or clinical application). The selection of a suitable disinfectant cannot be derived from genetic safety classification, as this indicates potential risks to other organisms and the environment but does not consider the tolerance or sensitivity of the GMOs or native oncolytic viruses to disinfectants.

The aim of this overview is to make the various aspects of efficacy testing and the spectrum of antiviral activity in the selection of disinfectants transparent for supervisory and monitoring authorities, biosafety officers, and operators of genetic engineering facilities. In this context, we primarily consider activity against viruses and address the anticipated changes caused by the European biocide legislation.

2. Efficacy of Disinfectants Depends on Viral Structure

The infectivity of virus particles is essentially determined by the nature of the viral envelope or the viral capsid (in the case of non-enveloped viruses). Their basic biological structure can be derived from their viral families or from the respective "donor and recipient organisms" that determine their biochemical and physical properties. Since disinfectants initially attack the viral envelope and/or the capsid of virus particles, the selection of the disinfectant for recombinant viruses or viral vectors (such as chimeric viruses) depends on the origin of the viral envelope or the viral capsid. Enveloped recombinant viruses such as viral vectors (GMOs) or "infectious" nucleic acids are unique cases since GMOs often contain modified viral proteins or other additionally expressed proteins in their viral membrane. Nevertheless, these genetically modified proteins are not expected to significantly alter the stability of the lipid envelope when exposed to disinfectants.

Comparative studies with a large number of enveloped viruses have shown that vaccinia virus—the model virus for enveloped viruses—still has the highest stability and is thus a good benchmark [11–14] and a well-suited model virus for enveloped viruses. Its stability might be higher than that of others, as the cell-associated enveloped virus (CEV) and the extracellular enveloped virus (EEV) are both surrounded by two membrane layers, while the intracellular enveloped virus (IEV) has three envelopes [15]. Aside from the structural proteins and lipids, viruses contain nucleic acid that can still be considered as infectious. However, not all disinfectants destroy or inactivate the nucleic acids (DNA or RNA) contained in the virus particles. Certain virus groups (e.g., vaccinia viruses, herpes viruses, adenoviruses, simian virus 40 (SV40), picornaviruses, or parvoviruses) possess so-called "infectious" virus genomes. In principle, progeny viruses can develop from these genomes if they enter or are introduced into suitable host cells. However, eukaryotic cells do not have natural mechanisms for the uptake of free (non-encapsulated) viral genomes. Therefore, extremely high concentrations and special technical procedures are required to introduce free nucleic acids into cells in sufficient quantities for productive replication. Therefore, the infectivity of isolated viral genomes plays only a minor role compared to the infectivity of complete viral particles in the context of the potential risk of infection they pose.

According to Klein and DeForest [16], the activity level for the declaration of disinfectants is derived from the virus structure [1,17]. Disinfectants that are "active against enveloped viruses" are sufficient for use against all enveloped viruses. Within the group of non-enveloped viruses, adenoviruses, noroviruses, and rotaviruses form the group of lipophilic non-enveloped viruses, for which products with the activity level *"limited spectrum of virucidal activity"* are used. For the more hydrophilic non-enveloped virus agents, the activity level *"virucidal activity"* is required. Correspondingly, the same applies mutatis mutandis to GMOs: GMOs or native oncolytic viruses based on enveloped viruses can be inactivated using disinfectants that are *"active against enveloped viruses"*. In the case of more lipophilic non-enveloped viruses—rotaviruses, noroviruses, adenoviruses or GMOs derived from them—disinfectants are required with at least the "limited spectrum of viruci-dal activity" activity level [2]. However, some GMOs or native oncolytic viruses based on hydrophilic non-enveloped viruses such as parvoviruses are not inactivated by products with a "virucidal activity" level. Thus, we require a disinfectant with activity against parvoviruses [18,19]. Therefore, the German Committee of Virus Disinfection (DVV/GfK) and the Disinfectant Commission of the VAH (Association of Applied Hygiene) [20] have implemented a new virucidal activity level *"virucidal activity PLUS"*, requiring a practical test against parvoviridae such as those described in the German guideline of 2012 for the quantitative evaluation of the virucidal activity of chemical disinfectants on non-porous surfaces [21].

3. Test Methods for the Declaration of Activity against Viruses

For the evaluation of the virucidal efficacy of disinfectants, a two-step procedure is provided at a national [22–25] and a European level [26–31]. In this model, the efficiency must be confirmed in quantitative suspension tests (phase 2, step 1) followed by tests under practical conditions (phase 2, step 2). In suspension tests, virus particles in solution are mixed with disinfectant plus interfering substances to simulate organic solutions and the remaining titres determined via endpoint titration. This generally determines whether a substance reduces viral infectivity and is thus active against viruses. Suspension tests are well suited for use in preliminary investigations but are carried out with a considerable excess of disinfectants, thus simulating conditions that are usually not present in practice. Therefore, the use of suspension tests often results in activities at lower concentrations and/or shorter exposure times compared to the tests performed under practical conditions (phase 2, step 2) and thus might overestimate efficacy.

Therefore, the tests performed under practical conditions must correspond to the situation in which the disinfectants are to be used, e.g., for hand hygiene, spraying surfaces, and immersing instruments. However, to ensure the efficacy of disinfectants, the test methods must not only simulate the type of application but should also consider possible organic contamination. At present, coordinated phase 2, step 2 tests are only used at the European level to assess antiviral efficacy for surface disinfection via the spray method (application without mechanical action, EN 16777) [28] and for chemical and chemo-thermal instrument disinfection via the immersion method (EN 17111) [29]. However, this is not the case for surface disinfection via mechanical action or hand disinfection. For this reason, viral test methods are currently being developed for application to the various areas of application, which also include the different areas of effect, by specifying the respective test viruses.

Specifically, hand disinfectants must be tested on the hands of volunteers and the achieved reduction must be compared to that achieved using a reference product tested in parallel on the same study subjects [32]. For surface disinfectants, the tests must con-sider whether the surfaces are wiped (i.e., disinfection is carried out with a mechanical component) or whether the product is sprayed on (without mechanical action). Instrument disinfectants are normally analyzed for their efficacy with immersion procedures.

The desired efficacy of products can only be guaranteed if they are also used as speci-fied in the respective application-specific test. To date, test methods have been developed for the medical and veterinary fields and for the food sector. Special test methods have not yet been developed for the laboratory sector. If surfaces or the hands of employees are to be disinfected in laboratories, however, products with proven efficacy can be used in accordance with the test methods appropriate to the medical or veterinary field. But if the products are to be used in a way that deviates from the established applications, additional tests should be performed that largely take into account the desired use.

If there are no official national lists for disinfectants with proven virucidal activity, or if the existing lists do not include virus-active products (i.e., [33–35]), the user must rely on the label from the manufacturer.

In conclusion, the claim or labelling of products as being *"active against enveloped viruses"*, or as displaying *"limited spectrum of virucidal activity"* and *"virucidal activity"*, does not mean that products have undergone uniform tests and have actually met the requirements for the confirmation of efficacy, especially in the application setting.

4. Efficacy Level of Disinfectants and Surrogate Viruses for Disinfectant Testing in Germany and Europe

In Germany, virucidal claims such as *"active against enveloped viruses"*, *"limited spectrum of virucidal activity"*, or *"virucidal activity"* are in principle harmonized with those in the rest of Europe. However, the proof of disinfectant efficacy in Germany is based on more extensive tests than those used in Europe as additional surrogate viruses (see tables) and reproducibility (i.e., more test runs) are required. Table 1a–c list the requirements for virus-active products in the medical field.

Table 1. (**a**) Overview of disinfectant activity level *"active against enveloped viruses"* and associated test methods for the human medical area. (**b**) Overview of disinfectant activity level *"limited spectrum of virucidal activity"* and associated test methods for the human medical area. (**c**) Overview of disinfectant activity level *"virucidal activity"* and associated test methods for the human medical area.

(a)		
Active against	Enveloped viruses (e.g., HBV, HCV, HIV, influenza viruses, herpes viruses)	
Basis of the declaration	**Test methods and associated test viruses**	
	DVV/VAH	European standards
Quantitative suspension test (Phase 2/Step 1)	DVV/RKI Guideline [22,23] EN 14476 [27]	EN 14476 [27]
	Vaccinia virus * BVDV (oxidative products)	Vaccinia virus *
Practical test: Surface disinfection (Phase 2/Step 2)	DVV Guideline [21] EN 16777 [28]	EN 16777 [28]
	Vaccinia virus *	Vaccinia virus *
Practical test: Instrument disinfectio (Phase 2/Step 2)	EN 17111 [29] (products for pre-cleaning with a combined cleaner/disinfectant)	
	Vaccinia virus *	Vaccinia virus *

(b)		
Active against	Enveloped viruses and lipophilic non-enveloped viruses (adenoviruses, noroviruses and rotaviruses)	
Basis of the declaration	**Test methods and associated test viruses**	
	DVV/VAH	European standards
Quantitative suspension test (Phase 2/Step 1)	DVV/RKI Guideline [22,23] EN 14476 [27]	EN 14476 [27]
	Adenovirus MNV	Adenovirus MNV
Practical test: Surface disinfection (Phase 2/Step 2)	DVV Guideline [21] EN 16777 [28]	EN 16777 [28]
	Adenovirus MNV	Adenovirus MNV
Practical test: Instrument disinfection (Phase 2/Step 2)	Claim cannot be issued	

Table 1. *Cont.*

	(c)	
Active against	Enveloped and non-enveloped viruses (except parvoviridae such as AAV, HAV, HEV)	Enveloped and non-enveloped viruses (except parvoviridae such as AAV, HAV, HEV, papillomaviruses)
Basis of the declaration	**Test methods and associated test viruses**	
	DVV/VAH	European standards
Quantitative suspension test (Phase 2/Step 1)	DVV/RKI Guideline [22,23] EN 14476 [27]	EN 14476 [27]
	Adenovirus MNV Poliovirus [†] SV40 [36] MVM (instrument disinfection $\geq 40\,°C$)	Adenovirus MNV Poliovirus MVM (instrument disinfection $\geq 40\,°C$)
Practical test: Surface disinfection (Phase 2/Step 2)	DVV Guideline [21] EN 16777 [28]	EN 16777 [28]
	Adenovirus MNV	Adenovirus MNV
Practical test: Instrument disinfection (Phase 2/Step 2)	EN 17111 [29]	
	Adenovirus (<40 °C) MNV (<40 °C) SV40 (<40 °C) MVM ($\geq$40 °C)	Adenovirus (<40 °C) MNV (<40 °C) MVM ($\geq$40 °C)

* Modified vaccinia virus Ankara (MVA) or vaccinia virus strain Elstree. HBV, hepatitis B virus; HCV, hepatitis C virus; HIV, human immunodeficiency virus; BVDV, bovine viral diarrhoea virus. MNV, murine norovirus. AAV, adeno-associated virus; HAV, hepatitis A virus; HEV, hepatitis E virus; SV40, simian virus 40; MVM, murine parvovirus (minute virus of mice; rodent protoparvovirus 1). [†] Although poliovirus is usually the most resistant test virus in the suspension test, it cannot be used for practical tests due to its loss in titre after drying [19,37].

As shown in Table 1a (*active against enveloped viruses*) and Table 1b (*limited spectrum of virucidal activity*), there is little to no difference in the test viruses. While the European method [27] uses vaccinia virus as the test virus in the quantitative suspension test (phase 2/step 1), the German methods [22,23] additionally use bovine viral diarrhoea virus (BVDV) when testing oxidative products with the activity level *"active against enveloped viruses"*.

In terms of *"virucidal activity"* (Table 1c), enteroviruses (EV) such as EV71 and EVD68 and poliovirus are covered by both the German and the European methods. However, this claim does not cover the highly resistant outbreak viruses hepatitis A (HAV) and hepatitis E (HEV). Also not included are parvoviridae, including adeno-associated viruses (AAV) used in gene therapy. Recently, clusters of AAV type 2 were globally reported to occur in children with acute severe hepatitis in 2022 [38,39]. An additional test virus simian virus (SV) 40 (as surrogate for human papillomavirus (HPV)) in the DVV/RKI test method makes another relevant clinical difference as this oncogenic virus is not included in the European standards.

Products with activity against enveloped viruses remain the correct choice for all indications where viruses are enveloped. Additionally, the most common non-enveloped viruses need to be adequately targeted in the area of the application of surface disinfection, including the disinfection of medical devices.

Since parvoviruses are particularly resistant to disinfectants, products that are to be used against parvoviridae, including AAV, native oncolytic parvoviruses, HAV and HEV, must also have proof of activity against parvoviruses. This requires that additional tests be performed with parvoviruses. However, the current European standards for testing under practical conditions [28,29] and drafts of new standards published in recent years do not include murine parvovirus (MVM) as a test organism mandated for testing under the practical conditions of surface and instrument disinfectants (with the exception of chemo-thermal instrument disinfection).

The routine use of murine parvovirus as a test organism is problematic because the active ingredient concentration or exposure time must be significantly increased to demonstrate sufficient efficacy against parvoviruses and those active ingredients are limited.

Therefore, for routine disinfection, the previous virucidal claim remains valid, but a new claim needs to be implemented in the EU to define virucidal activity against highly resistant viruses. As in the case of Germany [17], this could be the implementation of either *"virucidal activity PLUS"* or *"high level virucidal active"* for highly resistant viruses.

Table 2 presents a proposal for a new virucidal activity claim (*"virucidal activity PLUS"*) as already introduced in Germany.

Table 2. Proposal for a disinfectant activity level *"virucidal activity PLUS"* and associated test methods for the human medical area.

Active against	enveloped and non-enveloped viruses (incl. parvoviridae such as AAV, HAV, HEV, papillomaviruses)	Claim cannot be issued yet
Basis of the declaration	**Test methods and associated test viruses**	
	DVV/VAH	European standards
Quantitative suspension test (phase 2/step 1)	DVV/RKI Guideline [22,23] EN 14476 [27]	n.a.
	Adenovirus MNV Poliovirus SV40	n.a.
Practical test: Surface disinfection (phase 2/step 2)	DVV Guideline [21]	n.a.
	Adenovirus MNV MVM	n.a.

AAV, adeno-associated virus; HAV, hepatitis A virus; HEV, hepatitis E virus; MNV, murine norovirus; SV40, simian virus 40; MVM, murine parvovirus (minute virus of mice; rodent protoparvovirus 1). n.a. not applicable.

5. Disinfectant for Oncolytic Viruses or Viral Vectors in Therapy

For users of disinfectants, it is not easy to select suitable conditions or products based on the information available regarding the various test methods. For example, parvoviruses are intrinsically resistant to alcoholic formulations [18,37,40]. This high tolerance to alcohols was also found for HEV [37] and HAV [41,42]. Therefore, hand disinfectants are not available against parvovirus-based GMOs or native oncolytic parvoviruses, nor are they produced for HAV. For hand hygiene, wearing protective gloves and hand washing is recommended such as in the case of bacterial spores [43,44]. Note that hand washing can only mechanically reduce the viral load from the hands as soaps are not active against these resilient viruses. In contrast, for products based on peracetic acid, aldehydes, oxygen scavengers, and chloramine-T, activity against murine parvovirus was confirmed in practical tests in various testing laboratories, leading some to claim *"virucidal activity PLUS"* for surface disinfection without mechanical action [19,45].

For users of disinfectants, it is not easy to select suitable conditions or products based on the information available regarding the various test methods. For example, parvoviruses are intrinsically resistant to alcoholic formulations [18,37,40]. This high tolerance to alcohols was also found for HEV [37] and HAV [41,42]. Therefore, hand disinfectants are not available against parvovirus-based GMOs or native oncolytic parvoviruses, nor are they produced for HAV. For hand hygiene, wearing protective gloves and hand washing is recommended such as in the case of bacterial spores [43,44]. Note that hand washing can only mechanically reduce the viral load from the hands as soaps are not active against

6. Future Aspects Due to European Biocide Regulation

The implementation of the European Biocide Regulation No. 528/2012 [46] is unsettling the monitoring authorities and genetic engineering facilities in Europe, as many well-established products may or will no longer be available on the market. Nowadays, all biocidal products require authorization before they can be placed on the market, and the active substances contained in that biocidal product must be previously approved. The Biocidal Products Regulation (BPR) classifies the biocidal products into 22 biocidal product-types, with surface disinfectants belonging to product type (PT) 2 "Disinfectants and algaecides not intended for direct use on humans or animals". This product type contains, in addition to disinfectants used in the medical field or in laboratories, products for completely different applications such as the treatment of bath water, air conditioning systems, or soils (i.e., sand, soil). However, such a collective group makes it difficult to clearly delineate the areas of application and thus to select suitable products, e.g., for laboratories.

At present, the majority of disinfectants are marketable on the basis of transitional regulations, but they do not yet have authorization as biocidal products. The authorization of biocidal products includes not only efficacy testing but also the assessment of risks to humans and the environment and is carried out in a two-step process. First, the active substances contained in the biocidal products are evaluated using a European procedure. Active substances that have been positively evaluated ("approved") are included in the so-called Union list. This list results in the deadlines during which the authorization of the product must be applied for. If no application is submitted, product marketability ends after a fixed period. Against this background, impacts on the availability of disinfectants and potentially on their operational conditions are already present or expected as the majority of disinfectants are subject to this regulation. As an example, Greek authorities have proposed to classify ethanol, which has been reliable as a disinfectant in everyday use, as a "CMR" (carcinogenic, mutagenic, and toxic for reproduction) [47], which would severely affect its use and thus eliminate a safe, available, and inexpensive ingredient.

7. Conclusions

As has been shown, a variety of test methods exist at both the national and European level that can be used for the evaluation of disinfectants with declared virucidal activity. In addition to the methods and claims described here for use in the human medical area, there are also existing national and European methods that can be applied to the veterinary field or food, industrial, domestic, and institutional issues. However, there are still no methods or specifications for laboratories working with viral GMOs or oncolytic viruses available. Additionally, highly resistant parvoviridae, HAV and HEV, are not yet considered for surface disinfection in European standards. Also, practical test methods for virucidal efficacy are currently not available for all applications such as surface, instrument, textile, or hand disinfection, although much has been implemented in recent years [48]. However, as specified in EN 14885 [26], further tests under additional conditions such as test organisms, temperature, contact time, and interfering substances should be carried out according to the claimed use of a disinfectant. Therefore, EN 14885 enables tests with a highly resistant test virus such as parvovirus, using a modified standard to provide an opportunity to test disinfectants against resistant GMO or oncolytic parvoviruses. This could be the basis for the new claim *"virucidal PLUS"*.

In Europe, the labelling of the virucidal efficacy of products by the manufacturer is not uniformly regulated for medicinal products and biocides [49]. In both cases, however, only the area of application (e.g., food, medical, institutional) and the instructions for use must be indicated on the product label, but not the test methods (if available) on which the virucidal activity level is based. Unfortunately, as long as no practical tests for surface disinfection with mechanical action (four field test) or hand disinfection are approved at the European level, a manufacturer can state and advertise a *"virucidal activity"* of such products even though it has only undergone a quantitative suspension test.

Taken together, the recommendation for and selection of suitable disinfectants with proven efficacy for GMO and viral vectors is complex. Since the evaluation of laboratory test reports for the selection or comparison of different possible disinfectants with proven virucidal efficacy requires special knowledge and extensive experience, access to accordingly reviewed lists of disinfectants with proven efficacy for the different levels may be helpful for both the user and for supervisory or monitoring authorities.

Author Contributions: M.E. and M.S. conceptualize, wrote and edited the manuscript, I.S., J.B., F.B., J.G., N.H., J.A.M., H.F.R., I.R., S.R., E.S., J.S., H.F. and P.Z. conceptualize and critically revised the final version of the manuscript. All authors have read and agreed to the published version of the manuscript.

Funding: This research received no external funding.

Institutional Review Board Statement: Not applicable.

Informed Consent Statement: Not applicable.

Data Availability Statement: Not applicable.

Conflicts of Interest: The author I.R. is an employee of the company Boehringer Ingelheim Therapeutics GmbH and declares that no conflicts of interest exist regarding the data presented here. All other authors declare no conflict of interest.

References

1. Tarka, P.; Nitsch-Osuch, A. Evaluating the Virucidal Activity of Disinfectants According to European Union Standards. *Viruses* **2021**, *13*, 534. [CrossRef]
2. Eggers, M.; Schwebke, I.; Suchomel, M.; Fotheringham, V.; Gebel, J.; Meyer, B.; Morace, G.; Roedger, H.J.; Roques, C.; Visa, P.; et al. The European tiered approach for virucidal efficacy testing—Rationale for rapidly selecting disinfectants against emerging and re-emerging viral diseases. *Eurosurveillance* **2021**, *26*, 2000708. [CrossRef]
3. Hajda, J.; Leuchs, B.; Angelova, A.L.; Frehtman, V.; Rommelaere, J.; Mertens, M.; Pilz, M.; Kieser, M.; Krebs, O.; Dahm, M.; et al. Phase 2 Trial of Oncolytic H-1 Parvovirus Therapy Shows Safety and Signs of Immune System Activation in Patients With Metastatic Pancreatic Ductal Adenocarcinoma. *Clin. Cancer Res.* **2021**, *27*, 5546–5556. [CrossRef]
4. Chulpanova, D.S.; Solovyeva, V.V.; Kitaeva, K.V.; Dunham, S.P.; Khaiboullina, S.F.; Rizvanov, A.A. Recombinant Viruses for Cancer Therapy. *Biomedicines* **2018**, *6*, 94. [CrossRef]
5. Jhawar, S.R.; Thandoni, A.; Bommareddy, P.K.; Hassan, S.; Kohlhapp, F.J.; Goyal, S.; Schenkel, J.M.; Silk, A.W.; Zloza, A. Oncolytic Viruses—Natural and Genetically Engineered Cancer Immunotherapies. *Front. Oncol.* **2017**, *7*, 202. [CrossRef]
6. Bezeljak, U. Cancer gene therapy goes viral: Viral vector platforms come of age. *Radiol. Oncol.* **2022**, *56*, 1–13. [CrossRef]
7. Suerth, J.D.; Schambach, A.; Baum, C. Genetic modification of lymphocytes by retrovirus-based vectors. *Curr. Opin. Immunol.* **2012**, *24*, 598–608. [CrossRef]
8. Long, B.R.; Veron, P.; Kuranda, K.; Hardet, R.; Mitchell, N.; Hayes, G.M.; Wong, W.Y.; Lau, K.; Li, M.; Hock, M.B.; et al. Early Phase Clinical Immunogenicity of Valoctocogene Roxaparvovec, an AAV5-Mediated Gene Therapy for Hemophilia A. *Mol. Ther.* **2021**, *29*, 597–610. [CrossRef]
9. Keeler, A.M.; Flotte, T.R. Recombinant Adeno-Associated Virus Gene Therapy in Light of Luxturna (and Zolgensma and Glybera): Where Are We, and How Did We Get Here? *Annu. Rev. Virol.* **2019**, *6*, 601–621. [CrossRef]
10. Available online: https://zag.bvl.bund.de/organismen/index.jsf?dswid=8616&dsrid=287 (accessed on 20 June 2023).
11. Sauerbrei, A.; Schacke, M.; Glück, B.; Bust, U.; Rabenau, H.F.; Wutzler, P. Does limited virucidal activity of biocides include duck hepatitis B virucidal action? *BMC Infect. Dis.* **2012**, *12*, 276. [CrossRef]
12. Eggers, M.; Eickmann, M.; Kowalski, K.; Zorn, J.; Reimer, K. Povidone-iodine hand wash and hand rub products demonstrated excellent in vitro virucidal efficacy against Ebola virus and modified vaccinia virus Ankara, the new European test virus for enveloped viruses. *BMC Infect. Dis.* **2015**, *15*, 375. [CrossRef] [PubMed]
13. Eggers, M.; Eickmann, M.; Zorn, J. Rapid and effective virucidal activity of povidone-iodine products against Middle East Respiratory Syndrome Coronavirus (MERS-CoV) and Modified Vaccinia Virus Ankara (MVA). *Infect. Dis. Ther.* **2015**, *4*, 491–501. [CrossRef] [PubMed]
14. Siddharta, A.; Pfaender, S.; Vielle, N.J.; Dijkman, R.; Friesland, M.; Becker, B.; Yang, J.; Engelmann, M.; Todt, D.; Windisch, M.P.; et al. Virucidal activity of World Health Organization-recommended formulations against enveloped viruses, including Zika, Ebola, and emerging coronaviruses. *J. Infect. Dis.* **2017**, *215*, 902–906. [CrossRef] [PubMed]
15. Smith, G.L.; Vanderplasschen, A.; Law, M. The formation and function of extracellular enveloped vaccinia virus. *J. Gen. Virol.* **2002**, *83 Pt 12*, 2915–2931. [CrossRef] [PubMed]
16. Klein, M.; De Forest, A. Principles of viral inactivation. In *Disinfection, Sterilisation and Preservation*, 3rd ed.; Block, S.S., Ed.; Lea & Febiger: Philadelphia, PA, USA, 1983; pp. 422–434.

17. Schwebke, I.; Eggers, M.; Gebel, J.; Geisel, B.; Glebe, D.; Rapp, I.; Steinmann, J.; Rabenau, H.F. Prüfung und Deklaration der Wirksamkeit von Desinfektionsmitteln gegen Viren zur Anwendung im human-medizinischen Bereich Stellungnahme des Arbeitskreises Viruzidie beim Robert Koch-Institut. *Bundesgesundheitsblatt Gesundheitsforschung Gesundheitsschutz* **2017**, *60*, 353–363. [CrossRef] [PubMed]

18. Eterpi, M.; McDonnell, G.; Thomas, V. Disinfection efficacy against parvoviruses compared with reference viruses. *J. Hosp. Infect.* **2009**, *73*, 64–70. [CrossRef]

19. Rabenau, H.F.; Steinmann, J.; Rapp, I.; Schwebke, I.; Eggers, M. Evaluation of a Virucidal Quantitative Carrier Test for Surface Disinfectants. *PLoS ONE* **2014**, *9*, e86128. [CrossRef]

20. Gemeinsame Mitteilung der Desinfektionsmittel-Kommission im VAH und der Kommission Virusdesinfektion von DVV und GfV. Harmonisierung der Anforderungen an die viruzide Wirksamkeit von chemischen Flächendesinfektionsverfahren im praxisnahen Test. *Hyg. Med.* **2023**, *48*, 69–72.

21. Rabenau, H.F.; Schwebke, I.; Steinmann, J.; Eggers, M.; Rapp, I.; Neumann-Haefelin, D. Quantitative test for the evaluation of virucidal activity of chemical disinfectants on non-porous surfaces. *Hyg. Med.* **2012**, *37*, 459–466.

22. Rabenau, H.F.; Schwebke, I.; Blümel, J.; Eggers, M.; Glebe, D.; Rapp, I.; Sauerbrei, A.; Steinmann, E.; Steinmann, J.; Willkommen, H.; et al. Guideline of the German Association for Combating Viral Diseases (DVV) e.V. and the Robert Koch Institute (RKI) on testing chemical disinfectants for efficacy against viruses in human medicine, version of 1 December 2014. *Bundesgesundheitsblatt Gesundheitsforschung Gesundheitsschutz* **2015**, *58*, 493–504. [CrossRef]

23. 2nd Communication of the Expert Committee on Virus Disinfection of the DVV/GfV and the RKI on the examination temperature when testing chemical or chemothermal instrument disinfection procedures in accordance with the DVV/RKI Guideline in the version of 01.12.2014. *Bundesgesundheitsblatt Gesundheitsforschung Gesundheitsschutz* **2015**, *58*, 888.

24. DVG Test Methods. Available online: http://www.desinfektion-dvg.de/index.php?id=1810 (accessed on 20 October 2023).

25. Disinfectants Commission in the Association for Applied Hygiene (VAH). Requirements and methods for VAH certification of chemical disinfection procedures. *Loose-Leaf Collection*, 1 September 2022.

26. *EN 14885:2023*; Chemical Disinfectants and Antiseptics. Application of European Standards for Chemical Disinfectants and Antiseptics. CEN—Comité Européen de Normalisation: Brussels, Belgium, 2023.

27. *EN 14476:2013 + A2:2019*; Chemical Disinfectants and Antiseptics—Quantitative Suspension Test for the Determination of Virucidal Activity in Human Medicine—Test Method and Requirements (Phase 2, Step 1). CEN—Comité Européen de Normalisation: Brussels, Belgium, 2019.

28. *EN 16777:2019*; Chemical Disinfectants and Antiseptics—Quantitative Test on Non-Porous Surfaces without Mechanical Action to Determine Virucidal Activity in Human Medicine—Test Method and Requirements (Phase 2, Step 2). CEN—Comité Européen de Normalisation: Brussels, Belgium, 2019.

29. *EN 17111:2018*; Chemical Disinfectants and Antiseptics—Quantitative Carrier Test for Virucidal Activity for Instruments Used in the Field of Human Medicine—Test Method and Requirements (Phase 2, Step 2). CEN—Comité Européen de Normalisation: Brussels, Belgium, 2018.

30. *EN 14675: 2015*; Chemical Disinfectants and Antiseptics—Quantitative Suspension Test for the Determination of Virucidal Activity of Chemical Disinfectants and Antiseptics for Veterinary Use—Test Method and Requirements (Phase 2, Step 1). CEN—Comité Européen de Normalisation: Brussels, Belgium, 2015.

31. *EN 17122: 2020*; Chemical Disinfectants and Antiseptics—Quantitative Surface Test for the Determination of Virucidal Activity of Chemical Disinfectants and Antiseptics for Veterinary Use on Non-Porous Surfaces—Test Method and Requirements (Phase 2, Step 2). CEN—Comité Européen de Normalisation: Brussels, Belgium, 2020.

32. Eggers, M.; Suchomel, M. In-vivo efficacy of alcohol-based hand rubs against noroviruses: A novel standardized European test method simulating practical conditions. *J. Hosp. Infect.* **2023**, *135*, 186–192. [CrossRef] [PubMed]

33. Disinfectant List of the VAH. Available online: https://vah-online.de/en/desinfektionsmittel-liste (accessed on 20 October 2023).

34. Expertisenverzeichnis der Österreichischen Gesellschaft für Hygiene, Mikrobiologie und Präventivmedizin (ÖGHMP). Available online: https://expertisen.oeghmp.at/ (accessed on 20 October 2023).

35. Liste der vom BAG zugelassenen Desinfektionsmittel zur Bekämpfung von Influenza und Coronaviren(BAG), Schweiz. Available online: https://echa.europa.eu/documents/10162/29202019/switzerland_list_auth_disinfectant_products_en.pdf/f96e52fc-6dc6-9a2b-c770-3833b5000043 (accessed on 20 October 2023).

36. Miyazawa, D. Potential mechanisms by which adeno-associated virus type 2 causes unexplained hepatitis in children. *J. Med Virol.* **2022**, *94*, 5623–5624. [CrossRef] [PubMed]

37. Servellita, V.; Gonzalez, A.S.; Lamson, D.M.; Foresythe, A.; Huh, H.J.; Bazinet, A.L.; Bergman, N.H.; Bull, R.L.; Garcia, K.Y.; Goodrich, J.S.; et al. Adeno-associated virus type 2 in US children with acute severe hepatitis. *Nature* **2023**, *617*, 574–580. [CrossRef]

38. Hufbauer, M.; Wieland, U.; Gebel, J.; Steinmann, J.; Akgül, B.; Eggers, M. Inactivation of Polyomavirus SV40 as Surrogate for Human Papillomaviruses by Chemical Disinfectants. *Viruses* **2021**, *13*, 2207. [CrossRef]

39. Steinmann, J.; Eggers, M.; Rapp, I.; Todt, D.; Steinmann, E.; Brill, F.H.H.; Schwebke, I. Evaluation of the substitution of poliomyelitis virus for testing virucidal activities of instrument and surface disinfection. *J. Hosp. Infect.* **2022**, *122*, 60–63. [CrossRef]

40. Behrendt, P.; Friesland, M.; Wißmann, J.-E.; Kinast, V.; Stahl, Y.; Praditya, D.; Hueffner, L.; Nörenberg, P.M.; Bremer, B.; Maasoumy, B.; et al. Hepatitis E virus is highly resistant to alcohol-based disinfectants. *J. Hepatol.* **2022**, *76*, 1062–1069. [CrossRef]
41. Wolff, M.; Schmitt, J.; Rahaus, M.; König, A. Hepatitis A virus: A test method for virucidal activity. *J. Hosp. Infect.* **2001**, *48* (Suppl. SA), S18–S22. [CrossRef]
42. Kampf, G. Efficacy of ethanol against viruses in hand disinfection. *J. Hosp. Infect.* **2018**, *98*, 331–338. [CrossRef]
43. Recommendation of the Commission for Hospital Hygiene and Infection Prevention (KRINKO) at the Robert Koch Institute (RKI): Hygienemaßnahmen bei Clostridioides difficile-Infektion (CDI). *Bundesgesundheitsblatt Gesundheitsforschung Gesundheitsschutz* **2019**, *62*, 906–923. [CrossRef]
44. Recommendation of the Commission for Hospital Hygiene and Infection Prevention (KRINKO) at the Robert Koch Institute (RKI): Händehygiene in Einrichtungen des Gesundheitswesens. *Bundesgesundheitsblatt Gesundheitsforschung Gesundheitsschutz* **2016**, *59*, 1189–1220. [CrossRef] [PubMed]
45. Robert Koch Institute. List of disinfectants and disinfection methods tested and approved by the Robert Koch Institute. *Bundesgesundheitsblatt* **2017**, *60*, 1274–1297.
46. Regulation (EU) No 528/2012 of the European Parliament and of the Council of 22 May 2012 Concerning the Making Available on the Market and Use of Biocidal Products. Official Journal of the European Union L; 167, pp. 1–123. Available online: https://eur-lex.europa.eu/LexUriServ/LexUriServ.do?uri=OJ:L:2012:167:0001:0123:en:PDF (accessed on 20 October 2023).
47. Kramer, A.; Arvand, M.; Christiansen, B.; Dancer, S.; Eggers, M.; Exner, M.; Müller, D.; Mutters, N.T.; Schwebke, I.; Pittet, D. Ethanol is indispensable for virucidal hand antisepsis: Memorandum from the alcohol-based hand rub (ABHR) Task Force, WHO Collaborating Centre on Patient Safety, and the Commission for Hospital Hygiene and Infection Prevention (KRINKO), Robert Koch Institute, Berlin, Germany. *Antimicrob. Resist. Infect. Control.* **2022**, *11*, 93. [CrossRef]
48. *prEN 17430:2022*; Chemical Disinfectants and Antiseptics—Hygienic Handrub Virucidal—Test Method and Requirements (Phase 2/Step 2). CEN—Comité Européen de Normalisation: Brussels, Belgium, 2022.
49. Lilienthal, N.; Baumann, A.; Respondek, V.; Hübner, N.-O.; Schwebke, I. Quo vadis Händedesinfektionsmittel: Arzneimittel oder Biozidprodukt? *Epidemiol. Bull.* **2021**, *17*, 5–13. [CrossRef]

Review

Viral Vectors in Gene Therapy: Where Do We Stand in 2023?

Kenneth Lundstrom

PanTherapeutics, CH1095 Lutry, Switzerland; lundstromkenneth@gmail.com

Abstract: Viral vectors have been used for a broad spectrum of gene therapy for both acute and chronic diseases. In the context of cancer gene therapy, viral vectors expressing anti-tumor, toxic, suicide and immunostimulatory genes, such as cytokines and chemokines, have been applied. Oncolytic viruses, which specifically replicate in and kill tumor cells, have provided tumor eradication, and even cure of cancers in animal models. In a broader meaning, vaccine development against infectious diseases and various cancers has been considered as a type of gene therapy. Especially in the case of COVID-19 vaccines, adenovirus-based vaccines such as ChAdOx1 nCoV-19 and Ad26.COV2.S have demonstrated excellent safety and vaccine efficacy in clinical trials, leading to Emergency Use Authorization in many countries. Viral vectors have shown great promise in the treatment of chronic diseases such as severe combined immunodeficiency (SCID), muscular dystrophy, hemophilia, β-thalassemia, and sickle cell disease (SCD). Proof-of-concept has been established in preclinical studies in various animal models. Clinical gene therapy trials have confirmed good safety, tolerability, and therapeutic efficacy. Viral-based drugs have been approved for cancer, hematological, metabolic, neurological, and ophthalmological diseases as well as for vaccines. For example, the adenovirus-based drug Gendicine® for non-small-cell lung cancer, the reovirus-based drug Reolysin® for ovarian cancer, the oncolytic HSV T-VEC for melanoma, lentivirus-based treatment of ADA-SCID disease, and the rhabdovirus-based vaccine Ervebo against Ebola virus disease have been approved for human use.

Keywords: viral vector; gene therapy; cancer; chronic disease; vaccines; preclinical models; clinical trials; approved drugs

Citation: Lundstrom, K. Viral Vectors in Gene Therapy: Where Do We Stand in 2023? *Viruses* **2023**, *15*, 698. https://doi.org/10.3390/v15030698

Academic Editors: Ottmar Herchenröder and Brigitte Pützer

Received: 18 January 2023
Revised: 23 February 2023
Accepted: 2 March 2023
Published: 7 March 2023

1. Introduction

Gene therapy has been defined as the supplementation, correction, or modification of malfunctional genes by functional equivalents for therapeutic correction of the absence or reduced levels of gene expression activity [1]. A broader definition considers oligonucleotide- [2] and RNA interference (RNAi)-based gene silencing [3], immunotherapy, especially cancer immunotherapy, and vaccine development as gene therapy [4]. More recently, stem cell technologies [5], chimeric antigen receptor (CAR) T-cell therapy [6], and Clustered Regularly Interspaced Short Palindromic (CRISPR) [7], providing unprecedent possibilities for gene replacement and gene editing, have also received gene therapy status.

Viral vectors have played a central role in gene therapy because of their superior gene delivery capacity compared to non-viral vectors. Moreover, the virus-based transgene expression, depending on the needs, for both short-term and long-term duration can be achieved. For example, for cancer gene therapy, short-term high-level transgene expression is advantageous, whereas for chronic diseases such as hemophilia, long-term transgene expression is necessary. However, the application of viral vectors requires a higher biosafety level compared to non-viral vectors due to the risk of spread of virus progeny, especially in the case of using replication-competent and oncolytic viruses. Other factors of importance are the regulation and termination of transgene expression. The history of gene therapy has been impacted by some tragic events, which was a setback for its proclaimed status as "the medicine of the future". Although the retrovirus-based treatment of children with X-linked severe combined immunodeficiency (SCID-X1) was successful, the insertion of the therapeutic gene into the LMO2 proto-oncogene region of the genome led to the development of leukemia in a few patients [8]. In another case, inadequate planning and execution of clinical protocols for adenovirus-based treatment of the non-life-threatening ornithine transcarbamylase (OTC) deficiency resulted in the death of an 18-year-old patient [9]. These two incidents in the 1990s had a dramatic impact on the field of gene therapy, which put it more or less on hold until a renaissance occurred in recent years. However, during the years, efforts to develop more efficient and safer viral vectors continued, which has facilitated the return of gene therapy to the front of innovative drug and vaccine development. Although enormous progress has also been made in the area of non-viral vectors and their applications in gene therapy, the focus in this review is uniquely on viral vector systems and their utilization in preclinical studies and clinical trials.

2. Viral Vector-Based Delivery

Different types of viral vectors based on both DNA and RNA viruses have been engineered for gene therapy applications (Figure 1). The choice of vector is, to a large extent, affected by the disease indication, and need of short-term expression for acute diseases such as infectious diseases and cancers, and long-term expression required for chronic diseases. In the former case, high-level transient expression from replication-deficient viral vectors can provide therapeutic efficacy [10]. In the latter case, long-term expression is often achieved by extrachromosomal presence or chromosomal integration of the viral vector/transgene for extended therapeutic activity. Typically, replication-deficient and non-integrating vector systems are only capable of providing long-term transgene expression in post-mitotic tissues. In any viral vector-based gene therapy application, safety is of utmost importance [11]. Obviously, long-term treatment and presence of viral vector and/or transgene sequences in the host genome demands special requirements related to integration site, control of expression levels, and pharmacokinetics of the therapeutic product. In the context of cancer gene therapy, oncolytic viruses, which specifically replicate in tumor cells leading to their killing, have been evaluated as such, or as delivery vectors for anti-tumor genes both in vitro and in vivo [12]. A comprehensive description of various types of viral vectors is presented below and summarized in Table 1.

2.1. Adenovirus Vectors

Since the advent of gene transfer in mammalian cells, adenoviruses (Ad) vectors have been commonly used as viral delivery vehicles [13]. They are non-enveloped viruses possessing a double-stranded DNA (dsDNA) genome with a packaging capacity of up to 7.5 kb foreign DNA. However, Ad shuttle vectors have been engineered for accommodation of up to 14 kb of foreign DNA [14]. The first-generation Ad vectors were hampered by strong immunogenicity despite removal of the E1/E3 genes from the genome [15]. However, the immunogenicity has been reduced significantly in replication-deficient second- and third generation Ad vectors [16]. High-capacity third-generation adenovirus (HC-Adv) vectors, also known as helper-dependent gutless vectors, have the capacity to accommodate up to 37 kb of foreign DNA [17]. Moreover, replication-competent oncolytic adenoviruses have been developed for specific replication in tumor cells, resulting in the killing of tumor cells [18]. The engineering of packaging cell lines has facilitated large-scale GMP-grade Ad vector production [19]. Ad vectors provide persistent extrachromosomal transgene expression lasting for at least one year despite no integration into the host genome [20]. Moreover, a follow-up study in non-human primates showed transgene expression, although reduced to 10% of peak values, up to 7 years without any long-term adverse effects [21].

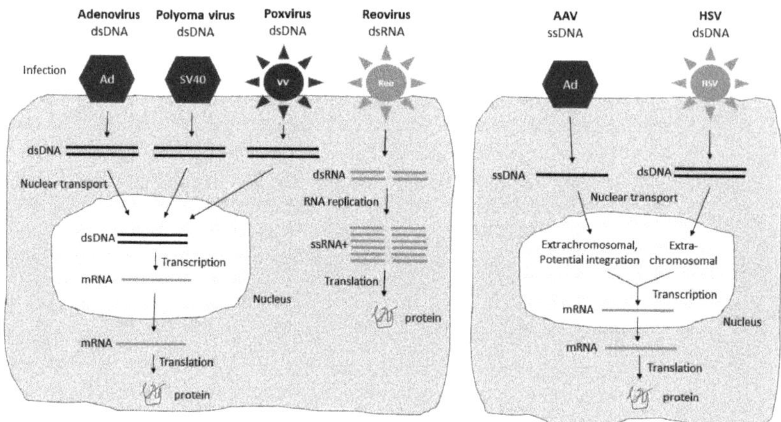

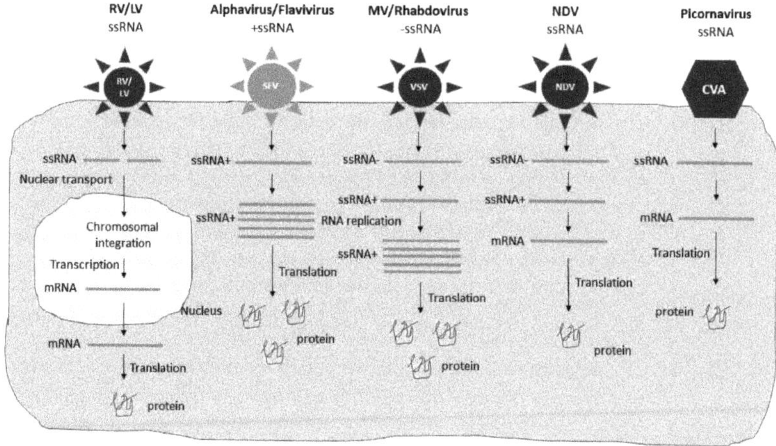

Figure 1. Viral vector expression systems. (**A**) Expression systems engineered for adenoviruses (Ad), simian virus 40 (SV40), vaccinia virus (VV), reoviruses, adeno-associated viruses (AAV), and herpes simplex viruses (HSV). (**B**) Viral vector systems for retro-and lentiviruses (RV), Semliki Forest virus (SFV), measles viruses and vesicular stomatitis virus (VSV), Newcastle disease virus (NDV), and coxsackieviruses A (CVA). dsDNA, double-stranded DNA, dsRNA, double-stranded RNA, ssDNA, single-stranded, DNA, ssRNA, single-stranded RNA, ssRNA+, ssRNA of positive polarity, ssRNA-, ssRNA of negative polarity.

2.2. Adeno-Associated Virus Vectors

The small non-enveloped single-stranded DNA (ssDNA) adeno-associated virus (AAV) can only accommodate 4 kb of foreign DNA [22], although, the packaging capacity has been improved by constructing fragmented, overlapping, or trans-splicing Dual AAV vectors [23,24]. AAV vectors generally do not cause toxic or pathogenic responses. However, repeated administration of AAV vectors has generated strong immune responses, reducing the efficacy of delivery and transgene expression [25]. This problem has been addressed by applying different AAV serotypes for each AAV re-administration. An alternative approach has been to utilize exosome-associated AAV (Exo-AAV), which has supported the application of reduced AAV doses resulting in reduced immune responses against the AAV capsid protein [26]. Moreover, Exo-AAV8 vectors have demonstrated long-term liver-directed gene transfer [27]. AAV vectors can transduce both dividing and non-dividing cells and usually remain in an extrachromosomal state, although integration of AAV-delivered genes into the host genome has been reported [28]. In fact, 30-fold higher AAV integration frequency was obtained by the introduction of 28S ribosomal DNA homology sequences in AAV vectors, which might contribute to superior treatment of genetic diseases [29].

2.3. Herpes Simplex Virus Vectors

The large herpes simplex viruses (HSV) are enveloped dsDNA viruses, which cause latent infection in neural ganglia [30]. The engineering of HSV expression vectors has resulted in long-lasting transgene expression [31]. The linear HSV forms a circularized viral episome in the nucleus and remains extrachromosomal without integration [32]. HSV vectors have an excellent capacity of accommodating more than 30 kb of foreign DNA [33]. Engineered HSV amplicons are able to package 150 kb of foreign genetic material [34]. However, HSV vectors have been associated with relatively strong cytopathogenicity, which has been addressed by the deletion of non-essential genes in the HSV genome [35]. Furthermore, the introduction of micro-RNA sequences (miR-145) in the HSV ICP27 gene has generated oncolytic HSV vectors, which can selectively reduce cell proliferation in non-small cell lung cancer (NSCLCs) cells [36]. Efficient HSV packaging systems have been engineered, such as the helper virus-free system for the HSV amplicon using an ICP27-deleted, oversized HSV-1 DNA in a bacterial artificial chromosome (BAC) [37].

2.4. Retrovirus and Lentivirus Vectors

The enveloped single-stranded RNA (ssRNA) retroviruses (RVs) possess a packaging capacity of 8 kb of foreign sequences [38]. The special feature of RVs comprises their reverse transcriptase activity, which allows the production of dsDNA copies of the RNA genome for integration into the host genome [39]. The chromosomal integration is advantageous for long-term transgene expression, although random integration has been of concern, even resulting in leukemia development in treated SCID-X1 patients [8]. For this reason, self-inactivating γRV (SIN-γRV) vectors have been engineered, which have proven safe with no cases of adverse integration or leukemia observed in clinical trials [40]. However, adenosine deaminase deficient severe combined immunodeficiency (ADA-SCID) seems to differ from other inherited immunodeficiencies, as insertional oncogenesis is rare after γRV treatment [41]. For example, none of the 10 patients in a clinical study developed leukemia [42], and among a total of 50 ADA-SCID patients treated with γRV, only one showed clinical evidence of leukemia [43]. Packaging cell lines have also been engineered for RV vectors to support large-scale production of GMP-grade materials [44]. One serious limitation of gene therapy applications for RV vectors is their capability to only transduce dividing cells and not non-dividing cells.

In contrast, lentivirus (LV) vectors, which also belong to the family of retroviruses, can transduce both dividing and non-dividing mammalian cells [45]. Otherwise, LV vectors share the same features with RVs of an ssRNA genome and a capacity of carrying 8 kb of foreign genetic material. Importantly, LV vectors show low cell cytotoxicity and due to their "semi-random" chromosomal integration provide improved biosafety for clinical applications, although some adverse events and insertional oncogenesis have been reported [46]. For example, modification of integration of human immunodeficiency virus-1 (HIV-1) by the fusion of the C-terminal HIV integrase-binding region of the LEDGF/p75 protein to the N-terminal chromodomain of heterochromatin protein-1 alpha (HP1 alpha) reduced the number of integration events [47]. Expression systems for non-human LV vectors such as simian immunodeficiency virus (SIV) [48], feline immunodeficiency virus (FIV) [49], and equine infectious anemia virus (EIAV) [50] have been engineered. LV producer cell lines have been designed to support large-scale production [51]. However, the low titers obtained, and residual toxicity have compromised their utilization [51].

2.5. Alphavirus Vectors

Alphaviruses are enveloped viruses with an ssRNA genome of positive polarity and a packaging capacity of 8 kb of foreign genetic material [52,53]. The positive polarity of alphaviruses allows the direct translation of viral RNA in the host cell cytoplasm. Alphaviruses possess a special feature of RNA self-replication, which generates extreme levels of transgene expression. The nature of expression is transient due to the rapid degradation of the alphavirus ssRNA. Alphavirus vectors can be used as recombinant particles, naked or liposome encapsulated RNA replicons, or plasmid DNA-based replicons [54]. Expression systems have been developed for Semliki Forest virus (SFV) [55], Sindbis virus (SIN) [56], and Venezuelan equine encephalitis virus (VEE) [57]. Moreover, naturally occurring oncolytic M1 viruses [58] and engineered oncolytic SFV vectors [59] have been used for cancer therapy.

2.6. Flavivirus Vectors

Similar to alphaviruses, flaviviruses are enveloped ssRNA viruses of positive polarity and therefore possess the feature of self-replicating RNA, providing high levels of transient transgene expression and the flexibility of using recombinant viral particles, RNA replicons and DNA replicons [60]. The packaging capacity of flaviviruses is approximately 6 kb. Kunjin virus (KUN) [60], West Nile virus (WNV) [61], Dengue virus (DENV) [62], tick-borne encephalitis virus (TBEV) [63], yellow fever virus (YFV) [64], and Zika virus (ZIKV) [65] have been subjected to the engineering of expression systems. In support of large-scale KUN [66] and TEBV [63] vector production, packaging cell lines have been engineered.

2.7. Measles Virus Vectors

The enveloped measles viruses (MVs) carry an ssRNA genome of negative polarity [67]. For this reason, the MV RNA first needs to be copied as a positive strand RNA template for self-replication of RNA in the host cytoplasm before being translated [68]. Approximately 6 kb of foreign genetic material can be introduced into MV vectors. Technologies for reverse genetics [69] and packaging cell lines [70] have been established. Oncolytic MV strains such as MV Hu-191 [71] and MV Schwartz [72] have also been used for cancer therapy.

Table 1. Examples of viral vectors used for gene therapy applications.

Virus	Genome	Insert Size	Advantages and Limitations
Adenovirus			
Ad5	dsDNA	<7.5 kb	Broad host range (dividing and non-dividing cells) [13]
Ad26			Excellent packaging capacity of HC-Adv [17]
ChAd			Persistent expression, no chromosomal integration [20]
HC-AdV		37 kb	Strong immunogenicity [14], reduced for gutless Ad [16]
			Oncolytic Ad vectors for tumor targeting and killing [18]
			Pre-existing immunity in humans [13]
			Packaging cell lines for large-scale GMP production [19]
AAV			
AAV2, 3	ssDNA	4 kb	Relatively broad host range [22]
AAV5, 6			Limited packaging capacity [22] improved by Dual AAV vectors [23,24]
AAV8, 9			Strong immune response after AAV re-administration, which could be reduced by re-administration with different AAV serotypes [25]
Dual AAV			Exo-AAV vectors have reduced immunogenicity, providing liver-targeted transgene expression [26,27]
Exo-AAV			Generally, AAV remains in an extrachromosomal state [28]

Table 1. *Cont.*

Virus	Genome	Insert Size	Advantages and Limitations
HSV			
HSV-1	dsDNA	>30 kb	Broad host cell range [31], excellent [33], extreme for HSV amplicons [34] foreign DNA packaging capacity
HSV-2			Long-lasting transgene expression from extrachromosomal circular HSV DNA [32]
HSV amplicons		150 kb	Deletion of non-essential HSV genome reduces cytotoxicity [35] Engineering of oncolytic HSV by introduction of miR145 [36] Engineering of helper virus-free packaging system [37]
γ-Retrovirus MMSV MSCV SIN-γRV	ssRNA	8 kb	Restricted host range, only dividing cells [38] Good packaging capacity of foreign genetic material [38] Chromosomal integration due to reverse transcriptase activity [39] Random integration causing leukemia [8] Targeted integration with self-inactivating vector [40] Packaging cell lines for large-scale production [44]
Lentivirus HIV-1 HIV-2 SIV FIV EIAV	ssRNA	8kb	Broad host range, including non-dividing cells [45] Good capacity to accommodate foreign genetic material [45] Non-random chromosomal integration [46] Non-human LV vectors available [47–50] Producer cell lines engineered for LV vectors [51]
Alphavirus SFV, SIN, VEE, M1	ssRNA	8 kb	Extremely broad host range, risk of neurovirulence [52] Good packaging capacity [53] RNA self-replication leading to extreme transgene expression [52] Low immunogenicity of alphaviruses [52] Transient expression not applicable for chronic diseases, but good for acute diseases and vaccines [52] Flexibility to use viral particles, RNA and DNA replicons for delivery [54] Oncolytic alphaviruses for cancer therapy [58,59]
Flavivirus KUN, WNV, DENV, TBEV YFV, ZIKV	ssRNA	6 kb	Broad host range, relatively good packaging capacity [60] RNA self-replication leading to high transgene expression [60] Flexibility to use viral particles, RNA and DNA replicons for delivery [60] Efficient packaging cell lines for KUN [66] and TBEV [63]
Measles virus MV	ssRNA	6 kb	Broad host range, relatively good packaging capacity [67] Positive strand RNA template needed for translation [68] Development of reverse genetics [69] and packaging cell lines [70] Oncolytic MV strains for cancer therapy [71,72]
Rhabdovirus VSV RABV Maraba	ssRNA	6 kb	Broad host range, relatively good packaging capacity [73] Positive strand RNA template needed for translation [73] Reverse genetics systems [74] Oncolytic rhabdoviruses for cancer therapy [75,76] Vaccinia-free packaging cell lines [77]
NDV NDV	ssRNA	4 kb	Broad host range, modest packaging capacity [78] Reverse genetics systems available [79] Oncolytic NDV for killing of tumor cells [79]

Table 1. *Cont.*

Virus	Genome	Insert Size	Advantages and Limitations
Poxvirus VV Avipox	dsDNA	>30 kb	Broad host range [80] Excellent packaging capacity [80] Tumor-selective replication-competent VV [81]
Picornavirus CVA21 CVB3 PV-1	ssRNA	6 kb	Relatively broad host range [82] Relatively good packacking capacity despite the small size [82] No chromosomal integration [82] Applications for gene therapy and vaccines [83,84]
Reovirus Reovirus-3	dsRNA	ND	Oncolytic activity in different types of cancer cells [85] Reoviruses replicate preferentially in Ras activated tumor cells [86] Combination therapy with radio-, chemo-, and immunotherapy [87] Endoplasmic reticular stress-mediated apoptosis in cancer cells [88]
Polyoma virus SV40	dsDNA	17.7 kb	Superb packaging capacity of 17.7 kb for SV40 with small genome [89] Vero cell-based SV40 packaging system [90] Inhibition of tumor cell progression [91]

AAV, adeno-associated virus; Ad, adenovirus; CVA21, coxsackievirus A21; CVB3, coxsackievirus B3; DENV, Dengue virus; dsDNA, double-stranded DNA; dsRNA, double-stranded RNA; Exo-AAV, exosome-associated AAV; FIV, feline immunodeficiency virus; HC-Adv, high-capacity Ad gutless vector; HIV, human immunodeficiency virus; HSV, herpes simplex virus; KUN, Kunjin virus; M1, oncolytic alphavirus; MMSV, Moloney murine sarcoma virus; MSCV, murine stem cell virus; ND, not determined; NDV. Newcastle disease virus; PV-I, poliovirus-1; SFV, Semliki Forest virus; SIN, Sindbis virus; SINγRV, self-inactivating gamma retrovirus; SIV, simian immunodeficiency virus; ssDNA, single-stranded DNA; ssRNA, single-stranded RNA; TBEV, tick-borne encephalitis virus; VEE, Venezuelan equine encephalitis virus; VV, vaccinia virus; WNV, West Nile virus; YFV, yellow fever virus; ZIKV, Zika virus.

2.8. Rhabdovirus Vectors

Also, rhabdoviruses are enveloped ssRNA viruses with a negative-stranded genome [73]. Reverse genetics methods have been applied for the generation of rhabdovirus expression systems [74]. Generally, 6 kb of foreign sequences can be accommodated in rhabdovirus vectors [73]. Expression systems have been engineered for rabies virus (RABV) [92], vesicular stomatitis virus (VSV) [93], and Maraba virus [94]. The majority of the oncolytic rhabdovirus vectors are based on VSV [75] and Maraba virus [76]. Moreover, vaccinia-free packaging cell lines have been established for VSV [77].

2.9. Newcastle Disease Virus Vectors

The enveloped negative-stranded ssRNA Newcastle disease virus (NDV) has a limited packaging capacity of only 4 kb of foreign genetic material [78]. However, this has not been a major issue as NDV vectors possess oncolytic activity and specifically replicate in tumor cells, resulting in efficient cell killing and tumor eradication [95]. Oncolytic NDV vectors have been used for cancer therapy in both preclinical animal models and clinical trials [76]. Reverse genetics have also been used for the NDV-73 T strain to modify the cleavage site of the fusion (F) protein, which decreased the pathogenicity in chicken without reducing the potency of tumor cell killing [96].

2.10. Poxvirus Vectors

Poxviruses are large, enveloped dsDNA viruses [80], which show an outstanding packaging capacity of more than 30 kb of foreign DNA. Among poxviruses, vaccinia virus (VV) vectors have been frequently used for prophylactic and therapeutic applications in the fields of infectious diseases and cancers [97]. Engineering of tumor-selective replication-proficient VV vectors has proven an attractive approach for cancer therapy [81]. In the context of avian poxviruses such as the non-replicating avipox virus, good biosafety standards have been achieved for non-avian species [98].

2.11. Picornavirus Vectors

The small non-enveloped picornaviruses contain an ssRNA genome and are capable of introducing up to 6 kb of foreign genetic material despite their small size [81]. Both coxsackievirus A21 (CVA21) [82] and the attenuated coxsackievirus B3 (CVB3) [83] have proven useful for gene therapy and vaccine development. Expression systems have also been engineered for the PV-1 poliovirus [84].

2.12. Reovirus Vectors

The enveloped dsRNA reoviruses possess oncolytic activity, showing killing of different types of cancer cells [85]. It has been documented that reoviruses replicate preferentially in tumor cells with activated genes of the Ras family or Ras-signaling pathway, which can be found in 60–80% of human malignancies [86]. Reovirus vectors have been demonstrated to invoke immune stimulation for reversing tumor-induced immunosuppression and promotion of anti-tumor immune responses [99]. Reoviruses have also been combined with radiotherapy, chemotherapy, immunotherapy, and surgery for cancer treatment [87]. Moreover, reovirus serotype 3 (Reolysin®) induces endoplasmic reticular stress-mediated apoptosis in in vivo models of pancreatic cancer [88].

2.13. Polyoma Virus Vectors

Although, the small non-enveloped dsDNA viruses possess a genome of only 5 kb, for example, the simian virus 40 (SV40) can package 17.7 kb of foreign DNA [89]. Packaging of virus-like particles (VLPs) containing no SV40 wildtype sequences can be carried out in vitro. Additionally, Vero cell-based packaging systems have been engineered for SV40 [90]. SV40 vectors have demonstrated successful delivery of anti-viral agents, DNA vaccines, suicide, chemoprotective, and anti-angiogenic genes for successful inhibition of tumor cell progression [91].

3. Gene Therapy Applications

Due to the many gene therapy studies conducted with viral vectors, it is only possible to provide an overview here through examples from preclinical studies and clinical trials for various diseases. The examples are selected to cover most disease indications using different types of viral vectors without indicating any preference of vector choice. The findings are also summarized in Tables 2–5.

3.1. Cancer

Different types of cancers have been frequently targeted by viral vector-based gene therapy and immunotherapy, and the potentially straightforward tumor killing with no need for long-term transgene expression. In addition to intratumoral administration, tumor targeting by specifically designed vectors [100], utilization of tumor-specific promoters [101], and application of oncolytic viruses [9] have been tested (Table 2).

Table 2. Preclinical and clinical examples of viral vectors applied for cancer therapy.

Viral Vector	Phase	Findings
Breast cancer		
HSV-HF10	Pre	Substantial tumor regression, prolonged survival in mice [102]
Reolysin + anti-PD1	Pre	Superior tumor regression, prolonged survival in mice after combination [103]
M1	Pre	Targeting and killing of 4T1 mammary tumors in mice [104]
PANVAC	Phase I	SD in 4 patients, complete response in 1 patient [105]

Table 2. *Cont.*

Viral Vector	Phase	Findings
Gliomas		
M1	Pre	Specific targeting of C6 glioma cells [106]
M1	Pre	Replication of M1 in gliomas in mice, rats, and macaques [107]
SFV-IL-12	Pre	87% reduction of RG2 glioma size in rats [108]
SFV-VA	Pre	100% eradication of small, 50% eradication of large tumors in mice [59]
RRV Toca 511-CD	Pre	Prolonged survival in mice with orthotopic gliomas [109]
m-ZIKV	Pre	Prolonged survival in mice with implanted glioblastomas [65]
MV-CEA	Phase I	Trial in progress in patients with recurrent glioblastoma [110]
RRV Toca 511	Phase I	Prolonged survival of 13.6 months in HGG patients [111]
RRV Toca 511	Phase II/III	No improvement in overall survival in HGG patients [112]
Colon cancer		
KUN-GM-CSF	Pre	Cure of >50% of mice with CT26 colon tumors [113]
VSV(M51R)	Pre	Reduced luciferase expression in tumors, prolonged survival in mice [114]
M-LPO	Pre	Superior oncolytic activity in mice [115]
SFV-LacZ RNA	Pre	Protection against tumor challenges in mice after a single injection of RNA [116]
VEE-CEA	Phase I	Antigen-specific responses and prolonged survival in colorectal cancer patients [108,117]
vvDD	Phase I	Th1-biased immune responses against vvDD and tumors in patients [118]
Melanoma		
KUN-GM-CSF	Pre	Significant tumor regression, 67% of mice tumor-free [113]
NDV-IL12/IL15	Pre	Superior survival after NDV-IL15 compared to NDV-IL12 in mice [119]
VSV-LCMV-GP	Pre	Significant tumor regression, prolonged survival in melanoma models [120]
VSV-XN2-ΔG	Pre	Strong tumor regression in mice [121]
CVA21-ICAM-DAF	Pre	Tumor regression, reduced tumor burden in mouse melanoma model [122]
MG1-hDCT + Ad	Pre	Ad-hDCT prime-Maraba MG1-hDCT booster elicited immune responses [94]
HSV-HF10 + CTLA4	Phase III	Good safety and antitumor activity in patients [102]
HSV T-VEC	Phase II/III	Good tolerance, promising therapeutic effect in melanoma patients [123]
HSV T-VEC	Approval	Approved for treatment of advanced melanoma in the US, Europe, Australia [124]
Pancreatic cancer		
Adsur-SYE	Pre	Complete tumor regression in mice [125]
PANVAC	Pre	Superior immune response in pancreatic mouse cancer models [126]
SV40-hRT-SST2	Pre	Long-term inhibition of tumors in Capan-1 mouse tumor model [91]
HSV-HF10	Phase I	PR in 3 patients, SD in 4 patients, PD in 9 patients [127]
Ovarian		
VSV-LCMP-GP	Pre	Tumor regression in ovarian cancer mouse models [128]
VSV-LCMP-GP + Rux	Pre	Superior therapeutic activity after combination therapy [128]
VSVMP-p DNA	Pre	87–98% tumor regression in ovarian mouse cancer models [129]
SIN AR339	Pre	Ovarian cancer cell killing, tumor regression in mice [130]
MV-CEA	Phase I	SD in all 9 patients, overall survival twice to the expected time [131]
Prostate		
MV-CEA	Pre	Delay of tumor growth, prolonged survival [132]
MV-sc-Fv-PSMA	Pre	Specific killing of prostate tumors, enhanced by radiation [133]
MV + MuV	Pre	Superior antitumor activity, survival after combination therapy [134]
VEE-PSMA	Pre	Strong Th1-biased immune responses in mice [135]
VEE-mSTEAP	Pre	Prime immunization with DNA, booster with VEE specific immunogenicity [136]
VEE-PSCA	Pre	Long-term survival in 90% of TRAMP mice [137]
VV-GLV-1h123-NIS	Pre	Inhibition of tumor growth, prolonged survival in prostate cancer models [138]
VSV-PSMA	Phase I	Good safety, disappointingly weak immune responses [139]

Adsur-SYE, adenovirus vector with survivin promoter, pancreatic cell-targeting ligand SYENFSA; CD, yeast cytosine deaminase; CTLA4, anti-CTLA-4 antibody; DAF, decay accelerating factor; HGG, high grade glioma; HSV, herpes simplex virus; ICAM-1, intercellular adhesion molecule-1; KUN. Kunjin virus; LCMP-GP, lymphocytic choriomeningitis virus-glycoprotein; M1, oncolytic alphavirus; M-LPO, liposome-encapsulated M1 alphavirus; mSTEAP, mouse six-transmembrane epithelial antigen of the prostate; m-ZIKV, mouse-adapted Zika virus; MV, measles virus; NDV, Newcastle disease virus; PD, progressive disease; PR, partial response; Pre, preclinical studies; PSCA, prostate stem cell antigen; PSMA, prostate-specific membrane antigen; RRV replicating retrovirus; Rux, ruxolitinib; SD, stable disease; SFV, Semliki Forest virus; SIN, Sindbis virus; TRAMP, transgenic adenocarcinoma of the prostate; VSV, vesicular stomatitis virus; vvDD, oncolytic vaccinia virus.

For example, substantial tumor regression and prolonged survival were observed in mouse breast tumor models after treatment with an HSV-HF10 vector [102], and the co-treatment with a reovirus vector and checkpoint inhibitor PD-1 antibody [103]. The oncolytic M1 alphavirus efficiently targeted and killed 4T1 mammary tumors in mice [104]. In one example of a clinical study, the PANVAC vaccine based on VV and fowl pox was subjected to a phase I trial in heavily pre-treated breast cancer patients [105]. Stable disease (SD) was observed in four patients and one patient showed a complete response [105]. In the context of gliomas, the M1 alphavirus showed specific targeting of C6 glioma cells [106] and replication in gliomas in mice, rats, and macaques [107]. Moreover, SFV particles expressing interleukin-12 (SFV-IL-12) were administered via an implanted canula, which reduced RG2 gliomas by 87% in rats [108]. In another study, the replication competent SFV VA7 showed strong killing of human glioma cells, and intravenous administration in BALB/c mice completely eradicated 100% of small and 50% of large subcutaneous U87 tumors [59]. In the case of RV vectors, the replicating retroviral vector (RRV) Toca 511 carrying the yeast cytosine deaminase (CD) provided extended survival in mice implanted with orthotopic gliomas [109]. ZIKV has demonstrated specific targeting and killing of glioblastoma stem cells (GSCs), and administration of the mouse-adapted ZIKV (m-ZIKV) strain prolonged survival substantially in mice with implanted glioblastomas [65]. Moreover, MV particles expressing the carcinoembryonic antigen (CEA) [140] have been subjected to a phase I trial in patients with recurrent glioblastoma multiforme [110]. In a phase I trial, patients with recurrent or progressive high-grade glioma (HGG) who received the RRV Toca 511 vector showed a statistically relevant extended survival of 13.6 months [111]. In contrast, the overall survival was not prolonged in phase II/III trials in HGG patients [112].

KUN-based expression of the granulocyte macrophage-colony stimulating factor (GM-CSF) resulted in cure in more than 50% of CT26 colon tumor-bearing mice [113]. In another approach, the oncolytic VSV(M51R) strain was administered intraperitoneally into BALB/c mice carrying luciferase-expressing CT26 tumors, which resulted in eradication of tumors demonstrated by reduced luciferase expression and prolonged survival of mice [114]. M1 alphavirus particles encapsulated in liposomes (M-LPO) were able to inhibit the growth of colorectal LoVo and liver Hep3B cancer cells [115]. Moreover, intravenous administration of M-LPO reduced the production of M1-specific neutralizing antibodies in mice, resulting in superior oncolytic activity [106]. In an interesting approach, only a single injection of 0.1 μg of naked SFV-LacZ replicon RNA provided protection in mice with implanted CT26 colon tumors against tumor challenges [116]. Additionally, therapeutic activity and prolonged survival were found in mice with pre-existing tumors [116]. In the case of clinical trials, VEE-CEA particles were administered to stage III and IV colorectal cancer patients in a phase I trial [117]. It was found that antigen-specific immune responses were detected in both stage III and IV patients, and the overall survival was extended. In another phase I trial, patients with advanced colorectal cancer were subjected to oncolytic vvDD poxvirus particles, which elicited potent Th1-biased immune responses against vvDD and tumors [118].

Melanoma has been frequently targeted for gene therapy applications of viral vectors. For example, KUN-GM-CSF particles generated significant tumor regression and cured 67% of mice with B16-OVA melanomas [113]. In another approach NDV vectors were applied for the expression of IL-12 and IL-15 [119]. Intratumoral administration of NDV-IL12 and NDV-IL15 into a mouse melanoma model suppressed tumor growth. NDV-IL15 was superior, showing 26.6% higher survival rate compared to NDV-IL12 [119]. In another study, chimeric VSV particles expressing the lymphocytic choriomeningitis virus glycoprotein (LCMV-GP) showed significant tumor regression and prolonged survival in syngeneic melanoma tumor models [120]. In another study on VSV, strong tumor regression was seen in C57BL/6 mice implanted with B16-OVA melanomas after subcutaneous injection of an oncolytic VSV vector [121]. In the context of picornaviruses, a single subcutaneous injection of CVA21 particles expressing the intercellular adhesion molecule-1 (ICAM-1) and the decay-accelerating factor (DAF) resulted in tumor regression and reduced tumor burden in a mouse melanoma model [122]. The oncolytic Maraba MG1 strain expressing the human dopachrome tautomerase (hDCT) neither elicited antitumor immune responses nor therapeutic activity in mice with B16-F10 metastases [94]. However, prime immunization with Ad-hDCT followed by a booster immunization with Maraba MG1-hDCT elicited strong immune responses [94]. In contrast, the Maraba MG1 strain provided a long-lasting cure in sarcoma-bearing mice, and protection against challenges with sarcoma tumors [79]. In a phase III study, HSV-HF10 was combined with the checkpoint inhibitor anti-CTLA-4 antibody, demonstrating a good safety profile and antitumor activity in patients with non-resectable or metastatic melanoma [102]. HSV vectors, especially the oncolytic talimogene laherparevec (HSV T-VEC) vector expressing GM-CSF, have been assessed in Phase II and III clinical trials, showing a tolerable adverse event profile and promising therapeutic efficacy superior to GM-CSF therapy [123]. However, responses in visceral metastases have been modest. HSV T-VEC has been approved for the treatment of advanced melanoma in the US, Europe, and Australia [124].

Due to the aggressive nature and difficulty to treat pancreatic cancer, gene therapy efforts have been welcomed as an alternative strategy. For example, administration of Ad vectors containing the survivin promoter and the pancreatic cancer cell-targeting ligand SYENFSA (SYE) resulted in complete regression of pancreatic neuroendocrine tumors (PNETs) in mice [125]. Related to poxviruses, a heterogenous prime-boost strategy applying the PANVAC system for VV and fowl pox vectors elicited enhanced immune responses in pancreatic mouse cancer models [126]. A replication-competent SV40 vector carrying the tumor-specific human telomerase (hTR) RNA promoter and the somatostatin receptor tumor-suppressor 2 (SST2) gene showed long-term inhibition of tumor growth in the Capan-1 pancreatic mouse tumor model [91]. In a phase I trial, the oncolytic HSV-HF10 was administered intratumorally to patients with non-resectable locally advanced pancreatic cancer, showing partial response (PR) in three patients, SD in four patients, and progressive disease (PD) in nine patients [127].

In the case of ovarian cancer, the VSV-LCMV-GP showed tumor regression in subcutaneous and orthotopic ovarian cancer mouse models [128]. Moreover, the therapeutic efficacy was improved by co-administration of VSV-LCMV-GP and the JAK1/2 inhibitor ruxolitinib [128]. Application of the liposome-encapsulated VSVMP-p DNA vector expressing the VSV membrane (M) protein for intraperitoneal injection in mice reduced the tumor weight by 90%, and prolonged survival of mice with implanted ovarian tumors [141]. Moreover, the ovarian tumor growth was inhibited by 87–98% [129]. In another study, intraperitoneal administration of the oncolytic SIN AR339 vector resulted in ovarian cancer cell killing and tumor regression in mice [130]. In a clinical setting, MV-CEA particles were evaluated in a phase I trial in patients with recurrent ovarian cancer [131]. No dose-limiting toxicity was associated with the treatment, and SD was achieved in all nine treated patients. Moreover, the median overall survival was 12.15 months, which is twice the expected time.

Related to prostate cancer, intratumoral administration of MV-CEA particles delayed tumor growth and prolonged survival in PC-3 prostate tumor-bearing mice [132]. In another study, an MV vector expressing a single-chain antibody (sc-Fv) specific for the extracellular domain of the prostate-specific membrane antigen (PSMA) was administered to mice with LNCaP and PC3-PSMA prostate tumors [133]. MV-sc-Fv-PSMA provided specific infection and killing of PSMA-positive prostate cancer cells, which was further enhanced by radiation therapy. Co-administration of oncolytic MV and mumps virus (MuV) vectors showed superior antitumor activity, and prolonged survival in mice with PC-3 prostate tumors compared to administration of either MV or MuV vectors alone [134]. In another approach, VEE-based expression of the prostate-specific membrane antigen (PSMA) elicited strong PSMA-specific immune responses in BALB/c and C57BL/6 mice [135]. A single immunization induced strong T- and B-cell responses, which were Th1-biased. Moreover, a booster immunization with VEE particles expressing the mouse six-transmembrane epithelial antigen of the prostate (mSTEAP) 15 days after a prime immunization with gold-coated conventional pcDNA-3-mSTEAP plasmids elicited specific immune responses against mSTEAP, a modest but significant delay of tumor growth, and prolonged the overall survival of mice [136]. Moreover, administration of VEE particles expressing the prostate stem cell antigen (PSCA) resulted in long-term survival in 90% of transgenic adenocarcinoma of the prostate (TRAMP) mice [137]. In addition, administration of the VV GLV-1h123 vector expressing the sodium iodide symporter (NIS) gene provided significant inhibition of tumor growth, and extended survival time in prostate cancer mouse models [138]. In the context of clinical evaluation, a phase I trial was conducted in patients with castration resistant metastatic prostate cancer (CRPC) with VEE-PSMA particles [139]. Although the procedure showed good safety standards, the PSMA-specific immune responses were disappointingly weak.

3.2. Cardiovascular Diseases

Gene therapy-based applications for cardiovascular diseases have mainly focused on Ad and AAV vectors (Table 3). For example, expression of the sarcoplasmic reticulum Ca^{2+} ATPase (SERCa2a) by an Ad vector restored both systolic and diastolic heart functions to normal levels in a rat model of heart failure [142]. Ad-SERCa2a also managed to improve coronary blood flow, and reduced cardiomyocyte size in a rat model for type 2 diabetes [143]. SERCa2a has also been expressed from AAV-1 vectors leading to increased coronary blood flow in a pig model [144]. Moreover, LV-based expression of SERCa2a provided protection against left ventricular dilation, improved systolic and diastolic functions, and reduced mortality rates in an ischemic rat heart failure model [145]. Moreover, expression of the hepatocyte growth factor (HGF) led to improved heart function in a postinfarct pig heart model [146]. In other approaches, cardiac arrythmia has been treated with Ad vectors expressing Connexin 43 (Cx43) or the I(Kr) potassium channel alpha subunit, resulting in increased conduction velocity, prevention of atrial fibrillation, and reduced tachycardia after myocardial infarction in pigs [147] and prevention of fibrillation in a swine model [148], respectively. The pMX5 retrovirus has been applied for the expression of the transcription factors GATA4, MEF2C, and TBX5 for the reprogramming of non-myocytes in the mouse heart to cardiomyocyte-like cells to reduce infarct size and to attenuate cardiac dysfunction [149].

Table 3. Preclinical and clinical examples of viral vectors applied for cardiovascular, metabolic, and hematological diseases.

Viral Vector	Phase	Findings
Cardiovascular		
Ad-SERCa2a	Pre	Restoration of systolic/diastolic heart function in rat heart model [142]
Ad-SERCa2a	Pre	Improved coronary blood flow, reduced cardiomyocyte size in rats [143]
AAV1-SERCa2a	Pre	Increased coronary blood flow in pig model [144]
LV-SERCa2a	Pre	Protection against dilation, improved systolic and diastolic functions [145]
Ad-HGF	Pre	Improved heart function in a post-infarct pig model [146]
Ad-Cx43	Pre	Prevention of atrial fibrillation, reduced tachycardia in post-infarct pigs [147]
Ad-KCNH2	Pre	Prevention of fibrillation in swine model [148]
pMX5-GATA4/TBX5	Pre	Reprogramming cells to reduce infarct size, attenuated cardiac dysfunction [149]
Ad-VEGF	Phase I	Improved myocardial perfusion reserve, relief in symptoms in angina patients [150]
Ad-VEGF	Phase II	Improved treadmill exercise, no improvement in myocardial perfusion [151]
Ad-FGF4	Phase I/II	Improved treadmill exercise [152,153], stress-induced myocardial perfusion [154]
AAVI-SERCa2a	Phase I	Improved in functional, symptomatic, ventricular/remodeling parameters [155]
AAV1-SERCa2a	Phase II	Improved walking, oxygen consumption, ventricular endosystolic volume [156]
AAV1-SERCa2a	Phase IIa	Reduced number of cardiovascular events and deaths [157]

Table 3. *Cont.*

Viral Vector	Phase	Findings
Metabolic		
AAV-GUS	Pre	Single injection reversed mucopolysaccharidosis phenotype in mice [158]
AAV-LDL-R	Pre	Nearly normal lipid levels, prevention of severe atherosclerosis in mice [159]
AAV-FGF21	Pre	Therapeutic efficacy in transgenic mice as model for T2DM [160]
AAV8-PAL	Pre	Long-term correction of hyperphenylalaninemia in mice [161]
AAV8-GAA	Pre	Therapeutic activity and attenuated Pompe disease phenotype in mice [162]
MSCV-Insulin	Pre	Decreased blood glucose, increased insulin, reversal of diabetes in mice [163]
MMTV-Ad36 E4orf1	Pre	Improved glucose excursion in mice [164]
AAV-PBGD	Phase I	Unable to correct AIP phenotype, but reduced hospitalization [165]
AAV-hAAT	Phase I	Above background levels of hAAT in patients [166]
AAV-hAAT	Phase II	Strong immunostaining of AAT in muscle biopsies [167]
Hematology		
Ad-FVIII	Pre	Physiological levels of FVIII in mice [168]
Ad-FIX	Pre	Long-term expression of FIX in nude mice [169]
Ad-cFIX	Pre	Correction of hemophilia B in dogs, but only 1–2% FIX after 3 weeks [170]
Ad-cFIX + CsA	Pre	CsA restored therapeutic FIX levels for at least 6 months [171]
AAV6/AAV8-FVIII	Pre	Therapeutic levels of FVIII lasting for >3 years in dogs [172]
AAV8-FVIII	Pre	1–2% of normal FVIII levels, prevention of 90% of bleeding episodes in dogs [173]
AAV8/AAV9-FVIII	Pre	1.9–11.3% of normal FVIII, no effect on chromosomal integration in dogs [174]
AAV8-FIX	Pre	25–40% of normal FIX levels in hemophilic dogs [175]
AAV-FVIII	Phase I/II	8–60% of normal FVIII levels in hemophilia A patients [176]
AAV5-hFVIII-SQ	Phase I	Clinical benefits, reduced bleeding events in hemophilia A patients [177]
AAV8-FIX	Phase I	1–6% of normal FIX levels in hemophilia B patients for 3.2 years [178]
scAAV2-FIX	Phase I	Stable expression of FIX for 7 years, reduced bleedings in patients [176]
AAVS3-FIX	Phase I/II	Stable expression for 27 months required immunosuppression in patients [179]
AAV5-FVIII	Approval	Conditional marketing approval for severe hemophilia A by EMA [180].
2bF8 LV	Pre	Sustained FVIII expression. correction of hemophilia A phenotype in mice [181]
SIN-LV-cFIX	Pre	Long-term stable expression of FIX in dogs [182]
2bF9/MGMT LV	Pre	2.9-fold increase in FIX expression, reduced blood clotting time [183]
LV-PKDL/R	Pre	LV-transduced HSCs corrected hemolytic anemia phenotype in mice [184]
MSCV-FANCA	CR	Transient gene correction in 2 Fanconi anemia patients [185]
LV-RPS19	Pre	Cure of DBA in an RPS19 DBA-deficient mouse model [186]
LentiGlobin BB305	Phase I	Stop of transfusion of red blood cells in β-thalassemia patients [187]
LentiGlobin BB305	Phase III	Sustained HbA^{T87Q}, non-β^0/β^0 genotype patients independent of transfusions [188]
GLOBE LV	Pre	In utero gene therapy providing normalized hematological phenotype in mice [189]
GLOBE LV	Phase I/II	Transfusion discontinued or reduced in β-thalassemia patients [190]
LV-HSCs	Pre	Anti-sickling protein expression in mice [191]
LentiGlobin BB305	CR	Transfusions in the SCD patient could be discontinued [192]
LentiGlobin BB305	Phase I/II	Clinical remission or reduced frequency of transfusions in SCD patients [193]
HIV-HSV-TK	Pre	Prolonged survival of mice with acute T-cell leukemia (ATL) [194]
SIN-GALV.fus	Pre	Antitumor activity against acute myeloid leukemia (AML) xenografts in mice [195]
AAV6-CD33-iCasp9	Pre	Antitumor and apoptotic activity, prolonged survival in zebrafish [196]
LOAd703 + CAR T	Pre	Lymphoma killing in cell lines and in xenograft mouse models [197]
HSVrantes/HSVB7.1	Pre	Complete tumor regression after combination therapy in mice [198]
HSV-1 T-01	Pre	Intratumoral and contralateral tumor regression in mice [199]
AAV8-h1567 mAb	Pre	Strong antitumor activity, prolonged survival in mice [200]
SIN + α4-IBB Ab	Pre	Complete lymphoma eradication, long-lasting immunity in mice [201]
CVA21 RNA	Pre	Rapid tumor regression in mice, comparable to CVA21 particles [202]

Table 3. *Cont.*

Viral Vector	Phase	Findings
VSV-IFN-β	Pre	Eradication of tumors, prolonged survival in mice [203]
Reolysin	Pre	Reduced tumor burden in xenograft and syngeneic myeloma mouse models [204]

2bF8 LV, LV vector with integrin alpha-2b promoter; 2bF9/MGMT LV, LV vector with alpha-2b promoter; FVIII gene; hAAV, adeno-associated virus; AAVS3, AAV3 with synthetic capsid protein; Ad, adenovirus; AIP, acute intermittent porphyria; CR, case report; CsA, cyclosporin A; Cx43, connexin 43; DBA, Diamond-Blackfan anemia; EMA, European Medicines Agency; FANCA, Fanconi anemia complementation group A; FGF4, fibroblast growth factor 4; FGF21, fibroblast growth factor 21; FIX, factor IX; FVIII, factor VIII; hAAT, human alpha-1-antitrypsin; GAA, acid α-glucosidase; GALV.fus, gibbon ape leukemia virus fusion protein; GUS, β-glucuronidase; h1567 mAb, anti-CCR4 monoclonal antibody; HbA, hemoglobin; HSCs, hematopoietic stem cells; HSV, herpes simplex virus; HSV-TK, herpes simplex virus-thymidine kinase; KCNH2, I(Kr) potassium channel alpha subunit; LDL-R, low density lipoprotein receptor; LentGlobin BB305, LV vector expressing HbA^{T87Q}; MMTV, mouse mammary tumor virus; MSCV, murine stem cell virus; PAL, phenylalanine amino lyase; PBGD, porphobilinogen deaminase; pMX5, retrovirus; Pre, preclinical studies; RPS19, ribosomal protein S19; scAAV8, self-complimentary AAV8; SERCa2a, sarcoplasmic reticulum Ca^{2+} ATPase; SIN-LV, self-inactivating LV; SIN, Sindbis virus; T2DM, type 2 diabetes mellitus; VEGF, vascular endothelial growth factor; VSV, vesicular stomatitis virus.

Related to clinical evaluation, in a phase I trial, intramyocardial administration of the vascular endothelial growth factor (VEGF) expressed from Ad vectors generated improvement in myocardial perfusion reserve and relief of symptoms in refractory angina patients [150]. In a phase II study in patients with severely symptomatic coronary artery disease, the Ad-VEGF vector showed significant improvement in treadmill exercise, although, no improvement in myocardial perfusion was observed [151]. In a series of phase I-II AGENT (Angiogenic GENe Therapy) trials, the fibroblast growth factor 4 (FGF4) was expressed from Ad vectors in patients with chronic stable angina [152–154]. The studies demonstrated symptomatic improvement in exercise time [152], sex-specific benefits for treadmill exercise [153], and improvement in stress-induced myocardial perfusion [154]. AAV1-SERCa2a has been evaluated in a phase I study in patients with heart failure, which led to an improvement in functional, symptomatic, and ventricular/remodeling parameters [155]. In a phase II study, improvements in a walking test, peak maximum oxygen consumption, and left ventricular endosystolic volume were seen in patients with class III/IV heart failure after AAV1-SERCa2a treatment [156]. In another phase IIa trial, AAV1-SERCa2a treatment reduced the number of cardiovascular events and deaths [157].

3.3. Metabolic Diseases

More than 30 metabolic diseases have been subjected to viral vector-based gene therapy studies [205] (Table 3). AAV vectors have been used in the majority of studies. For example, AAV-based expression of β-glucuronidase (GUS) has been used for treatment of the lysosomal storage disease mucopolysaccharidosis [158]. Intramuscular injection of AAV-GUS generated high levels of local GUS. In contrast, only low GUS activity was detected after intravenous administration in mice [158]. However, even low levels of GUS reduced the glycosaminoglycan levels to normal in the liver and reduced storage granules substantially, and a single administration of AAV-GUS was sufficient to reverse the disease phenotype in mice [158]. AAV vectors have also been used for the expression of the low-density lipoprotein receptor (LDL-R) in the liver, which provided nearly complete normalization of serum lipid levels and prevention of severe atherosclerosis in mice [159]. Related to type 2 diabetes mellitus (T2DM), expression of the fibroblast growth factor 21 (FGF21) from AAV vectors provided substantial reduction in body weight, adipose tissue hypertrophy and inflammation, and insulin resistance for more than one year in transgenic ob/ob mice or wildtype mice receiving a high-fat diet [160]. In the context of phenylketonuria (PKU), a single injection of an AAV8 vector, containing the human antitrypsin (hAAT) promoter for the liver-specific expression of phenylalanine amino lyase (PAL), generated long-term correction of hyperphenylalaninemia in mice [161]. Moreover, AAV8 vectors expressing the acid α-glucosidase (GAA) gene have been evaluated for the treatment of Pompe disease, a glycogen storage disease [162,206]. Liver-specific GAA expression led to therapeutic activity and attenuated the disease phenotype in mice. RVs such as murine stem cell virus (MSCV) have been used for expression of the human insulin gene in diabetic mice, showing decrease in blood glucose levels, increase in secreted insulin, and reversal of diabetes for up to 6 weeks [163]. Moreover, the hyperglycemic Ad36 E4orf1 protein was expressed from an murine mammary tumor virus (MMTV) vector generating improved glucose excursion in C57BL/6 mice despite their high fat diet, and enhanced glucose levels without increasing insulin sensitivity [164].

In the case of clinical trials, intravenous administration of AAV particles expressing the porphobilinogen deaminase (PBGD) gene in a phase I trial in patients with acute intermittent porphyria (AIP) did not correct the AIP phenotype but suggested a trend towards a reduction in hospitalization and heme treatment [165]. In another approach, a phase I trial on patients with alpha-1-antitrypsin (AAT) deficiency was conducted with AAV vectors expressing the human AAT gene [166]. The safe intramuscular administration of AAV-hAAT generated AAT expression above background levels, which was sustained for at least one year. A follow-up phase II trial demonstrated antibody responses in all patients, however, not against AAT [167]. Despite that, strong immunostaining of AAT was detected in muscle biopsies.

3.4. Hematological Diseases

Among hematological diseases, hemophilias have been successful targets for gene therapy to correct the mutated factor VIII (FVIII) [207] and factor IX (FIX) [208] genes causing hemophilia A and B, respectively (Table 3). Originally, Ad vectors were applied showing sustained expression of the full-length FVIII at physiological levels in mice [168]. Furthermore, Ad-based long-term FIX expression of more than 300 days could be established in nude mice [169]. Ad-based expression of the canine FIX (cFIX) provided complete correction of the hemophilic phenotype in FIX-deficient hemophilia B dogs [170]. However, the cFIX levels decreased to only 1–2% of normal FIX levels in three weeks, but co-administration of the immunosuppressive cyclosporin A (CsA) restored therapeutic FIX levels and correction of hemophilia B for at least 6 months [171].

The limited packaging capacity of AAV vectors has presented some difficulties related to hemophilia therapy due to the large size of the FVIII gene [209]. For this reason, the B-domain deleted (BDD) FVIII has been expressed from AAV vectors [210]. In addition, the choice of AAV serotype is important as AAV8 provided much higher FVIII activity than AAV2, 3, 5, and 7 serotypes [211]. For example, AAV2-based expression of the canine BDD FVIII was only transient, while AAV6 and AAV8 vectors provided persistent therapeutic levels of FVIII, lasting for more than 3 years [172]. In another canine study on AAV8-FVIII, 1–2% of normal FVIII levels were achieved, which prevented 90% of bleeding episodes [173]. Moreover, a study with AAV8 and AAV9 in nine dogs showed 1.9–11.3% of normal levels monitored for 10 years [174]. Liver samples from six dogs identified 1741 unique integration sites in the genome, none of which induced tumors or altered liver function. Related to hemophilia B, AAV8-based FIX delivery increased FIX expression by 8–12-fold, with 25–40% of normal FIX levels in hemophilic dogs [175].

In clinical trials, interim results from a phase I/II study in six hemophilia A patients treated with a single injection of AAV-FVIII generated 8–60% of normal FVIII levels [176]. Moreover, a single infusion of the AAV-FVIII SQ variant (AAV5-hFVIII-SQ) showed sustained clinically relevant benefits with a decrease in bleeding events, and no need for prophylactic FVIII use in severe hemophilia A patients in a multiyear follow-up study [177]. AAV8-FIX particles were evaluated in a phase I trial in hemophilia B patients, which provided 1–6% of normal FIX levels for at least 3.2 years [178]. In another approach, self-complementary AAV2 vectors expressing FIX (scAAV2-FIX) showed stable FIX production for 7 years, contributing to substantial reduction in bleeding in hemophilia B patients [176]. In a phase I/II study, the AAVS3 vector, containing a synthetic capsid protein, was subjected to expression of FIX (FLT180a), which resulted in dose-dependent increase in FIX levels with five patients showing 51–78%, three patients 23–43%, and one patient 260% of the normal FIX levels [179]. Although sustained FIX expression was detected for 27 months, immunosuppression with glucocorticoids was required in all patients. Approval for conditional marketing of an AAV5 vector expressing the BDD FVIII cDNA for the treatment of severe hemophilia A has been granted by the European Medicines Agency (EMA) [180].

LV vectors have also been evaluated for hemophilia gene therapy. For example, the FVIII gene was expressed from a platelet-specific integrin alpha 2b promoter engineered into an LV vector (2bF8 LV) and transduced into mouse bone marrow [181]. Mice transplanted with 2bF8 LV-transduce bone marrow generated functional FVIII activity, survival of tail clipping, and correction of the hemophilia A phenotype. In the case of hemophilia B, expression of cFIX from a self-inactivating LV (SIN-LV) vector, carrying a hepatocyte-specific promoter, generated long-term stable FIX expression in dogs [182]. In another approach, the 2bF9/MGMT LV vector, which contains the alpha-2b promoter, the FIX, and methylguanine-DNA-methyltransferase (MGMT) 140K genes, provided a 2.9-fold higher FIX expression and 3.7-fold higher FIX activity in platelets after hematopoietic stem cell (HSC) transduction [183]. In transplanted mice, the blood clotting time was significantly reduced while the expression of therapeutic platelet-FIX was enhanced in mice.

Hemolytic anemia has been approached by transduction of HSCs by LV expressing the pyruvate kinase L/R (PKL/R) to compensate for pyruvate kinase deficiency (PKD), which corrected the hematological phenotype in mice [184]. The oncoretroviral MSCV vector has been used for ex vivo transfer of the Fanconi anemia complementation group A (FANCA) gene to treat Fanconi anemia (FA) [185]. Despite good safety and tolerability, the gene correction was transient due to the low dose of infused gene-corrected cells. In the context of Diamond-Blackfan anemia (DBA), LV-based expression of the ribosomal protein S19 (RPS19) provided cure of DBA and lethal bone marrow in an RPS19-deficient DPA mouse model [186].

In addition, β-thalassemia caused by more than 200 mutations in the β-globin gene [212] has been the target for viral-based gene therapy. For example, ex vivo transduced LentiGlobin BB305, an LV vector expressing the adult human hemoglobin T87Q mutant gene (HbA^{T87Q}), allowed 12 β-thalassemia patients with the β^0/β^0 genotype to stop red blood cell transfusions and in 9 other patients, the transfusion volume could be reduced by 73% in a phase I study [187]. Interim results from a phase III trial with LentiGlobin BB305 confirmed the expression of sustained levels of HbA^{T87Q} and for patients with the non-β^0/β^0 genotype to become independent of transfusions [188]. The GLOBE LV vector has been subjected to intrahepatic administration in utero in a humanized mouse model, which resulted in a normalized hematological phenotype at 12–32 weeks of age [189]. In a phase I/II trial, rapid recovery was achieved in three adult and six pediatric β-thalassemia patients treated with GLOBE LV vector-transduced stem cells [190]. The transfusion could be completely discontinued in children and reduced in adults.

In the context of sickle cell disease (SCD), which is caused by a single mutation in the β-globin chain of the adult $\alpha2\beta2$ hemoglobin tetramer [213], HSCs transduced with LV vectors expressing a βA-globin variant have demonstrated long-term expression for 10 months and accumulation of anti-sickling protein up to 52% of total hemoglobin in mouse models [191]. In a case report, LentiGlobin BB305-transduced bone marrow cells showed no SCD-related clinical events and the patient's transfusions could be discontinued [192]. In a phase I/II trial, autologous CD34$^+$ cells were transduced with LentiGlobin BB305 expressing the anti-sickling β^{A-T87Q} globin gene, which caused no adverse events in three SCD patients [193]. Clinical remission was observed in two patients, and the frequency of transfusions could be reduced in one patient.

Among hematological diseases, leukemias, lymphomas, and myelomas have also been subjected to gene therapy applications using viral vectors, as described previously in more detail [214]. Briefly, LV (HIV) vectors expressing herpes simplex virus-thymidine kinase (HSV-TK) were administered intraperitoneally to adult T-cell leukemia (ATL)-NOD-SCID mice, which generated significantly lower levels of secreted IL-2 and prolonged survival of mice compared to administration of an HIV vector expressing GFP [194]. Expression of a hyperfusogenic gibbon ape leukemia virus envelope glycoprotein (GALV.fus) from a SIN vector resulted in antitumor activity against human acute myeloid leukemia (AML) xenografts in mice [195]. In another approach, the AAV6-CD33 vector carrying an antibody-binding CD33 epitope targeting leukemia cells was utilized for the expression of the inducible caspase 9 (iCasp9) suicide gene in an AML xenotransplantation model in zebrafish [196]. AAV6-CD33-iCasp9 treatment resulted in antileukemic activity, a higher number of apoptotic cells, and prolonged survival.

In the case of lymphomas, the oncolytic Ad vector LOAd703 expressing CD40L and 4-1BBL was combined with chimeric antigen receptor (CAR) T-cell therapy, demonstrating increased killing of lymphoma cell lines and lymphomas in xenograft mouse models [197]. HSV amplicon vectors have been used for the expression of RANTES (HSVrantes) and the T-cell costimulatory ligand B7.1 (HSVB7.1) [198]. Complete EL4 tumor regression was observed in mice after intratumoral co-administration of HSVrantes and HSVB7.1, and in contralateral tumors. Similarly, intratumoral injection of the third generation HSV-1 T-01 vector provided tumor regression not only in injected tumors but also in non-injected contralateral tumors in mice [199]. In another approach, AAV8 expressing the humanized single-chain variable fragment (scFV)-Fc fusion minibody of the anti-CCR4 monoclonal antibody h1567 showed strong antitumor activity and prolonged survival in mice after a single intravenous infusion [200]. The oncolytic SIN vector combined with the agonistic monoclonal antibody to the T-cell stimulatory receptor 4-1BB (α4-1BB Ab) showed complete eradication of a non-Hodgkin B cell lymphoma in an A20 mouse tumor model, and long-lasting antitumor immunity was established in surviving mice [201].

In the context of lymphomas, infectious oncolytic CVA21 RNA was intratumorally injected into KAS6/1 myeloma-bearing mice leading to rapid tumor regression, which was comparable to injection of fully infectious CVA21 particles [202]. Moreover, intravenous administration of the oncolytic VSV vector expressing interferon-β (IFN-β) eradicated myeloma cells and prolonged survival in immune-competent myeloma mice [203]. In addition, the oncolytic reovirus (Reolysin) showed selective replication and induced apoptosis in multiple myeloma cell lines and reduced the tumor burden in xenograft and syngeneic multiple myeloma mouse models [204].

3.5. Neurological Disorders

Several approaches have been explored for gene therapy of neurological disorders (Table 4). For instance, AAV-based expression of the glutamic acid decarboxylase 65 (GAD65) gene improved symptoms related to Parkinson's disease in a rat model, and relieved pain in a rat pain model [215]. In a comparative study, the glial cell-derived neurotrophic factor (GDNF) was expressed from Ad, AAV, and LV vectors resulting in regionally restricted GDNF expression in the striatum and substantia nigra, inhibition of toxin-induced degeneration of nigral dopamine neurons, and functional striatal dopamine innervation in a rat Parkinson's disease model [215]. Moreover, administration of AAV-GDNF or LV-GDNF to 6-hydroxydopamine (6-OHDA)-lesioned rats and 1-methyl-4-phenyl-1,2,3,6-tetrahydropyridine (MTTP)-lesioned primates generated sustained GDNF delivery for 3–6 months, which contributed to regeneration and functional recovery [216]. In another study, it was demonstrated that LV-GDNF administration to the striatum and substantia nigra reversed functional and motor deficits and completely prevented nigrostriatal degradation in MPTP-lesioned rhesus macaques [217]. In clinical settings, in a phase I clinical trial, the human aromatic-l-amino acid decarboxylase (hAAD) expressed from an AAV vector showed good tolerance, only minor adverse events, and a significant improvement in the Parkinson's Disease Rating Scale (UPDRS), which was sustained for at least 2 years in patients with moderate to advanced Parkinson's disease [218]. In a phase I/II clinical trial, tyrosine hydroxylase (TH), aromatic amino acid dopa decarboxylase (AADC), and GTP-cyclohydroxylase-1 (GCH-1) expressed from LV vectors (ProSavin) were subjected to intrastriatal administration in Parkinson's disease patients, which was safe, well tolerated, and provided significant improvement of motor function [219]. Moreover, a long-term phase I/II follow-up study with ProSavin showed a significant improvement in the UPDRS score 4 years after the treatment [220].

Table 4. Preclinical and clinical examples of viral vectors applied for neurological disorders, muscular diseases, and immunodeficiency.

Viral Vector	Phase	Findings
Neurological		
AAV-GAD65	Pre	Improved symptoms of Parkinson's disease in rats [215]
AAV-GAD65	Pre	Pain relief in rat pain model [215]
Ad-GDNF	Pre	Inhibition of toxin-induced degeneration of neurons in rat model [216]
AAV-GDNF	Pre	Inhibition of toxin-induced degeneration of neurons in rat model [216]
LV-GDNF	Pre	Inhibition of toxin-induced degeneration of neurons in rat model [216]
AAV-GDNF	Pre	GDNF for 3–6 months, regeneration, functional recovery in rats, primates [216]
LV-GDNF	Pre	GDNF for 3–6 months, regeneration, functional recovery in rats, primates [216]
LV-GDNF	Pre	Reversed functional and motor deficits, prevented degradation in primates [217]
AAV-hAAD	Phase I	Significant improvement in UPRDS in Parkinson's disease patients [218]
LV-ProSavin	Phase I/II	Safe, well tolerated, improved motor function in Parkinson's disease patients [219]
LV-ProSavin	Phase I/II	Significantly improved 4-year UPRDS score in Parkinson's disease patients [220]
AAV2/5-NGF	Pre	Long-term neuroprotection in rat Alzheimer's disease model [221]
AAV2/5-NGF	Phase I	Inconclusive results in Alzheimer's disease patients [222]
AAV-APPsα	Pre	Functional special memory, mitigated synaptic and cognitive deficits in mice [223]
LV-GDNF	Pre	Preserved learning and memory in mouse Alzheimer's disease model [224]
LV-Klotho	Pre	Less cognitive deficits and Alzheimer's disease-like pathologies in mice [225]
AAV5-miHTT	Pre	Prevention of ATT aggregate formation, neuronal dysfunction in HD rat model [226]
AAV-miHTT	Pre	Reduced mutant HTT mRNA and protein in transgenic HD minipig brain [227]
AAV-miHTT	Phase I/II	Study in progress on disease progression in Huntington's disease patients [228]
AAV9-MeCP2	Pre	Prolonged survival in a mouse Rett syndrome model [229]
AAV9-SMN	Phase I	Improved motor function, prolonged survival in SMA patients [230]
AAV9-SMN	Phase I	Improved motor function, prolonged survival in SMA patients [231]
AAV9-SMN	Approval	Approved for treatment of SMA patients in the US, the EU, and Canada [232]
Muscular		
Ad-ΔDys	Pre	Restored dystrophin protein levels in mice [233]
AAV-μDys	Pre	Amelioration of dystrophin phenotype in transgenic mtx mice [234]
AAV6-μDys	Pre	Reduced skeletal muscle pathology, prolonged lifespan in dystrophic mice [235]
AAV6-μDys	Pre	Efficient delivery of dystrophin in canine dystrophin model for 2 years [236]
AAV6-μDys	Phase I/II	Therapeutic levels of μDys, improved NSAA score in all DMD patients [237]
AAV9-μDys	Phase I	Study in progress in 4–12-year-old DMD patients [238]
AAV-PABPN1	Pre	Decreased muscle fibrosis, normal muscle strength in OPMD mouse model [239]
LV-PABPN1	Pre	Efficient ex vivo transduction and rescue of myoblasts from OPMD patients [240]
Immunodeficiency		
γRV-IL2RG	CR	Long-lasting clinical benefits in 8 out of 10 SCID-X1 patients [241]
γRV-IL2RG	CR	Normal growth, protection against infections in SCID-X1 patients after 18 years [242]
γRV-IL2RG	CR	Sustained clinical benefits in 10 SCID-X1 patients [243]
γRV-IL2RG	CR	T-ALL in SCID-X1 patients after unfavorable integration of the γRV vector [8,244]
SIN-γRV	CR	Successful treatment of 9 SCID-X1 patients without leukemia development [245]
SIN-LV	CR	Successful treatment of 44 SCID-X1 patients without leukemia development [245]
SIN-LV-ABCD1	CR	Prevention of progressive demyelination, clinical benefits in ALD patients [40]
SIN-γRV/LV-ADA	CR	Sustained ADA expression, metabolic correction in >100 SCID-ADA patients [246]

AAV, adeno-associated virus; ABCD1, adenosine triphosphate.binding cassette transporter; Ad, adenovirus; ADA, adenosine deaminase; ALD, adrenoleukodystrophy; ΔDys, truncated dystrophin; DMD, Duchenne muscular dystrophy; GAD65; glutamic acid decarboxylase; GDNF, glial-derived neurotrophic factor; hAAD, human aromatic-l-amino decarboxylase; HD, Huntington's disease; IL2RG, interleukin-2 receptor gamma subunit; HTT, huntingtin; LV, lentivirus; MeCP2, methyl CpG binding protein 2; μDys, mircro-dystrophin; miHTT, micro-RNA targeting HTT; NSAA, North Star Ambulatory Assessment; OPMD, oculopharyngeal muscular dystrophy; PABPN1, poly A-binding protein nuclear 1; Pre, preclinical studies; ProSavin, LV vector expressing tyrosine hydroxylase, aromatic amino acid dopa decarboxylase, and GTP-cyclohydroxylase-1; γRV, gamma retrovirus; SCID-X1, X-linked severe combined immunodeficiency; SIN-LV, self-inactivating LV; SMA, spinal muscular atrophy; SMN, survival motor neuron; T-ALL, T-cell acute lymphoblastic leukemia; UPDRS, United Parkinson's Disease Rating Scale.

In the case of Alzheimer's disease, a chimeric AAV2/5 vector with the AAV2 genome and the AAV5 capsid structure has been applied for the expression of the nerve growth factor (NGF) [221]. In comparison to AAV2-NGF, the AAV2/5-NGF showed superior transduction of septal cholinergic neurons in rats, which provided long-term neuroprotection. Although preclinical studies have shown promising results regarding neuroprotection, the results from a phase I trial with AAV2/5-NGF were inconclusive [222]. In another approach, the secreted amyloid precursor protein (AAPsα) was expressed from AAV vectors, which resulted in functional rescue of spatial memory and mitigated synaptic and cognitive deficits in mice [223]. Moreover, LV-GDNF administration preserved learning and memory in mice; although, the amyloid and tau pathologies were not reduced [224]. However, the upregulation of the brain-derived neurotrophic factor (BDNF) was induced, which can contribute to neuronal protection against atrophy and degeneration. In another study, LV-based expression of the anti-aging gene Klotho efficiently ameliorated cognitive deficits and Alzheimer's disease-like pathologies in the brains of APP/presenilin-1 transgenic mice [225].

Huntington's disease, caused by a mutation in the huntingtin (HTT) gene, has been explored for AAV-based gene silencing with miRNAs targeting HTT [226]. Administration of AAV5-miHTT suppressed mutant HTT mRNA, resulting in almost complete prevention of mutant HTT aggregate formation and suppression of DARPP-32-associated neuronal dysfunction in a rat model for Huntington's disease [226]. Moreover, AAV5-miHTT significantly decreased mutant HTT mRNA and protein levels in the brain of transgenic HD minipigs [227]. A phase I/II clinical trial is in progress for the evaluation of safety, tolerability, and proof-of-concept of a single-time bilateral injection of AAV-miHTT (AMT-1309) in adults with early-stage Huntington's disease compared with control individuals for disease progression [228]. In the context of the X-linked Rett syndrome (RTT), the transcription regulator methyl CpG-binding protein 2 (MeCP2) was expressed from an AAV9 vector showing prolonged survival in an RTT mouse model [229]. In attempts to treat spinal muscular atrophy (SMA), which is associated with muscle weakness and atrophy, but caused by deterioration of motor neurons in the brainstem and spinal cord, an AAV9 vector has been employed for the expression of the survival motor neuron (SMN) gene [230]. In a phase I trial, AAV9-SMN delivery generated remarkable improvements in motor function and survival rates [230]. In another phase I study, a single intravenous AAV9-SMN injection improved motor function and extended survival in SMA patients [231]. AAV9-SMN1 has been approved in the US for treatment of children with SMA up to the age of two years, and in the EU and Canada in SMA patients under the brand name Zolgensma [232].

3.6. Muscular Diseases

Several gene therapy applications targeting muscular diseases, particularly various muscular dystrophies, have been successful [233]. For example, related to Duchenne muscular dystrophy (DMD), Ad-based expression of a truncated form of dystrophin restored dystrophin-related protein levels in mouse skeletal muscle [234]. The large size of dystrophin has been a major issue for AAV-based expression due to its limited packaging capacity, which has led to the engineering of "micro-dystrophin" cassettes (μDys) [235]. AAV-μDys were used for the production of transgenic mtx mice, which ameliorated the dystrophin phenotype with restored levels of normal C57BL/10 mice [235]. Moreover, AAV6-μDys restored dystrophin levels in respiratory, cardiac, and limb musculature, reducing the skeletal muscle pathology, and substantially prolonging the lifespan of severely dystrophic mice [236]. Additionally, the AAV6-μDys resulted in efficient delivery of dystrophin throughout different skeletal muscles in a canine dystrophin model, which lasted for at least two years [237]. In the context of clinical trials, AAV6-μDys has been subjected to a phase I/II trial in DMD patients, in which, according to interim results, therapeutic levels of μDys, 81% dystrophin-positive fibers, and improvement in the North Star Ambulatory Assessment (NSAA) score were seen in all patients [238]. Moreover, a phase I trial with the AAV9-mini-dystrophin vector is in progress in 4-12-year-old DMD patients for the verification of safety, tolerability, dystrophin expression and distribution, and muscle strength [247]. Several other AAV-based phase I/II and phase III are in progress in DMD patients (NCT03368742, NCT03375164, and NCT04281485), showing minimal adverse events, good safety in four patients, robust expression of μDys, and functional muscle improvement based on interim results [238].

In the case of oculopharyngeal muscular dystrophy (OPMD), which is caused by trinucleotide repeat expansion in the poly A-binding protein nuclear 1 (PABPN1) gene, patients suffer from late onset of ptosis, swallowing difficulties, and formation of nuclear aggregates in skeletal muscles [239]. Significant reduction in insoluble aggregates, decrease in muscle fibrosis, and normalization of muscle strength was seen in an OPMD mouse model after AAV-PABPN1 administration [240]. For ex vivo studies in myoblasts from OPMD patients, LV-based delivery was utilized due to the low transduction efficacy of AAV in primary myoblasts [241]. In contrast, the LV-PABPN1 transduction was efficient and provided myoblast cell rescue.

3.7. Immunodeficiency

The area where gene therapy has seen the greatest progress is undoubtedly in immunodeficiency, and the treatment of SCID and other immunodeficiencies. Despite the great excitement due to successful defective Moloney γRV-based correction of SCID-X1 in children, a major setback was encountered as the therapeutic gene was inserted into the LMO2 proto-oncogene region of the genome leading to leukemia development in a few patients [8,228]. In this first clinical trial, CD34$^+$ cells were transduced with the RV vector expressing the interleukin-2 receptor gamma subunit (IL2RG) in 10 SCID patients, which established normal T-cell counts within 3–6 months and demonstrated long-lasting clinical benefits in 8 out of 10 patients [242]. Remarkably, in a follow-up study of 18 years, all but one patient presented normal growth and protection against infections associated with SCID-X1 disease [243]. In another study, sustained clinical benefits were obtained in 10 SCID-X1 patients [244], although 2-14 years after the therapeutic intervention, T-cell acute lymphoblastic leukemia (T-ALL) was discovered in patients where the γRV vector was integrated either into the LMO2 [8] or the CCDN2 locus [245]. For this reason, SIN-γRV vectors have been engineered, which has confirmed that no cases of leukemia developed in nine newly treated SCID-X1 patients [40]. Similarly, engineering of SIN-LV vectors allowed successful treatment of another 44 SCID-X1 patients without any leukemia development [40].

In the context of X-linked adrenoleukodystrophy (ALD), SIN-LV vector expressing the adenosine triphosphate-binding cassette transporter (ABCD1) were ex vivo transduced into patient-derived autologous CD34$^+$ cells [248]. When SIN-LV-ABCD1 transduced cells were reinfused in two ALD patients, progressive cerebral demyelination was prevented providing clear clinical benefits [248]. Related to adenosine deaminase-severe combined immunodeficiency (ADA-SCID), the defected adenosine deaminase (ADA) gene [246] has been replaced by delivery with SIN-γRV or SIN-LV vectors [246]. Today, more than 100 ADA-SCID patients have been treated, resulting in sustained ADA expression, metabolic correction, and high overall survival [249].

3.8. Other Diseases

In addition to the disease indications described above, other disease areas such as ophthalmologic and lung diseases have been subjected to gene therapy. Moreover, infectious diseases have been mainly subjected to vaccine development, which in a broad sense can be considered as gene therapy. As these areas have previously been described in detail elsewhere [250], only a short summary is included here (Table 5).

Ophthalmology has been considered as a favorable area for gene therapy due to the relatively easy access to treatable space, allowing topical administration of gene therapy vectors. For example, intravitreal administration AAV vectors expressing the brain-derived neurotrophic factor (BDNF) showed protection of retinal ganglion cells, and reduced the intraocular pressure in a rat glaucoma model [251]. The intraocular pressure could also be reduced in mice by overexpression of matrix metalloproteinase 3 (MMP-3) from an AAV vector [252]. Regarding macular dystrophy X-linked retinoschisis (XLRS), the loss of the extracellular matrix protein retinoschisis 1 (RS1) was compensated by AAV-based delivery of RS1 to the eye of RS1 knockout mice, which generated significant improvement in retinal structure and function [253]. AAV vectors have been utilized for gene therapy of color blindness, achromatopsia [254]. As mutations in the cyclic nucleotide gated channel (CNGC) and the guanine nucleotide α-transducin (GNAT) genes cause achromatopsia, AAV-GNAT2 expression under the control of a human red cone opsin promoter was used to restore color vision in mice [255]. Improved photopic electrophysiological responses and functional vision were obtained in dogs subjected to subretinal injection of AAV5-CNGB3 [256].

Table 5. Preclinical and clinical examples of viral vectors applied for ophthalmologic and lung diseases, and vaccine development against infectious diseases.

Viral Vector	Phase	Findings
Ophthalmologic		
AAV-BDNF	Pre	Retinal ganglion cell protection, reduced intraocular pressure in rat glaucoma [251]
AAV-MMP-3	Pre	Reduced intraocular pressure in mice [252]
AAV-RS1	Pre	Significant improvement in retinal structure, function in RS1 knockout mice [253]
AAV-GNAT	Pre	Restoration of color vision in mice [255]
AAV5-CNGB3	Pre	Improved photopic electrophysiological responses, functional vision in dogs [256]
AAV-sFLT01	Phase I	Good safety and tolerability in AMD patients [257]
AAV-sFLT01	Phase IIa	No serious adverse events, improved vision in AMD patients [258]
AAV2-ND4	Phase I	Significant improvement of visual acuity in LHON patients [259]
AAV2-ND4	Phase I	Enhanced visual acuity in LHON patients [260]
AAV2-RPE65	Phase III	Maximum vision improvement in patients with inherited retinal dystrophy [261]
AAV2-RPE65	Approval	Approved for treatment of visual loss in the US, Australia, and Canada [262]
Lung		
AAV-CFTR	Pre	Long-term (6 months) CFTR expression in rabbit airway epithelium [263]
AAV-CFTR	Pre	Safe delivery of CFTR DNA to rhesus macaque lung [264]
HD-Ad-CFTR	Pre	Transduction of airway basal cells from CF patients, restored CFTR activity [265]
HIV-CFTR	Pre	Partial recovery of CFTR function in CF knockout mice for 110 days [266]
FIV-CFTR	Pre	Restored CFTR activity in CF pigs [267]
SIV-CFTR	Pre	Functional CFTR in mouse lung, human air–liquid interface cultures [268]
Infectious		
ChAdOx1 nCoV-19	Phase III	Good safety and 62–90% vaccine efficacy [269]
ChAdOx1 nCoV-19	Approval	Granted EUA in the UK [270]
Ad5.S-nb2	Phase II	Strong immunogenicity, good safety in adults [271]
Ad5.S-nb2	Approval	Granted EUA in China [270]
rAd26-S/rAd5-S	Phase III	91.6% vaccine efficacy from interim results [272]
rAd26-S/rAd5-S	Approval	Granted EUA in Russia [273]
Ad26.COV2.S	Phase III	Vaccine efficacy after single dose [274]
Ad26.COV2.S	Approval	Granted EUA in the US [270]
VSV-ZEBOV	Phase III	Good vaccine efficacy in Guinea and Sierra Leone [275,276]
VSV-ZEBOV	Approval	Approval as Ervebo for vaccination against EVD [277]

AAV, adeno-associated virus; Ad, adenovirus; AMD, age-related macular degeneration; BDNF, brain-derived neurotrophic factor; CF, cystic fibrosis; CFTR, cystic fibrosis transmembrane conductance regulator; CNGB3, cyclic nucleotide gated channel B3; EUA, emergency use authorization; FIV, feline immunodeficiency virus; GNAT, guanine nucleotide transducing; HD-Ad, helper-dependent adenovirus; HIV, human immunodeficiency virus; LHON, Leber's hereditary optic neuropathy; MMP-3, matrix metalloproteinase 3; ND4, NADH dehydrogenase protein subunit 4; Pre, preclinical studies; RS1, retinoschisis 1; sFLOT01, fusion protein of VEGF and the Fc portion of the human IgG1; SIV, simian immunodeficiency virus.

Regarding clinical applications, AAV vectors expressing the sFLT01 fusion protein comprising the VEGF and the Fc portion of the human IgG1 showed good safety and tolerability in a phase I trial in 19 age-related macular degeneration (AMD) patients [257]. Furthermore, no treatment-related serious adverse events were recorded, but improved vision was registered in 11 AMD patients treated with AAV-sFLT01 in a phase IIa study [243,258]. Leber's hereditary optic neuropathy (LHON), characterized by rapid loss of vision, is caused by a mutation in the NADH dehydrogenase protein subunit 4 (ND4) [278]. AAV2-ND4 treatment resulted in significant improvement in visual acuity in six out of nine LHON patients [259]. In another phase I trial, modest but statistically significant improved visual acuity was seen for 14 LHON patients [260]. Moreover, patients with RPE65-mediated inherited retinal dystrophy were subjected to AAV2-based RPE65 gene replacement therapy in a phase III study, which provided maximum possible vision improvement [261]. AAV2-RPE65, Voretigene neparvovec, has been approved for treatment of visual loss due to inherited retinal dystrophy in patients in the US, Australia, and Canada under the brand name Luxturna [262].

Gene therapy for lung diseases has mainly focused on the potential of developing some breakthrough treatment for cystic fibrosis. As cystic fibrosis is caused by mutations in the cystic fibrosis transmembrane conductance regulator (CFTR) gene [263], viral vector based CFTR expression represents an attractive approach. For example, AAV2-CFTR administered via fiberoptic bronchoscopy to the rabbit lung provided CFTR expression for at least 6 months in the airway epithelium [279]. In another approach, AAV2-CFTR was administered to the right lower lung lobe of rhesus macaques resulting in safe long-term delivery of CFTR DNA [264]. A helper-dependent Ad (HD-Ad) vector has also been engineered for intranasal delivery in mice and bronchoscopic instillation in pigs [265]. The HD-Ad-CFTR also demonstrated transduction of human airway basal cells from cystic fibrosis patients and restoration of CFTR channel activity [265]. Among LV vectors, HIV-based expression of CFTR in the mouse epithelium resulted in a partial recovery of electrophysiological functions in cystic fibrosis knockout mice for at least 110 days [266]. Moreover, a FIV-CFTR based vector pseudotyped with the GP64 protein restored CFTR activity in pigs with cystic fibrosis [267]. SIV-based functional expression of CFTR was also established in mouse lung and in human air–liquid interface cultures as a preparation for the first in-human trial [268].

Finally, vaccine development against infectious diseases using viral vectors has been very successful, recently. Needless to say, the unprecedented rapid development of different Ad-based COVID-19 vaccines has strongly contributed to the downgrading of the COVID-19 pandemic to an endemic status. The ChAdOx1 nCoV-19 vaccine [269], based on the ChAdOx1 chimpanzee Ad, and the Ad5-S-nb2 vaccine [271], based on the human Ad5 serotype, carry the SARS-CoV-2 spike (S) protein as an antigen and have demonstrated high vaccine efficacy in phase III clinical trials after two immunization doses. In contrast, the rAd26-S/dAd5-S (Sputnik) vaccine [272] is based on a prime vaccination with the Ad26 serotype expressing the S protein, followed by a booster vaccination with the Ad5 serotype also expressing the S protein, showing good efficacy in phase II and III studies. The strategy of this vaccination regimen is to limit immune reactions against Ad and reduction in vaccine efficacy by using another Ad serotype for the booster vaccination. The Ad26 serotype-based Ad26.COV2.S vaccine [274] also expresses the S protein, but in contrast to the other Ad-based vaccines, a single immunization has shown efficacy in clinical trials. The positive results from clinical trials supported the granting of Emergency Use Authorization (EUA) for the ChAdOx1 nCoV-19 vaccine in the UK in December 2020, the Ad26.COV2.S vaccine in the US in February 2021, and the Ad5-S-nb2 vaccine in China in February 2021 [270]. Controversially, the rAd26-S/rAd5-S vaccine received approval in Russia already in August 2020, after only being preliminary evaluated in 76 Russian volunteers [273]. Although good safety and vaccine efficacy have been achieved, emerging SARS-CoV-2 variants and detection of rare serious adverse events due to mass vaccinations will require intelligent re-engineering of existing vaccines to meet the new demands.

In the context of other vaccines, the VSV-based Ebola virus vaccine (VSV-ZEBOV) has demonstrated good safety profiles and excellent efficacy in two phase III studies conducted in Guinea [275], and in Guinea and Sierra Leone [276]. In 2020, the VSV-ZEBOV vaccine was approved under the name Ervebo for vaccinations against Ebola virus disease (EVD) [277].

4. Conclusions

In summary, viral vectors have been successfully applied for a broad spectrum of disease indications. Encouraging results have been obtained in preclinical studies in animal models, and also in clinical trials in patient groups with long-lasting cure, confirmed especially in SCID-X1 patients [243]. Gendicine™, a replication-deficient Ad vector expressing the p53 gene, was approved in China [280], and more than 30,000 patients with head and neck cancer have been treated with it. Gendicine™ has demonstrated good safety and efficacy, especially in combination with chemo- and radiotherapy [281]. In the US and Europe, a second-generation oncolytic HSV vector expressing GM-CSF has been approved for melanoma therapy [124]. The AAV-based Onasemnogene aboparvovec (Zolgensma) has been approved for the treatment of SMA [232]. Furthermore, approval for the AAV-based Voretigene neparvovec (Luxturna) was received for the treatment of inherited retinal dystrophy [262]. As mentioned above, several Ad-based COVID-19 vaccines have been granted EUA [272], and the Ebola vaccine Ervebo has been approved [277]. However, it is important to keep in mind the case of Glybera™, the AAV-based treatment of lipoprotein lipase deficiency [282]. Despite its approval in Europe, the clinical use of Glybera™ was discontinued due to lack of demand for this rare monogenic inherited disease.

Moreover, several issues need still to be addressed to make viral vector-based gene therapy highly attractive. Despite the advantage of viral vector-based gene delivery compared to non-viral vector systems, the safety of using particularly oncolytic and replication-proficient vectors is of utmost importance. Safety issues have also surfaced related chromosomal integration, where random integration has caused severe adverse events. Another issue of concern has been the difficulties in transferring successful proof-of-concept findings from rodents to larger animals and especially to humans. A potential "bridge" to success, particularly in the field of cancer therapy, has been to target domestic animals. For example, canines develop natural tumors and in addition to developing veterinary drugs, they serve as a potential model for pre-evaluation of efficacy before conducting human trials, partly because they represent good models for delivery to a larger organism, and partly because the natural tumors in canines closely resemble human cancers in contrast to induced and implanted tumors in rodent models.

Furthermore, an often-asked question is which viral vector system, and which therapeutic target should be chosen. Based on all gene therapy examples described in this review, it is obvious that there is not a single vector suitable for all applications. For this reason, viral vector diversity is important in research and development of promoting gene therapy. Due to the extensive number of preclinical and clinical trials conducted with viral vectors, the goal has been to give an overview of which viral vectors are suitable for which indication. For example, self-replicating RNA viruses have proven excellent for high-level short-term transgene expression required for cancer therapy, and development of vaccines against infectious diseases and cancers. In contrast, inherited diseases and chronic diseases, such as immunodeficiency, hematological diseases, and muscular dystrophy, which require long-term expression of therapeutic genes albeit not necessarily at high levels, have favored the application of Ad, AAV, HSV, RV, and LV vectors. Both vectors providing extrachromosomal expression and chromosomal integration have proven useful for therapeutic efficacy, lasting for several years. As with any other method of drug development, the management of serious adverse events is important. Not unexpectedly, the delivery of viral vectors causes adverse events, as does generally any drug. For this reason, efforts have been made to reduce the risk of using viral vectors and to decrease the severity of adverse events by the deletion of non-essential genetic material from viral vectors, the use of attenuated or less cytopathogenic viral vectors, and monitoring the spread of viruses and establishing a control of their replication and expression capacity. As seen for long-term follow-up studies, treatments for several years have not revealed adverse events, including the extreme example of 18 years of therapeutic efficacy without any side effects in SCID-X1 patients treated with RV vectors. These positive findings have encouraged the transition to clinical applications. However, in the light of the ever-tightening requirements associated with clinical evaluation, it is important to, already at an early stage of vector development, include appropriate design and engineering steps to fully comply with the requirements for clinical studies, and to facilitate regulatory implementations.

Funding: No funding was received for writing this review.

Institutional Review Board Statement: Not applicable.

Informed Consent Statement: Not applicable.

Conflicts of Interest: The author declares no conflict of interest.

References

1. Lundstrom, K. New era in gene therapy. In *Novel Approaches and Strategies for Biologics, Vaccines and CancerTherapies*; Elsevier: San Diego, CA, USA, 2015; pp. 15–37.
2. Martínez, T.; Wright, N.; López-Fraga, M.; Jiménez, A.I.; Paneda, C. Silencing human genetic diseases with oligonucleotide-based therapies. *Hum. Genet.* **2013**, *132*, 481–493. [CrossRef]
3. Bobbin, M.L.; Rossi, J.J. RNA interference (RNAi)-based therapeutics: Delivering on the promise? *Annu. Rev. Pharmacol. Toxicol.* **2016**, *56*, 103–122. [CrossRef]
4. Ramirez-Montagut, T. Cancer vaccines. In *Novel Approaches and Strategies for Biologics, Vaccines and Cancer Therapies*; Elsevier: San Diego, CA, USA, 2015; pp. 365–388.
5. Anurogo, D.; Yuli Prasetyo Budi, N.; Thi Ngo, M.H.; Huang, Y.-H.; Pawitanm, J.A. Cell and Gene Therapy for Anemia: Hematopoietic Stem Cells and Gene Editing. *Int. J. Mol. Sci.* **2021**, *22*, 6275. [CrossRef]
6. Sermer, D.; Brentjens, R. CAR-T cell therapy: Full speed ahead. *Hematol. Oncol.* **2019**, *37*, 95–100. [CrossRef]
7. Lino, C.A.; Harper, J.C.; Carney, J.P.; Timlin, J.A. Delivering CRISPR: A review of the challenges and approaches. *Drug Delivery* **2018**, *25*, 1234–1257. [CrossRef]
8. Hacein-Bey-Abina, S.; Garrigue, A.; Wang, G.P.; Soulier, J.; Lim, A.; Morillon, E.; Clappier, E.; Caccavelli, L.; Delabesse, E.; Beldjord, K.; et al. Insertional oncogenesis in 4 patients after retrovirus-mediated gene therapy of SCID-X1. *J. Clin. Investig.* **2008**, *118*, 3132–3142. [CrossRef]
9. Raper, S.E.; Chirmule, N.; Lee, F.S.; Wivel, N.A.; Bagg, A.; Gao, G.-P.; Wilson, J.M.; Batshaw, M.L. Fatal systemic inflammatory response syndrome in an ornithine transcarbamylase deficient patient following adenoviral gene transfer. *Mol. Genet. Metab.* **2003**, *80*, 148–158. [CrossRef]
10. Bulcha, J.T.; Wang, Y.; Ma, H.; Tai, P.W.L.; Gao, G. Viral vector platforms within the gene therapy landscape. *Signal Transd. Targeted Ther.* **2021**, *6*, 53. [CrossRef]
11. Rothe, M.; Modlich, U.; Schambach, A. Biosafety challenges for use of lentiviral vectors in gene therapy. *Curr. Gene Ther.* **2013**, *13*, 453–468. [CrossRef]
12. Fukuhara, H.; Ino, Y.; Todo, T. Oncolytic virus therapy: A new era of cancer treatment at dawn. *Cancer Sci.* **2016**, *107*, 1373–1379. [CrossRef]
13. Tatsis, N.; Ertl, H.C.J. Adenoviruses as vaccine vectors. *Mol. Ther.* **2004**, *10*, 616–629. [CrossRef]
14. Ehrke-Schulz, E.; Zhang, W.; Schiwon, M.; Bergmann, T.; Solanki, M.; Liu, J.; Boehme, P.; Leitner, T.; Ehrhardt, A. Cloning and Large-Scale Production of High-Capacity Adenoviral Vectors Based on the Human Adenovirus Type 5. *J. Vis. Exp.* **2016**, *107*, e52894.
15. Crystal, R.G. Adenovirus: The First Effective in Vivo Gene Delivery Vector. *Hum. Gene Ther.* **2014**, *25*, 3–11. [CrossRef]
16. Wen, S.; Schneider, D.B.; Driscoll, R.M.; Vassalli, G.; Sassani, A.B.; Dichek, D.A. Second-generation adenovirus vectors do not prevent rapid loss of transgene expression and vector DNA from the arterial wall. *Arterioscler. Thromb. Vasc. Biol.* **2000**, *20*, 1452–1458. [CrossRef]
17. Ricobaraza, A.; Gonzalez-Aparicio, M.; Mora-Jimenez, L.; Lumbreras, S.; Hernadez-Alcoceba, R. High-Capacity Adenovirus Vectors: Expanding the Scope of Gene Therapy. *Int. J. Mol. Sci.* **2020**, *21*, 3643. [CrossRef]
18. Wang, F.; Wang, Z.; Tian, H.; Qi, M.; Zhai, Z.; Li, S.; Li, R.; Zhang, H.; Wang, W.; Fu, S.; et al. Biodistribution and safety assessment of bladder cancer specific oncolytic adenovirus in subcutaneous xenografts tumor model in nude mice. *Curr. Gene Ther.* **2012**, *12*, 67–76. [CrossRef]
19. Wei, Q.; Fan, J.; Liao, J.; Zou, Y.; Song, D.; Liu, J.; Cui, J.; Liu, F.; Ma, C.; Hu, X.; et al. Engineering the rapid adenovirus production and amplification (RAPA) cell line to expedite the generation of recombinant adenoviruses. *Cell Physiol. Biochem.* **2017**, *41*, 2383–2398. [CrossRef]
20. Ehrhardt, A.; Xu, H.; Kay, M.A. Episomal Persitence of Recombinant Adenoviral Vector Genomes during the Cell Cycle In Vivo. *J. Virol.* **2003**, *77*, 7689–7695. [CrossRef]
21. Brunetti-Pierri, N.; Ng, T.; Iannitti, D.; Cioffi, W.; Stapleton, G.; Law, M.; Breinholt, J.; Palmer, D.; Grove, N.; Rice, K.; et al. Transgene expression up to 7 years in nonhuman primates following hepatic transduction with helper-dependent adenoviral vectors. *Hum. Gene Ther.* **2013**, *24*, 761–765. [CrossRef]

22. Samulski, R.; Muzycka, N. AAV-mediated gene therapy for research and therapeutic purposes. *Annu. Rev. Virol.* **2014**, *1*, 427–451. [CrossRef]
23. Grieger, C.; Samulski, R.J. Packaging capacity of adeno-associated virus serotypes: Impact of larger genomes on infectivity and postentry steps. *J. Virol.* **2005**, *79*, 9933–9944. [CrossRef]
24. McClements, M.E.; MacLaren, R.R. Adeno-associated virus (AAV) dual vector strategies for gene therapy encoding large transgenes. *Yale J. Biol. Med.* **2017**, *90*, 611–623.
25. Park, K.; Kim, W.J.; Cho, Y.H.; Lee, Y.-I.; Lee, H.; Jeong, S.; Cho, E.-S.; Chang, S.-I.; Moon, S.-K.; Kang, B.-S.; et al. Cancer gene therapy using adeno-associated virus vectors. *Front. Biosci.* **2008**, *13*, 2653–2659. [CrossRef]
26. Mingozzi, F.; High, K.A. Immune responses to AAV vectors: Overcoming barriers to successful gene therapy. *Blood* **2013**, *122*, 23–36. [CrossRef] [PubMed]
27. Meliani, A.; Boisgerault, F.; Fitzpatrick, Z.; Marmier, S.; Leborgne, C.; Collaud, F.; Sola, M.S.; Charles, S.; Ronzitti, G.; Vignaud, A.; et al. Enhanced liver gene transfer and evasion of preexisting humoral immunity with exosome-enveloped AAV vectors. *Blood Adv.* **2017**, *1*, 2019–2031. [CrossRef]
28. Deyle, D.R.; Russell, D.W. Adeno-associated virus integration. *Curr. Opin. Mol. Ther.* **2009**, *11*, 442–447.
29. Wang, Z.; Lisowski, L.; Finegold, M.J.; Nakai, H.; Kay, M.A.; Grompe, M. AAV Vectors Containing rDNA Homology Increased Chromosomal Integration and Transgene Persistence. *Mol. Ther.* **2012**, *20*, 1902–1911. [CrossRef] [PubMed]
30. Singh, N.; Tscharke, D.C. Herpes Simplex Virus Latency Is Noisier The Closer We Look. *J. Virol.* **2020**, *94*, e01701-19. [CrossRef] [PubMed]
31. Epstein, A.L.; Marconi, P.; Argnani, R.; Manservigi, A. HSV-1 derived recombinant and amplicon vectors for gene transfer and gene therapy. *Curr. Gene Ther.* **2005**, *5*, 445–458. [CrossRef]
32. Morisette, G.; Flamand, L. Herpesviruses and chromosomal integration. *J. Virol.* **2010**, *84*, 12100–12109. [CrossRef]
33. Holmes, K.D.; Cassam, A.K.; Chan, B.; Peters, A.A.; Weaver, L.C.; Dekaban, G.A. A multi-mutant herpes simplex virus vector has minimal cytotoxic effects on the distribution of filamentous actin, alpha-actinin and a glutamate receptor in differentiated PC-12 cells. *J. Neurovirol.* **2000**, *6*, 33–45. [CrossRef] [PubMed]
34. Kasai, K.; Saeki, Y. DNA-based methods to prepare helper viurs-free herpes amplicon vectors and versatile design of amplicon vector plasmids. *Curr. Gene Ther.* **2006**, *6*, 303–314. [CrossRef]
35. Wu, N.; Watkins, S.C.; Schaffer, P.A.; De Luca, N.A. Prolonged gene expression and cell surviavl after unfection by a herpes simplex virus mutant defective in the immediate-early early genes encoding ICP4, ICP27, and ICP22. *J. Virol.* **1996**, *70*, 6358–6369. [CrossRef] [PubMed]
36. Li, J.M.; Kao, K.C.; Li, L.F. Micro-RNA-145 regulates oncolytic herpes simplex virus-1 for selective killing of human non-small lung cancer cells. *Virol. J.* **2013**, *10*, 241. [CrossRef] [PubMed]
37. Saeki, Y.; Fraefel, C.; Ichikawa, T.; Breakfield, X.O.; Chiocca, E.A. Improved helper virus-free packaging system for HSV amplicon vectors using an ICP27-deleted, oversized HSV-1 DNA in a bacterial artificial chromosome. *Mol. Ther.* **2001**, *3*, 591–601. [CrossRef]
38. Schambach, A.; Morgan, M. Retroviral vectors for cancer gene therapy. *Curr. Strat. Cancer Gene Ther.* **2016**, *209*, 17–35.
39. Lesbats, P.; Engelman, A.N.; Cherepanov, P. Retroviral DNA Integration. *Chem. Rev.* **2016**, *116*, 12730–12757. [CrossRef]
40. Hacein-Bey-Abina, S.; Pai, S.Y.; Gaspar, H.B.; Armant, M.; Berry, C.C.; Blanche, S.; Bleesing, J.; Blondeau, J.; de Boer, H.; Buckland, K.F.; et al. A modified γ-retrovirus vector for X-linked severe combined immunodeficiency. *N. Engl. J. Med.* **2014**, *371*, 1407–1417. [CrossRef]
41. Pai, S.-Y. Built to last: Gene therapy for ADA SCID. *Blood* **2021**, *138*, 1287–1288. [CrossRef]
42. Reinhardt, B.; Habib, O.; Shaw, K.L.; Garabedian, E.; Carbonaro-Saracino, D.A.; Terrazas, D.; Campo Fernandez, B.; De Oliveira, S.; Moore, T.B.; Ikeda, A.K.; et al. Long-term outcomes after gene therapy for adenosine deaminase severe combined immune deficiency. *Blood* **2021**, *138*, 1304–1316. [CrossRef]
43. Kohn, L.A.; Kohn, D.B. Gene therapies for primary immune deficiencies. *Front. Immunol.* **2021**, *120*, 3635–3646. [CrossRef]
44. Berg, K.; Schäfer, V.N.; Barthnicki, N.; Eggenschwiler, R.; Cantz, T.; Stitz, J. Rapid establishment of stable retroviral packaging cells and recombinant susceptible target cell lines employing novel transposon vectors derived from *Sleeping Beauty*. *Virology* **2019**, *531*, 192–202. [CrossRef]
45. Kay, M.A.; Glorioso, J.C.; Naldini, L. Viral vectors for gene therapy: The art of turning infectious agents into vehicles of therapeutics. *Nat. Med.* **2001**, *7*, 33–40. [CrossRef]
46. Ciuffi, A. Mechanisms governing lentivirus integration site selection. *Curr. Gene Ther.* **2008**, *8*, 419–429. [CrossRef]
47. Silvers, R.M.; Smith, J.A.; Schowalter, M.; Litwin, S.; Liang, Z.; Geary, K.; Daniel, R. Modification of integration site preferences of an HIV-1-based vector by expression of a novel synthetic protein. *Hum. Gene Ther.* **2010**, *2*, 337–349. [CrossRef]
48. Nakajima, T.; Nakamaru, K.; Ido, E.; Terao, K.; Hayami, M.; Hasegawa, M. Development of novel simian immunodeficiency virus vectors carrying a dual gene expression system. *Hum. Gene Ther.* **2000**, *11*, 1863–1874. [CrossRef] [PubMed]
49. Hartmann, K. Clinical aspects of feline retroviruses: A review. *Viruses* **2012**, *4*, 2684–2710. [CrossRef] [PubMed]
50. Olsen, J.C. Gene transfer vectors derived from equine infectious anemia virus. *Gene Ther.* **1998**, *5*, 1481–1487. [CrossRef] [PubMed]
51. Ferreira, M.V.; Cabral, E.T.; Coroadinha, A.S. Progress and Perspectives in the Development of Lentiviral Vector Producer Cells. *Biotechnol. J.* **2021**, *16*, e2000017. [CrossRef]
52. Strauss, J.H.; Strauss, E.G. The alphaviruses; gene expression, replication and evolution. *Microbiol. Rev.* **1994**, *58*, 491–562. [CrossRef]

53. Ehrengruber, M.U.; Schlesinger, S.; Lundstrom, K. Alphaviruses: Semliki Forest Virus and Sindbis Virus Vectors for Gene Transfer into Neurons. *Curr. Protocols Neurosci.* **2011**, *57*, 4–22. [CrossRef]

54. Lundstrom, K. Self-amplifying RNA viruses as RNA vaccines. *Int. J. Mol. Sci.* **2020**, *2020*, 5130. [CrossRef] [PubMed]

55. Liljeström, P.; Garoff, H. A new generation of animal cell expression vectors based on the Semliki Forest virus replicon. *Biotechnology* **1991**, *9*, 1356–1361. [CrossRef]

56. Xiong, C.; Levis, R.; Shen, P.; Schlesinger, S.; Rice, C.M.; Huang, H.V. Sindbis virus: An efficient, broad host range vector for gene expression in animal cells. *Science* **1989**, *243*, 1188–1191. [CrossRef]

57. Davis, N.L.; Willis, L.V.; Smith, J.F.; Johnston, R.E. In vitro synthesis of infectious Venezuelan equine encephalitis virus RNA from a cDNA clone: Analysis of a viable deletion mutant. *Virology* **1989**, *171*, 189–204. [CrossRef] [PubMed]

58. Zhang, J.; Liu, Y.; Tan, J.; Zhang, Y.; Wong, C.-W.; Lin, Z.; Liu, X.; Sander, M.; Yang, X.; Liang, L.; et al. Necroptotic virotherapy of oncolytic alphavirus M1 cooperated with Doxorubicin displays promising therapeutic efficacy in TNBC. *Oncogene* **2021**, *40*, 4783–4795. [CrossRef]

59. Heikkilä, J.E.; Vähä-Koskela, M.J.; Ruotsalainen, J.J.; Martikainen, M.W.; Stanford, M.M.; McCart, J.A.; Bell, J.C.; Hinkkanen, A.E. Intravenously administered alphavirus vector VA7 eradicates orthotopic human glioma xenografts in nude mice. *PLoS ONE* **2010**, *5*, e8603. [CrossRef]

60. Pijlman, G.P.; Suhrbier, A.; Khromykh, A.A. Kunjin virus replicons: An RNA-based, non-cytopathic viral vector system for protein production, vaccine and gene therapy applications. *Exp. Opin. Biol. Ther.* **2006**, *6*, 134–145. [CrossRef] [PubMed]

61. Scholle, I.; Girard, Y.A.; Zhao, Q.; Higgs, S.; Mason, P.W. Trans-packaged West Nile virus-like particles: Infectious properties in vitro and in infected mosquito vectors. *J. Virol.* **2004**, *78*, 11605–11614. [CrossRef]

62. Pang, X.; Zhang, M.; Dayton, A.I. Development of dengue virus type 2 replicons capable of prolonged expression in host cells. *BMC Microbiol.* **2001**, *1*, 18.

63. Gherke, R.; Ecker, M.; Aberle, S.W.; Allison, S.L.; Heinz, F.X.; Mandl, C.W. Incorporation of tick-borne encephalitis virus replicons into virus-like particles by a packaging cell line. *J. Virol.* **2003**, *77*, 8924–8933. [CrossRef]

64. Jones, C.T.; Patkar, C.G.; Kuhn, R.J. Construction and applications of yellow fever virus replicons. *Virology* **2005**, *331*, 247–259. [CrossRef]

65. Zhu, Z.; Gorman, M.J.; McKenzie, L.D.; Chai, J.N.; Hubert, C.G.; Prager, B.C.; Fernandez, E.; Richner, J.M.; Zhang, R.; Shan, C.; et al. Zika virus has oncolytic activity against glioblastoma stem cells. *J. Exp. Med.* **2017**, *214*, 2843–2857. [CrossRef]

66. Khromykh, A.A.; Varnavski, A.N.; Westaway, E.G. Encapsidation of the flavivirus Kunjin replicon RNA by using a complementation system providing Kunjin virus structural proteins in trans. *J. Virol.* **1998**, *72*, 5967–5977. [CrossRef] [PubMed]

67. Apostolopoulos, V. Vaccine delivery methods into the future. *Vaccine* **2016**, *4*, 9. [CrossRef]

68. Lundstrom, K. Self-replicating vehicles based on negative strand RNA viruses. *Cancer Gene Ther.* **2022**, *15*, 1–14.

69. Billeter, M.A.; Naim, H.Y.; Udem, S.A. Reverse genetics of measles virus and resulting multivalent recombinant vaccines: Applications of recombinant measles virus. *Curr. Top. Microbiol. Immunol.* **2009**, *329*, 129–162. [PubMed]

70. Lal, G.; Rajala, M. Engineering of measles virus to target cancer cells, an attempt. *Intl. J. Infect. Dis.* **2016**, *45*, 333–334. [CrossRef]

71. Zhao, D.; Chen, P.; Yang, H.; Wu, Y.; Zeng, X.; Zhao, Y.; Wen, Y.; Zhao, X.; Liu, X.; Wei, Y.; et al. Live attenuated measles virus vaccine induces apoptosis and promotes tumor regression in lung cancer. *Oncol. Rep.* **2013**, *29*, 199–204. [CrossRef]

72. Boisgerault, N.; Guillerme, J.-B.; Pouliquen, D.; Mesel-Lemoine, M.; Achard, C.; Combredet, C.; Fontenau, J.F.; Tangy, F.; Grégoire, M. Natural oncolytic activity of live-attenuated measles virus against human lung and colorectal adenocarcinomas. *Biomed. Res. Int.* **2013**, *2013*, 387362. [CrossRef]

73. Finke, S.; Conzelmann, K.K. Recombinant rhabdoviruses: Vectors for vaccine development and gene therapy. *Curr. Top. Microbiol. Immunol.* **2005**, *292*, 165–200.

74. Nolden, T.; Finke, S. Rapid Reverse Genetics Systems for Rhabdovirus From Forward to Reverse and Back Again. *Methods Mol. Biol.* **2017**, *1602*, 171–184. [PubMed]

75. Urbiola, C.; Santer, F.R.; Petersson, M.; van der Pluijm, G.; Horninger, W.; Erlmann, P.; Wollmann, G.; Kimpel, J.; Culig, Z.; von Laer, D. Oncolytic activity of the rhabdovirus VSV-GP against prostate cancer. *Int. J. Cancer* **2018**, *143*, 1786–1796. [CrossRef] [PubMed]

76. Le Boeuf, F.; Selman, M.; Son, H.H.; Bergeron, A.; Chen, A.; Tsang, J.; Butterwick, D.; Arulanandam, R.; Forbes, N.E.; Tzelepis, F.; et al. Oncolytic Maraba Virus MG1 as a Treatment for Sarcoma. *Int. J. Cancer* **2017**, *141*, 1257–1264. [CrossRef] [PubMed]

77. Ito, N.; Takayama-Ito, M.; Yamada, K.; Hosokawa, J.; Sugiyama, M.; Minamoto, N. Improved recovery of rabies virus from cloned cDNA using a vaccinia virus-free reverse genetics system. *Microbiol. Immunol.* **2003**, *47*, 613–677. [CrossRef] [PubMed]

78. Ganar, K.; Das, M.; Sinha, S.; Kumar, S. Newcastle disease virus: Current status and our understanding. *Virus Res.* **2014**, *184*, 71–81. [CrossRef]

79. Schirrmacher, V.; Griesbach, A.; Ahlert, T. Antitumor effects of Newcastle disease virus in vivo: Local versus systemic effects. *Int. J. Oncol.* **2001**, *18*, 945–952. [CrossRef]

80. Kwak, H.; Honig, H.; Kaufman, H.L. Poxviruses as vectors for cancer immunotherapy. *Curr. Opin. Drug Discov. Devel.* **2003**, *6*, 161–168. [PubMed]

81. Zeh, H.J.; Bartlett, D.L. Development of a replication-selective oncolytic proxvirus for the treatment of human cancers. *Cancer Gene Ther.* **2002**, *9*, 1001–1012. [CrossRef]

82. Bradley, S.; Jakes, A.D.; Harrington, K.; Pandha, H.; Melcher, A.; Errington-Mais, F. Applications of coxsackievirus A21 in oncology. *Oncol. Virother.* **2014**, *3*, 47–55. [CrossRef]
83. Kim, D.-S.; Nam, J.-H. Application of attenuated coxsackievirus B3 as viral vector system for vaccines and gene therapy. *Hum. Vaccin.* **2011**, *7*, 410–416. [CrossRef] [PubMed]
84. Jia, Q.; Liang, F.; Ohka, S.; Nomoto, A.; Hashikawa, T. Expression of brain-derived neurotrophic factor in the central nervous system of mice using a poliovirus-based vector. *J. Neurovirol.* **2002**, *8*, 14–23. [CrossRef] [PubMed]
85. Clements, D.; Helson, E.; Gujar, S.A.; Lee, P.W. Reovirus in cancer therapy: An evidence-based review. *Oncol. Virother.* **2014**, *3*, 69–82.
86. Figova, K.; Hrabeta, J.; Eckschlager, T. Reovirus—Possible therapy of cancer. *Neoplasma* **2006**, *53*, 457–462.
87. Zhao, X.; Chester, C.; Rajasekaran, N.; He, Z.; Kohrt, H.E. Strategic Combinations: The Future of Oncolytic Virotherapy with Reovirus. *Mol. Cancer Ther.* **2016**, *15*, 767–773. [CrossRef]
88. Carew, J.S.; Espita, C.M.; Zhao, W.; Kelly, K.R.; Coffey, M.; Freeman, J.W.; Nawrocki, S.T. Reolysin is a novel reovirus-based agent that induces endoplasmic reticular stress-mediated apoptosis in pancreatic cancer. *Cell Death Dis.* **2013**, *4*, e728. [CrossRef] [PubMed]
89. Kimchi-Sarfaty, C.; Gottesman, M.M. SV40 pseudovirions as highly efficient vectors for gene transfer and their potential application in cancer therapy. *Curr. Pharm. Biotechnol.* **2004**, *5*, 451–458. [CrossRef] [PubMed]
90. Toscano, M.G.; van der Velden, J.; van der Werf, S.; Odijk, M.; Roque, A.; Camacho-Garcia, R.J.; Herrera-Gomez, I.G.; Mancini, I.; de Haan, P. Generation of a Vero-based packaging cell line to produce SV40 gene delivery vectors for use in clinical gene therapy studies. *Mol. Ther. Methods Clin. Dev.* **2017**, *6*, 124–134. [CrossRef]
91. Cordelier, P.; Bienvenu, C.; Lulka, H.; Marrache, F.; Bouisson, M.; Openheim, A.; Strayer, D.S.; Vaysse, N.; Pradayrol, L.; Buscail, L. Replication-deficient rSV40 mediate pancreatic gene transfer and long-term inhibition of tumor growth. *Cancer Gene Ther.* **2007**, *14*, 19–29. [CrossRef]
92. Luo, J.; Zhao, J.; Tian, Q.; Mo, W.; Wang, Y.; Chen, H.; Guo, X. A recombinant rabies virus carrying GFP between N and P affects viral transcription in vitro. *Virus Genes* **2016**, *52*, 379–387. [CrossRef]
93. An, H.; Kim, G.N.; Kang, C.Y. Genetically modified VSV(NJ) vector is capable of accommodating a large foreign gene insert and allows high level gene expression. *Virus Res.* **2013**, *171*, 168–177. [CrossRef] [PubMed]
94. Pol, J.G.; Zhang, L.; Bridle, B.W.; Stephenson, K.B.; Resseguier, J.; Hanson, S.; Chen, L.; Kazdhan, N.; Bramson, J.L.; Stojdl, J.F.; et al. Maraba virus as a potent oncolytic vaccine vector. *Mol. Ther.* **2014**, *22*, 420–429. [CrossRef]
95. Reichard, K.; Lorence, R.M.; Cascino, C.J.; Peeples, M.E.; Walter, R.J.; Fernando, M.B.; Reyes, H.M.; Greager, J.A. Newcastle disease virus selectively kills human tumor cells. *J. Surg. Res.* **1992**, *52*, 448–453. [CrossRef] [PubMed]
96. Cheng, X.; Wang, W.; Xu, Q.; Harper, J.; Carroll, D.; Galinski, M.S.; Suzich, J.; Jin, H. Genetic modification of oncolytic Newcastle disease virus for cancer therapy. *J. Virol.* **2016**, *90*, 5343–5352. [CrossRef]
97. Lundstrom, K. Application of Viruses for Gene Therapy and Vaccine Development. In *The Biological Role of a Virus. Advances in Environmental Microbiology*; Springer Nature: Cham, Switzerland, 2022; pp. 285–341.
98. Pastoret, P.-P.; Vanderplasschen, A. Poxviruses as vaccine vectors. *Comp. Immunol. Microbiol. Infect. Dis.* **2003**, *26*, 343–355. [CrossRef] [PubMed]
99. Gujar, S.A.; Marcato, P.; Pan, D.; Lee, P.W. Reovirus virotherapy overrides tumor antigen presentation evasion and promotes protective antitumor immunity. *Mol. Cancer. Ther.* **2010**, *9*, 2924–2933. [CrossRef]
100. Liu, Y.; Deisseroth, A. Tumor vascular targeting therapy with viral vectors. *Blood* **2006**, *107*, 3027–3033. [CrossRef]
101. Montaño-Samaniego, M.; Bravo-Estupiñan, D.M.; Méndez-Guerrero, O.; Alarcon-Hernandez, E.; Ibanez-Hernandez, M. rategies for Targeting Gene Therapy in Cancer Cells with Tumor-Specific Promoters. *Front. Oncol.* **2020**, *10*, 605380. [CrossRef]
102. Eissa, I.R.; Naoe, Y.; Bustos-Villalobos, I.; Ichinose, T.; Tanaka, M.; Zhiwen, W.; Mukoyama, N.; Morimoto, T.; Miyajima, N.; Hitoki, H.; et al. Genomic signature of the natural oncolytic herpes simplex virus HF10 and its therapeutic role in preclinical and clinical trials. *Front. Oncol.* **2017**, *7*, 149. [CrossRef] [PubMed]
103. Mostafa, A.A.; Meyers, D.E.; Thirukkumaran, C.M.; Liu, P.J.; Gratton, K.; Spurrell, J.; Shi, Q.; Thakur, S.; Morris, D.G. Oncolytic Reovirus and immune checkpoint inhibitor as a novel immunotherapeutic strategy for breast cancer. *Cancers* **2018**, *10*, 205. [CrossRef] [PubMed]
104. Lin, Y.; Zhang, H.; Liang, J.; Li, K.; Zhu, W.; Fu, L.; Wang, F.; Zheng, X.; Shi, H.; Wu, S.; et al. Identification and characterization of alphavirus M1 as a selective oncolytic virus targeting ZAP-defective human cancers. *Proc. Natl. Acad. Sci. USA* **2014**, *111*, E4504–E4512. [CrossRef]
105. Mohebtash, M.; Tsang, K.Y.; Madan, R.A.; Huen, N.Y.; Poole, D.J.; Jochems, C.; Jones, J.; Ferrara, T.; Heery, C.R.; Arlen, P.M.; et al. A pilot study of MUC-1/CEA/TRICOM poxviral-based vaccine in patients with metastatic breast and ovarian cancer. *Clin. Cancer Res.* **2011**, *17*, 7164–7173. [CrossRef] [PubMed]
106. Hu, J.; Cai, X.F.; Yan, G. Alphavirus M1 induces apoptosis of malignant glioma cells via downregulation and nucleolar translocation of p21WAF1/CIP1 protein. *Cell Cycle* **2009**, *8*, 3328–3339. [CrossRef]
107. Cai, J.; Zhu, W.; Lin, Y.; Zhang, S.; Chen, X.; Gong, S.; He, S.; Hu, J.; Yan, G.; Liang, J. Systematic Characterization of the Biodistribution of the Oncolytic Virus M1. *Hum. Gene Ther.* **2020**, *31*, 1203–1213. [CrossRef] [PubMed]
108. Roche, F.P.; Sheahan, B.J.; O'Mara, S.M.; Atkins, G.J. Semliki Forest virus-mediated gene therapy of the RG2 rat glioma. *Neuropathol. Appl. Neurobiol.* **2010**, *36*, 648–660. [CrossRef]

109. Huang, T.T.; Parab, S.; Burnett, R.; Diago, O.; Ostertag, D.; Hofman, F.M.; Lopez Espinoza, F.; Martin, B.; Ibanez, C.E.; Kasahara, N.; et al. Intravenous administration of retroviral replicating vector, Toca 511, demonstrates efficacy in orthotopic immune-competent mouse glioma model. *Hum. Gene Ther.* **2015**, *26*, 82–93. [CrossRef] [PubMed]

110. Viral Therapy in Treating Patients with Recurrent Glioblastoma Multiforme. Available online: www.clinicaltrials.govNCT00390 299 (accessed on 11 January 2023).

111. Cloughesy, T.F.; Landolfi, J.; Hogan, D.J.; Bloomfield, S.; Carter, B.; Chen, C.C.; Elder, J.B.; Kalkanis, S.N.; Kesari, S.; Lai, A.; et al. Phase I trial of vocimagine amiroretrorepvec and 5-fluorocytosine for recurrent high-grade glioma. *Sci. Transl. Med.* **2016**, *8*, 341ra75. [CrossRef] [PubMed]

112. Tocagen Reports Results of Toca 5 Phase 3 Trial in Recurrent Brain Cancer. Tocagen. Published 12 September 2019. Available online: https://bit.ly/2lPm19v (accessed on 11 January 2022).

113. Hoang-Le, D.; Smeenk, L.; Anraku, I.; Pijlman, G.P.; Wang, X.J.; de Vrij, J.; Liu, W.J.; Le, T.T.; Schroder, W.A.; Krohmykh, A.A.; et al. A Kunjin replicon vector encoding granulocyte macrophage colony-stimulating factor for intra-tumoral gene therapy. *Gene Ther.* **2009**, *16*, 190–199. [CrossRef] [PubMed]

114. Day, G.L.; Bryan, M.L.; Northrup, S.A.; Lyles, D.S.; Westcott, M.M.; Sterwart, J.H., 4th. Immune effects of M51R vesicular stomatitis virus treatment of carcinomatosis from colon cancer. *J. Surg. Res.* **2020**, *245*, 127–135. [CrossRef]

115. Wang, Y.; Huang, H.; Zou, H.; Tian, X.; Hu, J.; Qiu, P.; Ferreira, L. Liposome Encapsulation of Oncolytic Virus M1 To Reduce Immunogenicity and Immune Clearance in Vivo. *Mol. Pharm.* **2019**, *16*, 779–785. [CrossRef]

116. Ying, H.; Zaks, T.Z.; Wang, R.-F.; Irvine, K.R.; Kammula, U.S.; Marincola, F.M.; Leitner, W.W.; Restifo, N. Cancer therapy using a self-replicating RNA vaccine. *Nat. Med.* **1999**, *5*, 823–827. [CrossRef]

117. Crosby, E.J.; Hobeika, A.C.; Niedzwiecki, D.; Rushing, C.; Hsu, D.; Berglund, P.; Smith, J.; Osada, T.; Iii, W.R.G.; Hartman, Z.C.; et al. Long-term Survival of Patients with Stage III colon Cancer Treated with VRP-CEA(6D), an Alphavirus Vector that Increases the CD8+ Effector Memory T Cell to Treg Ratio. *J. Immunother. Cancer* **2020**, *8*, e001662. [CrossRef] [PubMed]

118. Downs-Canner, S.; Guo, Z.S.; Ravindranathan, R.; Breitbach, C.J.; O'Malley, M.E.; Jones, H.L.; Moon, A.; McCart, J.A.; Shuai, Y.; Zeh, H.J.; et al. Phase I study of intravenous oncolytic poxvirus (vvDD) in patients with advanced solid cancers. *Mol. Ther.* **2016**, *24*, 1492–1501. [CrossRef] [PubMed]

119. Niu, Z.; Bai, F.; Sun, T.; Tian, H.; Yu, D.; Yin, J.; Li, S.; Li, T.; Cao, H.; Yu, Q.; et al. Recombinant Newcastle disease virus expressing IL15 demonstrates promising antitumor efficiency in melanoma model. *Technol. Cancer Res. Treat.* **2015**, *14*, 607–615.

120. Kimpel, J.; Urbiola, C.; Koske, I.; Tober, R.; Banki, Z.; Wollmann, G.; von Laer, D. The Oncolytic virus VSV-GP is effective against malignant melanoma. *Viruses* **2018**, *10*, 108. [CrossRef] [PubMed]

121. Galivo, F.; Diaz, R.M.; Wongthida, P.; Thompson, J.; Kottke, T.; Barber, G.; Melcher, A.; Vile, R. Single-cycle viral gene expression, rather than progressive replication and oncolysis, is required for VSV therapy of B16 melanoma. *Gene Ther.* **2010**, *17*, 158–170. [CrossRef]

122. Shafren, D.R.; Au, G.G.; Nguyen, T.; Newcombe, N.G.; Haley, E.S.; Beagley, L.; Johansson, E.S.; Hersey, P.; Barry, R.D. Systemic therapy of malignant human melanoma tumors by a common cold-producing enterovirus, coxsackievirus a21. *Clin. Cancer Res.* **2014**, *10*, 53–60. [CrossRef]

123. Johnson, D.B.; Puzanov, I.; Kelley, M.C. Talimogene laherparevec (T-VEC) for the treatment of advanced melanoma. *Immunotherapy* **2015**, *7*, 611–619. [CrossRef]

124. Rehman, H.; Silk, A.W.; Kane, M.P.; Kaufman, H.L. Into the clinic: Talimigene laherparevec (T-VEC), a first-in-class intratumoral oncolytic viral therapy. *J. Immunother. Ther. Cancer* **2016**, *4*, 53. [CrossRef]

125. Nagasato, M.; Rin, Y.; Yamamoto, Y.; Henmi, M.; Hiraoka, N.; Chiwaki, F.; Matussaki, K.; Tagawa, M.; Sasaki, H.; Aoki, K. A tumor-targeting adenovirus with high gene transduction efficiency for primary pancreatic cancer and ascites cells. *Anticancer Res.* **2017**, *37*, 3599–3605.

126. Petrulio, C.A.; Kaufman, H.L. Development of the panvac-vf vaccine for pancreatic cancer. *Expert Rev. Vaccines* **2006**, *5*, 9–19. [CrossRef] [PubMed]

127. Hirooka, Y.; Kasuya, H.; Ishikawa, T.; Hawashima, H.; Ohno, E.; Villalobos, I.B.; Naoe, Y.; Ichinose, T.; Koyoma, N.; Goto, H.; et al. A phase I clinical trial of EUS-guided intratumoral injection of the oncolytic virus, HF10 for unresectable locally advanced pancreatic cancer. *BMC Cancer* **2018**, *18*, 596. [CrossRef] [PubMed]

128. Long, J.; Yang, Y.; Kang, T.; Zhao, W.; Cheng, H.; Wu, Y.; Du, T.; Liu, B.; Yang, L.; Luo, F.; et al. Ovarian cancer therapy by VSVMP gene mediated by a paclitaxel-enhanced nanoparticle. *ACS Appl. Mater. Interfaces* **2017**, *9*, 39152–39164. [CrossRef]

129. Zhong, Q.; Wen, Y.J.; Yang, H.S.; Luo, H.; Fu, A.-F.; Yang, F.; Chen, L.-J.; Chen, X.; Qi, X.-R.; Lin, H.G.; et al. Efficient cisplatin-resistant human ovarian cancer growth and prolonged survival by gene transferred vesicular stomatitis virus matrix protein in nude mice. *Ann. Oncol.* **2008**, *19*, 1584–1591. [CrossRef]

130. Unno, Y.; Shino, Y.; Kondo, F.; Igarashi, N.; Wang, G.; Shimura, R.; Yamaguchi, T.; Asano, T.; Saisho, H.; Skeya, S.; et al. Oncolytic Viral Therapy for Cervical and Ovarian Cancer Cells by Sindbis Virus AR339 Strain. *Clin. Cancer Res.* **2005**, *11*, 4553–4560. [CrossRef]

131. Galanis, E.; Hartmann, L.C.; Cliby, W.A.; Long, H.J.; Peethambaram, P.P.; Barrette, B.A.; Kaur, J.S.; Haluska, P.J., Jr.; Aderca, I.; Zollman, P.J.; et al. Phase I trial of intraperitoneal administration of an oncolytic measles virus strain engineered to express carcinoembryonic antigen for recurrent ovarian cancer. *Cancer Res.* **2010**, *70*, 875–882. [CrossRef]

132. Msaouel, P.; Iankov, I.D.; Allen, C.; Morris, J.C.; von Messling, V.; Cattaneo, R.; Koutsilieris, M.; Russell, S.J.; Galanis, E. Engineered measles virus as a novel oncolytic therapy against prostate cancer. *Prostate* **2009**, *69*, 82–91. [CrossRef]
133. Liu, C.; Hasegawa, K.; Russell, S.J.; Sadelain, M.; Peng, K.-W. Prostate-specific membrane antigen retargeted measles virotherapy for the treatment of prostate cancer. *Prostate* **2009**, *69*, 1128–1141. [CrossRef]
134. Son, H.A.; Zhang, L.; Cuong, B.K.; Tong, H.V.; Cuong, L.D.; Hang, N.T.; Nhung, H.T.M.; Yamamoto, N.; Toan, N.L. Combination of vaccine-strain measles and mumps viruses enhances oncolytic activity against human solid malignancies. *Cancer Investig.* **2018**, *7*, 106–117. [CrossRef] [PubMed]
135. Durso, R.J.; Andjelic, S.; Gardner, J.P.; Margitich, D.J.; Donovan, G.P.; Arrigale, R.R.; Wang, X.; Maughan, M.F.; Donovan, G.P.; Arrigale, R.R.; et al. A Novel Alphavirus Vaccine Encoding Prostate-specific Membrane Antigen Elicits Potent Cellular and Humoral Immune Responses. *Clin. Cancer Res.* **2007**, *13*, 3999–4008. [CrossRef]
136. Garcia-Hernandez, M.L.; Gray, A.; Hubby, B.; Kast, W.M. In Vivo effects of Vaccination with Six-Transmembrane Epithelial Antigen of the Prostate: A Candidate Antigen for Treating Prostate Cancer. *Cancer Res.* **2007**, *67*, 1344–1351. [CrossRef] [PubMed]
137. Garcia-Hernandez, M.L.; Gray, A.; Hubby, B.; Klinger, O.J.; Kast, W.M. Prostate Stem Cell Antigen Vaccination Induces a Long-Term Protective Immune Response against Prostate Cancer in the Absence of Autoimmunity. *Cancer Res.* **2008**, *68*, 861–869. [CrossRef] [PubMed]
138. Mansfield, D.C.; Kyula, J.N.; Rosenfelder, N.; Chao-Chu, J.; Kramer-Marek, G.; Khan, A.A.; Roulstone, V.; McLaughlin, M.; Melcher, A.A.; Vile, R.G.; et al. Oncolytic vaccinia virus as a vector for therapeutic sodium iodide symporter gene therapy in prostate cancer. *Gene Ther.* **2016**, *23*, 357–368. [CrossRef] [PubMed]
139. Slovin, S.F.; Kehoe, M.; Durso, R.; Fernandez, C.; Olson, W.; Gao, J.P.; Israel, R.; Scher, H.I.; Morris, S. A Phase I Dose Escalation Trial of Vaccine Replicon Particles (VRP) Expressing Prostate-specific Membrane Antigen (PSMA) in Subjects with Prostate Cancer. *Vaccine* **2013**, *31*, 943–949. [CrossRef]
140. Msaouel, P.; Dispenzieri, A.; Galanis, E. Clinical testing of engineered oncolytic measles virus strains in the treatment of cancer: An overview. *Curr. Opin. Mol. Ther.* **2009**, *11*, 43–53.
141. Lin, X.; Chen, X.; Wei, Y.; Zhao, J.; Fan, L.; Wen, Y.; Wu, H.; Zhao, X. Efficient inhibition of intraperitoneal human ovarian cancer growth and prolonged survival by gene transfer of vesicular stomatitis virus matrix protein in nude mice. *Gynecol. Oncol.* **2007**, *104*, 540–546. [CrossRef]
142. Miyamoto, M.I.; del Monte, F.; Schmidt, U.; DiSalvo, T.S.; Kang, Z.B.; Matsui, T.; Guerrero, J.L.; Gwathmey, J.K.; Rosenzweig, A.; Hajjar, R.J. Adenoviral gene transfer of SERCa2a improves left-ventricular function in aortic-banded rats in transition to heart failure. *Proc. Natl. Acad. Sci. USA* **2000**, *97*, 793–798. [CrossRef]
143. Sakata, S.; Lebeche, D.; Sakata, Y.; Sakata, N.; Chemaly, E.R.; Liang, L.; Nakajima-Takenaka, C.; Tsuji, T.; Konishi, N.; del Monte, F.; et al. Transcoronary transfer of SERCa2a increases coronary blood flow and decreases cardiomyocyte size in a type 2 diabetic rat model. *Am. J. Physiol. Heart Circ. Physiol.* **2007**, *292*, H1204–H1207. [CrossRef]
144. Hadri, L.; Bobe, R.; Kawase, Y.; Ladage, D.; Ishikawa, K.; Atassi, F.; Lebeche, D.; Kranisa, E.G.; Leopold, J.A.; Lompre, A.-M.; et al. SERCA2a gene transfer enhances eNOS expression and activity in endothelial cells. *Mol. Ther.* **2010**, *18*, 1284–1292. [CrossRef]
145. Niwano, K.; Arai, M.; Koitabashi, N.; Watanabe, A.; Ikeda, Y.; Miyoshi, H.; Kurabayashi, M. Lentiviral vector–mediated SERCA2 gene transfer protects against heart failure and left ventricular remodeling after myocardial infarction in rats. *Mol. Ther.* **2008**, *16*, 1026–1032. [CrossRef]
146. Yang, Z.-J.; Chen, B.; Sheng, Z.; Zhang, D.G.; Jia, E.Z.; Wang, W.; Ma, D.C.; Zhu, T.B.; Wang, L.S.; Li, C.J.; et al. Improvement of heart function in postinfarct heart failure swine models after hepatocyte growth factor transfer: Comparison of low-, medium- and high-dose groups. *Mol. Biol. Rep.* **2010**, *37*, 2075–2081. [CrossRef] [PubMed]
147. Igarashi, T.; Finet, J.E.; Takeuchi, A.; Fujino, Y.; Strom, M.; Greener, I.D.; Rosenbaum, D.S.; Donahue, J.K. Connexin gene transfer preserves conduction velocity and prevents atrial fibrillation. *Circulation* **2012**, *125*, 216–225. [CrossRef] [PubMed]
148. Amit, G.; Kikuchi, K.; Greener, I.D.; Yang, L.; Novack, V.; Donahue, J.K. Selective molecular potassium channel blockade prevents atrial fibrillation. *Circulation* **2010**, *121*, 2263–2270. [CrossRef] [PubMed]
149. Qian, L.; Huang, Y.; Spencer, C.I.; Foley, A.; Vedantham, V.; Liu, L.; Conway, S.J.; Fu, J.D.; Srivastava, D. In vivo reprogramming of murine cardiac fibroblasts into induced cardiomyocytes. *Nature* **2012**, *485*, 593–598. [CrossRef]
150. Leikas, A.J.; Hassinen, I.; Hedman, A.; Kivelä, A.; Ylä-Herttuala, S.; Hartikainen, J.E.K. Long-term safety and efficacy of intramyocardial adenovirus-mediated VEGF-D$^{\Delta N \Delta C}$ gene therapy eight-year follow-up of phase I KAT301 study. *Gene Ther.* **2022**, *29*, 289–293. [CrossRef]
151. Stewart, D.J.; Hilton, J.D.; Arnold, J.M.O.; Gregoire, J.; Rivard, A.; Archer, S.L.; Charbonneau, F.; Cohen, E.; Curtis, M.; Buller, C.E.; et al. Angiogenic gene therapy in patients with nonrevascularizable ischemic heart disease: A phase 2 randomized, controlled trial of AdVEGF(121) (AdVEGF121) versus maximum medical treatment. *Gene Ther.* **2006**, *13*, 1503–1511. [CrossRef]
152. Grines, C.L.; Watkins, M.W.; Helmer, G.; Penny, W.; Brinker, J.; Marmur, J.D.; West, A.; Rade, J.J.; Marrott, P.; Hammond, H.K.; et al. Angiogenic Gene Therapy (AGENT) trial in patients with stable angina pectoris. *Circulation* **2002**, *105*, 1291–1297. [CrossRef]
153. Henry, T.D.; Grines, C.L.; Watkins, M.W.; Dib, N.; Barbeu, G.; Moreadith, R.; Andrasfay, T.; Engler, R.L. Effects of Ad5FGF-4 in patients with angina: An analysis of pooled data from the AGENT-3 and AGENT-4 trials. *J. Am. Coll. Cardiol.* **2007**, *50*, 1038–1046. [CrossRef] [PubMed]

154. Grines, C.L.; Watkins, M.W.; Mahmarian, J.J.; Iskandrian, A.E.; Rade, J.J.; Marrott, P.; Pratt, C.; Kleiman, N. A randomized, double-blind, placebo-controlled trial of Ad5FGF-4 gene therapy and its effect on myocardial perfusion in patients with stable angina. *J. Am. Coll. Cardiol.* **2003**, *42*, 1339–1347. [CrossRef]
155. Jaski, B.E.; Jessup, M.L.; Mancini, D.M.; Cappola, T.P.; Pauly, D.F.; Greenberg, B.; Borrow, K.; Dittrich, H.; Zsebo, K.M.; Hajjar, R.J. Calcium Up-Regulation by Percutaneous Administration of Gene Therapy In Cardiac Disease (CUPID) Trial Investigators. Calcium upregulation by percutaneous administration of gene therapy in cardiac disease (CUPID Trial), a first-in-human phase 1/2 clinical trial. *J. Card. Fail.* **2009**, *15*, 171–181.
156. Jessup, M.; Greenberg, B.; Mancini, D.; Cappola, T.; Pauly, D.F.; Jaski, B.; Yaroshinsky, A.; Zsebo, K.M.; Dittrich, H.; Haijjar, H. Calcium Upregulation by Percutaneous Administration of Gene Therapy in Cardiac Disease (CUPID) Investigators. Calcium Upregulation by Percutaneous Administration of Gene Therapy in Cardiac Disease (CUPID): A phase 2 trial of intracoronary gene therapy of sarcoplasmic reticulum Ca2+-ATPase in patients with advanced heart failure. *Circulation* **2011**, *124*, 304–313. [PubMed]
157. Zsebo, K.; Yaroshinsky, A.; Rudy, J.J.; Wagner, K.; Greenberg, B.; Jessup, M.; Hajjar, R.J. Long-term effects of AAV1/SERCA2a gene transfer in patients with severe heart failure: Analysis of recurrent cardiovascular events and mortality. *Circ. Res.* **2014**, *114*, 101–108. [CrossRef] [PubMed]
158. Watson, G.L.; Sayles, J.N.; Chen, C.; Elliger, S.S.; Elliger, C.A.; Raju, N.R.; Kurtzman, G.J.; Podsakoff, G.M. Treatment of lysosomal storage disease in MPS VII mice using a recombinant adeno-associated virus. *Gene Ther.* **1998**, *5*, 1642–1649. [CrossRef]
159. Lebherz, C.; Gao, G.; Louboutin, J.P.; Millar, J.; Rader, D.; Wilson, J.M. Gene therapy with novel adeno-associated virus vectors substantially diminishes atherosclerosis in a murine model of familial hypercholesterolemia. *J. Gene Med.* **2004**, *6*, 663–672. [CrossRef]
160. Jimenez, V.; Jambrina, C.; Casana, E.; Sacristan, V.; Munoz, S.; Rodo, J.; Grass, I.; Garcia, M.; Mallolo, C.; Leon, X.; et al. FGF21 gene therapy as treatment for obesity and insulin resistance. *EMBO Mol. Med.* **2018**, *10*, e8791. [CrossRef]
161. Tao, R.; Xiao, L.; Zhou, L.; Zheng, Z.; Long, J.; Zhou, L.; Tang, M.; Dong, B.; Yao, S. Long-Term Metabolic Correction of Phenylketonuria by AAV-Delivered Phenylalanine Amino Lyase. *Mol. Ther. Methods Clin. Dev.* **2020**, *19*, 507–517. [CrossRef] [PubMed]
162. Puzzo, F.; Colella, P.; Biferi, M.G.; Bali, D.; Paulk, N.K.; Vidal, P.; Collaud, F.; Simon-Sola, M.; Charles, S.; Hardet, R.; et al. Rescue of Pompe disease in mice by AAV-mediated liver delivery of secretable acid alpha-glucosidase. *Sci. Transl. Med.* **2017**, *9*, eeam6375. [CrossRef]
163. Xu, J.; Lu, Y.; Ding, F.; Zhan, X.; Zhu, M.; Wang, Z. Reversal of diabetes in mice by intrahepatic injection of bone-derived GFP-murine mesenchymal stem cells infected with the recombinant retrovirus-carrying human insulin gene. *World J. Surg.* **2007**, *31*, 1872–1882. [CrossRef]
164. Hegde, V.; Na, H.-N.; Dubuisson, O. An adenovirus-derived protein: A novel candidate for anti-diabetic drug development. *Biochemie* **2016**, *121*, 140–150. [CrossRef]
165. D'Avola, D.; López-Franco, E.; Sangro, B.; Paneda, A.; Grossios, N.; Gil-Farina, I.; Benito, A.; Twisk, J.; Paz, M.; Ruiz, J.; et al. Phase I open label liver-directed gene therapy clinical trial for acute intermittent porphyria. *J. Hepatol.* **2016**, *65*, 776–783. [CrossRef]
166. Brantly, M.L.; Chulay, J.D.; Wang, L.; Muller, C.; Humphries, M.; Spencer, L.T.; Rouhani, F.; Conlon, T.J.; Calcedo, R.; Betts, M.R.; et al. Sustained transgene expression despite Tlymphocyte responses in a clinical trial of rAAV1-AAT gene therapy. *Proc. Natl. Acad. Sci. USA* **2009**, *106*, 16363–16368. [CrossRef] [PubMed]
167. Flotte, T.R.; Trapnell, B.C.; Humphries, M.; Carey, B.; Calcedo, R.; Rouhani, F.; Campbell-Thompson, M.; Yachnis, A.T.; Sandhaus, R.A.; McElvaney, N.G.; et al. Phase 2 clinical trial of a recombinant adeno-associated viral vector expressing alphal-antitrypsin: Interim results. *Hum. Gene Ther.* **2011**, *22*, 1239–1247. [CrossRef] [PubMed]
168. Balagué, C.; Zhou, J.; Dai, Y.; Alemany, R.; Josephs, S.F.; Andreason, G.; Hariharan, M.; Sethi, E.; Prokopenko, E.; Jan, H.Y.; et al. Sustained high-level expression of full-length human factor VIII and a restoration of clotting activity in hemophilic mice using a minimal adenovirus vector. *Blood* **2000**, *95*, 820–828. [CrossRef] [PubMed]
169. Dai, Y.; Schwarz, E.M.; Gu, D.; Zhang, W.W.; Sarvetnick, N.; Verma, I.M. Cellular and humoral immune responses to adenoviral vectors containing factor IX gene: Tolerization of factor IX and vector antigens allows long-term expression. *Proc. Natl. Acad. Sci. USA* **1995**, *92*, 1401–1405. [CrossRef]
170. Kay, M.A.; Landen, C.N.; Rothenberg, S.R.; Taylor, L.A.; Leland, F.; Wiehle, S.; Fang, B.; Bellinger, D.; Finegold, M.; Thompson, A.R.; et al. In vivo hepatic gene therapy: Complete albeit transient correction of factor IX deficiency in hemophilia B dogs. *Proc. Natl. Acad. Sci. USA* **1994**, *91*, 2353–2357. [CrossRef] [PubMed]
171. Fang, B.; Eisensmith, R.C.; Wang, H.; Kay, M.A.; Cross, R.E.; Landen, C.N.; Gordon, G.; Bellinger, D.A.; Read, M.S.; Hu, P.C.; et al. Gene therapy for hemophilia B: Host immunosuppression prolongs the therapeutic effect of adenovirus-mediated factor IX expression. *Hum. Gene Ther.* **1995**, *6*, 1039–1044. [CrossRef]
172. Jiang, H.; Lillicrap, D.; Patarroyo-White, S.; Liu, T.; Qian, X.; Scallan, C.D.; Powell, S.; Keller, T.; McMurray, M.; Labelle, A.; et al. Multiyear therapeutic benefit of AAV serotypes 2, 6, and 8 delivering factor VIII to hemophilia A mice and dogs. *Blood* **2006**, *108*, 107–115. [CrossRef]
173. Callan, M.B.; Haskins, M.E.; Wang, P.; Zhou, S.; High, K.A.; Arruda, V.R. Successful Phenotype Improvement following Gene Therapy for Severe Hemophilia A in Privately Owned Dogs. *PLoS ONE* **2016**, *11*, e0151800. [CrossRef]

174. Nguyen, G.N.; Everett, J.K.; Kafle, S.; Roche, A.M.; Raymond, H.E.; Leiby, J.; Wood, C.; Assenmacher, C.-A.; Merricks, E.P.; Long, C.T.; et al. A long-term study of AAV gene therapy in dogs with hemophilia A identifies clonal expansions of transduced liver cells. *Nat. Biotechnol.* **2021**, *39*, 47–55. [CrossRef]
175. Crudele, J.M.; Finn, J.D.; Siner, J.I.; Martin, N.B.; Niemeyer, G.P.; Zhou, S.; Mingozzi, F.; Lothrop, C.D., Jr.; Arruda, V.R. AAV liver expression of FIXPadua prevents and eradicates FIX inhibitor without increasing thrombogenicity in hemophilia B dogs and mice. *Blood* **2015**, *125*, 1553–1561. [CrossRef]
176. Nathwani, A.C. Gene therapy for hemophilia. *Hematol. Am. Soc. Hematol. Educ. Program* **2019**, *2019*, 1–8. [CrossRef]
177. Pasi, K.J.; Rangarajan, S.; Mitchell, N.; Lester, W.; Symington, E.; Madan, B.; Laffan, M.; Russell, C.B.; Li, M.; Pierce, G.F.; et al. Multiyear Follow-up of AAV5-hFVIII-SQ Gene Therapy for Hemophilia A. *N. Engl. J. Med.* **2020**, *382*, 29–40. [CrossRef]
178. Nathwani, A.C.; Reiss, U.M.; Tuddenham, E.G.; Rosales, C.; Chowdary, P.; McIntosh, J.; Della Peruta, M.; Lheriteau, E.; Patel, N.; Raj, D.; et al. Long-term safety and efficacy of factor IX gene therapy in hemophilia B. *N. Engl. J. Med.* **2014**, *371*, 1994–2004. [CrossRef] [PubMed]
179. Chowdary, P.; Shapiro, S.; Makris, M.; Evans, G.; Boyce, S.; Talks, K.; Dolan, G.; Reiss, U.; Phillips, M.; Riddell, A.; et al. Phase 1-2 Trial of AAVS3 Gene Therapy in Patients with Hemophilia B. *N. Engl. J. Med.* **2022**, *387*, 237–247. [CrossRef] [PubMed]
180. VandenDriessche, T.; Pipe, S.W.; Pierce, G.F.; Kaczmarek, R. First conditional marketing authorization approval in the European Union for hemophilia "A" gene therapy. *Mol. Ther.* **2022**, *30*, 3335–3336. [CrossRef]
181. Shi, Q.; Wilcox, D.A.; Fahs, S.A.; Fang, J.; Johnson, B.D.; Du, L.M.; Desai, D.; Montgomery, R.R. Lentivirus-mediated platelet-derived factor VIII gene therapy in murine haemophilia A. *J. Tromb. Haemost.* **2007**, *5*, 352–361. [CrossRef] [PubMed]
182. Cantore, A.; Ranzani, M.; Bartholomae, C.C.; Volpin, M.; Della Valle, P.; Sanvito, F.; Sergi Sergi, L.; Gallina, P.; Benedicenti, F.; Bellinger, D.; et al. Liver-directed lentiviral gene therapy in a dog model of hemophilia B. *Sci. Transl. Med.* **2015**, *7*, 277ra28. [CrossRef]
183. Chen, Y.; Schroeder, J.A.; Gao, C.; Li, J.; Hu, J.; Shi, Q. In vivo enrichment of genetically manipulated platelets for murine hemophilia B gene therapy. *J. Cell Physiol.* **2021**, *236*, 354–365. [CrossRef]
184. Garcia-Gomez, M.; Calabria, A.; Garcia-Bravo, M.; Benedicenti, F.; Kosinski, P.; Lopez-Manzaneda, S.; Hill, C.; Del Mar Manu-Pereira, M.; Martin, M.A.; Orman, I.; et al. Safe and efficient gene therapy for pyruvate kinase deficiency. *Mol. Ther.* **2016**, *24*, 1187–1198. [CrossRef]
185. Kelly, P.F.; Radtke, S.; von Kalle, C.; Balcik, B.; Bohn, K.; Mueller, R.; Schuesler, T.; Haren, M.; Reeves, L.; Cancelas, J.A.; et al. Stem Cell Collection and Gene Transfer in Fanconi Anemia. *Mol. Ther.* **2007**, *15*, 211–219. [CrossRef]
186. Jaako, P.; Debnath, S.; Olsson, K.; Modlich, U.; Rothe, M.; Schambach, A.; Flygare, J.; Karlsson, S. Gene therapy cures the anemia and lethal bone marrow failure in a mouse model of RPS19-defificent Diamond-Blackfan anemia. *Haematologica* **2014**, *99*, 1792–1798. [CrossRef] [PubMed]
187. Cavazzana, M.; Mavilio, F. Gene Therapy for Hemoglobinopathies. *Hum. Gene Ther.* **2018**, *29*, 1106–1113. [CrossRef] [PubMed]
188. Lal, A.; Locatelli, F.; Kwiatkowski, J.L.; Porter, J.B.; Trasher, A.J.; Homgeng, S.; Sauer, M.G.; Thuret, I.; Lal, A.; Algeri, M.; et al. Northstar-3: Interim results from a phase 3 study evaluating lentiglobin gene therapy in patients with transfusion-dependent β-thalassemia and either a β0 or IVS-I-110 mutation at both alleles of the HBB gene. *Blood* **2019**, *134*, 815. [CrossRef]
189. Shangaris, P.; Loukogeorgakis, S.P.; Subramaniam, S.; Flouri, C.; Jackson, L.H.; Wang, W.; Blundell, M.P.; Liu, S.; Eaton, S.; Bakhamis, N.; et al. In Utero Gene Therapy (IUGT) Using GLOBE Lentiviral Vector Phenotypically Corrects the Heterozygous Humanised Mouse Model and Its Progress Can Be Monitored Using MRI Techniques. *Sci. Rep.* **2019**, *9*, 11592. [CrossRef]
190. Marktel, S.; Scaramuzza, S.; Cicalese, M.P.; Giglio, F.; Galimberti, S.; Lidonnici, M.R.; Calbi, V.; Assanelli, A.; Bernardo, M.E.; Rossi, C.; et al. Intrabone hematopoietic stem cell gene therapy for adult and pediatric patients affected by transfusion-dependent ss-thalassemia. *Nat. Med.* **2019**, *25*, 234–241. [CrossRef]
191. Pawliuk, R.; Westerman, K.A.; Fabry, M.E.; Payen, E.; Tighe, R.; Bouhassira, E.E.; Acharya, S.A.; Ellis, J.; London, I.M.; Eaves, C.J.; et al. Correction of sickle cell disease in transgenic mouse models by gene therapy. *Science* **2001**, *294*, 2368–2371. [CrossRef]
192. Ribeil, J.-A.; Hacein-Bey-Abina, S.; Payen, E.; Magnani, A.; Semeraro, M.; Magrin, E.; Caccavelli, L.; Neven, B.; Bourget, P.; El Nemer, W.; et al. Gene therapy in a patient with sickle cell disease. *N. Engl. J. Med.* **2017**, *376*, 848–855. [CrossRef]
193. Urbinati, F.; Campo Fernandez, B.; Masiuk, K.E.; Poletti, V.; Hollis, R.P.; Koziol, C.; Kaufman, M.L.; Brown, D.; Mavilio, F.; Kohn, D.B. Gene Therapy for Sickle Cell Disease: A Lentiviral Vector Comparison Study. *Hum. Gene Ther.* **2018**, *29*, 1153–1166. [CrossRef]
194. Miyake, K.; Inokuchi, K.; Miyake, N.; Dan, K.; Shimada, T. HIV vector-mediated targeted suicide gene therapy for adult T-cell leukemia. *Gene Ther.* **2007**, *14*, 1662–1667. [CrossRef]
195. Tan, L.; Xu, B.; Liu, R.; Liu, H.; Tan, H.; Huang, W. Gene therapy for acute myeloid leukemia using Sindbis vectors expressing a fusogenic membrane glycoprotein. *Cancer Biol. Ther.* **2010**, *9*, 350–357. [CrossRef]
196. Khan, N.; Bammidi, S.; Jayandharan, G.R. A CD33 Antigen-Targeted AAV6 Vector Expressing an Inducible Caspase-9 Suicide Gene Is Therapeutic in a Xenotransplantation Model of Acute Meyloid Leukemia. *Bioconjug. Chem.* **2019**, *30*, 2404–2416. [CrossRef]
197. Wenthe, J.; Naseri, S.; Labani-Motlagh, A.; Enblad, G.; Wikström, K.I.; Eriksson, E.; Loskog, A.; Lövgren, T. Boosting CAR T-cell responses in lymphoma by simultaneous targeting of CD40/4-1BB using oncolytic viral gene therapy. *Cancer Immunol. Immunother.* **2021**, *70*, 2851–2865. [CrossRef] [PubMed]
198. Kutubuddin, M.; Federoff, H.J.; Challita-Eid, P.M.; Halterman, M.; Day, B.; Atkinson, M.; Planelles, V.; Rosenblatt, J.D. Eradication of pre-established lymphoma using herpes simplex virus amplicon vectors. *Blood* **1999**, *93*, 643–654. [CrossRef] [PubMed]

199. Ishino, R.; Kawase, Y.; Kitawaki, T.; Sugimoto, N.; Oku, M.; Uchida, S.; Imataki, O.; Matsuoka, A.; Taoka, T.; Kawakami, K.; et al. Oncolytic Virus Therapy with HSV-1 for Hematological Malignancies. *Mol. Ther.* **2021**, *29*, 762–774. [CrossRef] [PubMed]
200. Han, T.; Abdel-Motal, U.M.; Chang, D.-K.; Sui, J.; Muvaffak, A.; Campbell, J.; Zhu, Q.; Kupper, T.S.; Marasco, W.A. Human anti-CCR4 minibody gene transfer for the treatment of cutaneous T-cell lymphoma. *PLoS ONE* **2012**, *7*, e44455. [CrossRef]
201. Yu, M.; Scherwitzl, I.; Opp, S.; Tsirigos, A.; Meruelo, D. Molecular and metabolic pathways mediating curative treatment of a non-Hodgkin B cell lymphoma by Sindbis viral vectors and anti-4-1BB monoclonal antibody. *J. Immunother. Cancer* **2019**, *7*, 185. [CrossRef]
202. Hadac, E.M.; Kelly, E.J.; Russell, S.J. Myeloma xenograft destruction by a nonviral vector delivering oncolytic infectious nucleic acid. *Mol. Ther.* **2011**, *19*, 1041–1047. [CrossRef]
203. Naik, S.; Nace, R.; Barber, G.N.; Russell, S.J. Potent systemic therapy of multiple myeloma utilizing oncolytic vesicular stomatitis virus coding for interferon-β. *Cancer Gene Ther.* **2012**, *19*, 443–450. [CrossRef]
204. Kelly, K.R.; Espitia, C.M.; Mahalingam, D.; Oyajobi, B.O.; Coffey, M.; Giles, F.J.; Carew, J.S.; Nawrocki, S.T. Reovirus therapy stimulated endoplasmic reticular stress, NOXA induction, and augments bortezomib-mediated apoptosis in multiple myeloma. *Oncogene* **2012**, *31*, 3023–3038. [CrossRef]
205. Alexander, I.E.; Cunningham, S.C.; Logan, G.J.; Christodoulou, J. Potential of AAV vectors in the treatment of metabolic disease. *Gene Ther.* **2008**, *15*, 831–839. [CrossRef]
206. Salabarria, S.M.; Nair, J.; Clement, N.; Smith, B.K.; Raben, N.; Fuller, D.D.; Byrne, B.J.; Corti, M. Advancements in AAV-mediated Gene Therapy for Pompe Disease. *J. Neuromusc. Dis.* **2020**, *7*, 15–31. [CrossRef] [PubMed]
207. Gale, A.J.; Pellequer, J.L.; Getzoff, E.D.; Griffin, J.H. Structural basis for hemophilia A caused by mutations in the C domains of blood coagulation factor VIII. *Thromb. Haemost.* **2000**, *83*, 78–85. [CrossRef]
208. Ludwig, M.; Sabharwal, A.K.; Brackmann, H.H.; Olek, K.; Smith, K.J.; Birktoft, J.J.; Bajaj, S.P. Hemophilia B caused by five different nondeletion mutations in the protease domain of factor IX. *Blood* **1992**, *79*, 1225–1232. [CrossRef] [PubMed]
209. Wang, L.; Herzog, R.W. AAV-mediated gene transfer for treatment of hemophilia. *Curr. Gene Ther.* **2005**, *5*, 349–360. [CrossRef]
210. Sarkar, R.; Xiao, W.; Kazazian, H.H., Jr. A single adenoassociated virus (AAV)-murine factor FVIII. *J. Thromb. Haemost.* **2003**, *1*, 220–226. [CrossRef] [PubMed]
211. Sarkar, R.; Tetreault, R.; Gao, G.; Wang, L.; Bell, P.; Chandler, R.; Wilson, J.M.; Kazazian, H.H., Jr. Total correction of hemophilia A mice with canine FVIII using an AAV 8 serotype. *Blood* **2004**, *103*, 1253–1260. [CrossRef] [PubMed]
212. Gambari, R. Alternative options for DNA-based experimental therapy of β-thalassemia. *Expert Opin. Biol. Ther.* **2012**, *12*, 443–462. [CrossRef]
213. Ingram, V. A specific chemical difference between the globins of normal human and sickle cell anemia hemoglobin. *Nature* **1956**, *178*, 792–794. [CrossRef]
214. Lundstrom, K. Gene Therapy in Hematology. In *Comprehensive Hematology and Stem Cell Research*; Elsevier: Chennai, India, 2023; *in press.*
215. Kim, J.; Yoon, Y.S.; Lee, H.; Chang, W. AAV-GAD gene for rat models of neuropathic pain and Parkinson's disease. *Acta Neurochir. Suppl.* **2008**, *101*, 99–105.
216. Björklund, A.; Kirik, D.; Rosenblad, C.; Georgievska, B.; Lundberg, C.; Mandel, R.J. Towards a neuroprotective gene therapy for Parkinson's disease: Use of adenovirus, AAV and lentivirus vectors for gene transfer of GDNF to the nigrostriatal system in the rat Parkinson model. *Brain Res.* **2000**, *886*, 82–98. [CrossRef]
217. Kordower, J.H.; Emborg, M.E.; Bloch, J.; Ma, S.Y.; Chu, Y.; Leventhal, L.; McBride, J.; Chen, E.Y.; Palfi, S.; Roitberg, B.Z.; et al. Neurodegeneration prevented by lentiviral vector delivery of GDNF in primate model of Parkinson's disease. *Science* **2000**, *290*, 767–773. [CrossRef]
218. Eberling, J.L.; Jagust, W.J.; Christine, C.W.; Starr, P.; Larson, P.; Bankiewicz, K.S.; Aminoff, M.J. Results from a phase I safety trial of hAADC gene therapy for Parkinson disease. *Neurology* **2008**, *70*, 1980–1983. [CrossRef]
219. Palfi, S.; Gurruchaga, J.M.; Ralph, G.S.; Lepetit, H.; Lavisse, S.; Buttery, P.C.; Watts, C.; Miskin, J.; Kelleher, M.; Deeley, S.; et al. Long-term safety and tolerability of ProSavin, a lentiviral vector-based gene therapy for Parkinson's disease: A dose escalation, open-label, phase 1/2 trial. *Lancet* **2014**, *383*, 1138–1146. [CrossRef]
220. Palfi, S.; Gurruchaga, J.M.; Lepetit, H.; Howard, K.; Ralph, G.S.; Mason, S.; Gouello, G.; Domenech, P.; Buttery, P.C.; Hantraye, P.; et al. Long-Term Follow-up of a Phase I/II of ProSavin, a Lentiviral Vector Gene Therapy for Parkinson's Disease. *Hum. Gene Ther. Clin. Dev.* **2018**, *29*, 148–155. [CrossRef]
221. Wu, K.; Meyer, E.M.; Bennett, J.A.; Meyers, C.A.; Hughes, J.A.; King, M.A. AAV2/5-mediated NGF gene delivery protects septal cholinergic neurons following axotomy. *Brain Res.* **2005**, *1061*, 107–113. [CrossRef] [PubMed]
222. Fol, R.; Braudeau, J.; Ludewig, S.; Abel, T.; Weyer, S.W.; Roederer, J.P.; Brod, F.; Audrain, M.; Bemelmans, A.P.; Buchholz, C.J.; et al. Viral gene transfer of APPsα rescues synaptic failure in an Alzheimer's disease model. *Acta Neuropathol.* **2016**, *131*, 247–266. [CrossRef] [PubMed]
223. Rafii, M.S.; Tuszynski, M.H.; Thomas, R.G.; Barbra, D.; Brewer, J.B.; Rissman, R.A.; Siffert, J.; Aisen, P.S.; AAV2-NGF Study Team. Adeno-associated viral vector (Serotype 2)–nerve growth factor for patients with Alzheimer Disease. *JAMA Neurol.* **2018**, *75*, 834–841. [CrossRef] [PubMed]

224. Revilla, S.; Ursulet, S.; Alvarez-Lopez, J.M.; Castro-Freire, M.; Peripina, U.; Garcia-Mesa, Y.; Bortolozzi, A.; Gimenez-Llort, L.; Kaliman, P.; Cristofol, R.; et al. Lenti-GDNF gene therapy protects against Alzheimer's disease-like neuropathology in 3xTg-AD mice and MC65 cells. *CNS Neurosci. Ther.* **2014**, *20*, 961–972. [CrossRef]
225. Zeng, C.-Y.; Yang, T.-T.; Zhou, H.-J.; Zhao, Y.; Kuang, X.; Duan, W.; Du, J.R. Lentiviral vector-mediated overexpression of klotho in the brain improves Alzheimer's disease-like pathology and cognitive deficits in mice. *Neurobiol. Aging* **2019**, *78*, 18–28. [CrossRef]
226. Miniarikova, J.; Zimmer, V.; Martier, R.; Bouwers, C.C.; Pythoud, C.; Richetin, K.; Rey, M.; Lubelski, J.; Evers, M.M.; van Deventer, S.J.; et al. AAV5-miHTT gene therapy demonstrates suppression of huntingtin aggregation and neuronal dysfunction in a rat model of Huntington's disease. *Gene Ther.* **2017**, *24*, 630–639. [CrossRef]
227. Evers, M.M.; Miniarikova, J.; Juhas, S.; Vallès, A.; Bohuslavova, B.; Juhasova, J.; Skalnikova, H.K.; Vodicka, P.; Valekova, I.; Brouwers, C.; et al. AAV5-miHTT gene therapy demonstrates broad distribution and strong human mutant huntingtin lowering in Huntington's disease minipig model. *Mol. Ther.* **2018**, *26*, 2163–2177. [CrossRef] [PubMed]
228. Rodrigues, F.B.; Wild, E.J. Huntington's Disease Clinical Trials Corner: April 2020. *J. Huntington's Dis.* **2020**, *9*, 185–197. [CrossRef]
229. Sinnett, S.E.; Gray, S.J. Recent endeavors in MECP2 gene transfer for gene therapy of Rett syndrome. *Discov. Med.* **2017**, *24*, 153–159.
230. Pattali, R.; Mou, Y.; Li, X.-J. AAV9 vector: A novel modality in gene therapy for spinal muscular atrophy. *Gene Ther.* **2019**, *26*, 287–295. [CrossRef]
231. Mendell, J.R.; Al-Zaidy, S.; Shell, R.; Arnold, W.D.; Rodino-Klapac, L.R.; Prior, T.W.; Lowes, L.; Alfano, L.; Berry, K.; Church, K.; et al. Single-dose gene-replacement therapy for spinal muscular atrophy. *N. Engl. J. Med.* **2017**, *377*, 1713–1722. [CrossRef]
232. Hoy, S.M. Onasemnogene Abeparvovec First Global Approval. *Drugs* **2019**, *79*, 1255–1262. [CrossRef]
233. Chamberlain, J.R.; Chamberlain, J.S. Progress toward gene therapy for Duchenne muscular dystrophy. *Mol. Ther.* **2017**, *25*, 1125–1131. [CrossRef]
234. Yuasa, K.; Miaygoe, Y.; Yamamoto, K.; Nabeshima, Y.; Dickson, G.; Takeda, S. Effective restoration of dystrophin-associated proteins in vivo by adenovirus-mediated transfer of truncated dystrophin cDNAs. *FEBS Lett.* **1998**, *425*, 329–336. [CrossRef]
235. Sakamoto, M.; Yuasa, K.; Yoshimura, M.; Yokota, T.; Ikemoto, T.; Suzuki, M.; Dickson, G.; Miyagoe-Suzuki, Y.; Takeda, S. Micro-dystrophin cDNA ameliorates dystrophic phenotypes when introduced into mdx mice as a transgene. *Biochem. Biophys. Res. Comm.* **2002**, *293*, 1265–1272. [CrossRef] [PubMed]
236. Gregorevic, P.; Allen, J.M.; Minami, E.; Blankinship, M.J.; Haraguchi, M.; Meuse, L.; Finn, E.; Adams, M.E.; Froehner, S.C.; Murry, C.E.; et al. rAAV6-microdystrophin preserves muscle function and extends lifespan in severely dystrophic mice. *Nat. Med.* **2006**, *12*, 787–789. [CrossRef] [PubMed]
237. Wang, Z.; Storb, R.; Halbert, C.L.; Banks, G.B.; Butts, T.M.; Finn, E.E.; Allen, J.M.; Miller, A.D.; Chamberlain, J.S.; Tapscott, S.J. Successful regional delivery and long-term expression of a dystrophin gene in canine muscular dystrophy: A preclinical model for human therapies. *Mol. Ther.* **2012**, *20*, 1501–1507. [CrossRef]
238. Mendell, J.R.; Sahenk, Z.; Lehman, K.; Nease, C.; Lowes, L.P.; Miller, N.F.; Iammarino, M.A.; Alfano, L.N.; Nicholl, A.; Al-Zaidy, S.; et al. Assessment of systemic delivery of rAAVrh74.MHCK7.micro-dystrophin in children with Duchenne muscular dystrophy. *JAMA Neurol.* **2020**, *77*, 1122–1131. [CrossRef]
239. Malerba, A.; Klein, P.; Bachtarzi, H.; Jarmin, S.A.; Cordova, G.; Ferry, A.; Strings, V.; Polay Spinoza, M.; Mamchaoui, K.; Lacau St Guily, J.; et al. PABPN1 gene therapy for oculopharyngeal muscular dystrophy. *Nat. Commun.* **2017**, *8*, 14848. [CrossRef]
240. Valori, C.F.; Ning, K.; Wyles, M.; Mead, R.J.; Grierson, A.J.; Shaw, P.J.; Azzouz, M. temic delivery of scAAV9 expressing SMN prolongs survival in a model of spinal muscular atrophy. *Sci. Transl. Med.* **2010**, *235*, 35ra42.
241. Duque, S.I.; Arnold, W.D.; Odermatt, P.; Li, X.; Porensky, P.N.; Schmelzer, L.; Meyer, K.; Kolb, S.J.; Schümperli, D.; Kaspar, B.K.; et al. A large animal model of spinal muscular atrophy and correction of phenotype. *Ann. Neurol.* **2015**, *77*, 399–414. [CrossRef] [PubMed]
242. Cavazzana-Calvo, M.; Hacein-Bey, S.; de Saint Basile, G.; Gross, F.; Yvon, E.; Nusbaum, P.; Selz, F.; Hue, C.; Certain, S.; Casanova, J.L.; et al. Gene therapy of human severe combined immunodeficiency (SCID)-X1 disease. *Science* **2000**, *28*, 669–672. [CrossRef]
243. Fischer, A.; Hacein-Bey-Abina, S. Gene therapy for severe combined immunodeficiencies and beyond. *J. Exp. Med.* **2020**, *217*, e20190607. [CrossRef] [PubMed]
244. Gaspar, H.B.; Parsley, K.L.; Howe, S.; King, D.; Gilmour, K.C.; Sinclair, J.; Brouns, G.; Schmidt, M.; Von Kalle, C.; Barington, T.; et al. Gene therapy of X-linked severe combined immunodeficiency by use of a pseudotyped gammaretroviral vector. *Lancet* **2004**, *364*, 2181–2187. [CrossRef]
245. Howe, S.J.; Mansour, M.R.; Schwarzwaelder, K.; Bartholomae, C.; Hubank, M.; Kempski, H.; Brugman, M.H.; Pike-Overzet, K.; Chatters, S.J.; de Ridder, D.; et al. Insertional mutagenesis combined with acquired somatic mutations causes leukemogenesis following gene therapy of SCID-X1 patients. *J. Clin. Investig.* **2008**, *118*, 3143–3150. [CrossRef]
246. Kohn, D.B.; Hershfield, M.S.; Puck, J.M.; Aiuti, A.; Blincoe, A.; Gaspar, H.B.; Notarangelo, L.D.; Grunebaum, E. Consensus approach for the management of severe combined immune deficiency caused by adenosine deaminase deficiency. *J. Allergy Clin. Immunol.* **2019**, *143*, 852–863. [CrossRef] [PubMed]

247. Butterfield, R.; Shieh, P.; Geffen, D.; Yong, F.; Binks, M.; McDonnell, T.G.; Ryan, K.A.; Belluscio, B.; Neelakanten, S.; Levy, D.; et al. One Year Data from Ambulatory Boys in a Phase 1b, Open-Label Study of Fordadistrogene Movaparvovec (PF-06939926) for Duchenne Muscular Dystrophy (DMD). MDA Conference, Poster 53. Available online: www.mdaconference.org/abstract-library/one-year-data-from-ambulatory-boys-in-a-phase-1b-open-label-study-of-fordadistrogene-movaparvovec-pf-069399 26-for-duchenne-muscular-dystrophy-dmd/ (accessed on 11 January 2023).

248. Cartier, N.; Hacein-Bey-Abina, S.; Bartholomae, C.C.; Veres, G.; Schmidt, M.; Kutschera, I.; Vidaud, M.; Abel, U.; Dal-Cortivo, L.; Caccavelli, L.; et al. Hematopoietic stem cell gene therapy with a lentiviral vector in X-linked adrenoleukodystrophy. *Science* **2009**, *326*, 818–823. [CrossRef] [PubMed]

249. Kohn, D.B.; Booth, C.; Shaw, K.L.; Xu-Bayford, J.; Garabedian, E.; Trevisan, V.; Carbonaro-Sarracino, D.A.; Soni, K.; Terrazas, D.; Snell, K.; et al. Autologous Ex Vivo Lentiviral Gene Therapy for Adenosine Deaminase Deficiency. *N. Engl. J. Med.* **2021**, *384*, 2002–2013. [CrossRef] [PubMed]

250. Lundstrom, K. Gene Therapy Cargoes Based on Viral Vector Delivery. *Curr Gene Ther.* **2023**, *23*, 111–134. [CrossRef]

251. Martin, K.R.; Quigley, H.A.; Zack, D.J.; Levkovitch-Verbin, H.; Kielczewski, J.; Valenta, D.; Baumrind, L.; Pease, M.E.; Klein, R.L.; Hauswirth, W.W. Gene therapy with brain-derived neurotrophic factor as a protection: Retinal ganglion cells in a rat glaucoma model. *Investig. Ophthalmol. Vis. Sci.* **2003**, *44*, 4357–4365. [CrossRef]

252. O'Callaghan, J.; Crosbie, D.E.; Cassidy, P.S.; Sherwood, J.M.; Flügel-Koch, C.; Lütjen-Drecoll, E.; Humphries, M.M.; Reina-Torres, E.; Wallace, D.; Kiang, A.S.; et al. Therapeutic potential of AAV-mediated MMP-3 secretion from corneal endothelium in treating glaucoma. *Hum. Mol. Genet.* **2017**, *26*, 1230–1246. [CrossRef]

253. Bush, R.A.; Zeng, Y.; Colosi, P.; Kjellstrom, S.; Hiriyanna, S.; Vijayasarathy, C.; Santos, M.; Li, J.; Wu, Z.; Sieving, P.A. Preclinical dose-escalation study of intravitreal AAV-RS1 gene therapy in a mouse model of X-linked retinoschisis: Dose-dependent expression and improved retinal structure and function. *Hum. Gene Ther.* **2016**, *27*, 376–389. [CrossRef] [PubMed]

254. Hassall, M.M.; Barnard, A.R.; MacLaren, R.E. Gene Therapy for Color Blindness. *Yale J. Biol. Med.* **2017**, *90*, 543–551.

255. Alexander, J.J.; Umino, Y.; Everhart, D.; Chang, B.; Min, S.H.; Li, Q.; Timmers, A.M.; Hawes, N.L.; Pang, J.-J.; Barlow, R.B.; et al. Restoration of cone vision in a mouse model of achromatopsia. *Nat. Med.* **2007**, *13*, 685–687. [CrossRef]

256. Komáromy, A.M.; Alexander, J.J.; Rowlan, J.S.; Garcia, M.M.; Chiodo, V.A.; Kaya, A.; Tanaka, J.C.; Acland, G.M.; Hauswirth, W.W.; Aguirre, G.D. Gene therapy rescues cone function in congenital achromatopsia. *Hum. Mol. Genet.* **2010**, *19*, 2581–2593. [CrossRef]

257. Heier, J.S.; Kherani, S.; Desai, S.; Dugel, P.; Kaushal, S.; Cheng, S.H.; Delacono, C.; Purvis, A.; Richards, S.; Le-Halpere, A.; et al. Intravitreous injection of AAV2-sFLT01 in patients with advanced neovascular age-related macular degeneration: A phase I, open-label trial. *Lancet* **2017**, *390*, 50–61. [CrossRef]

258. Constable, I.J.; Pierce, C.M.; Lai, C.-M.; Magno, A.L.; Degli-Esposti, M.A.; French, M.A.; McAllister, I.L.; Butler, S.; Barone, S.B.; Schwartz, S.D.; et al. Phase 2a randomized clinical trial: Safety and post hoc analysis of subretinal rAAV.sFLT-1 for wet age-related macular degeneration. *EBioMedicine* **2016**, *14*, 168–175. [CrossRef]

259. Guy, J.; Feuer, W.J.; Davis, J.L.; Porciatti, V.; Gonzalez, P.J.; Koilkonda, R.D.; Yuan, H.; Hauswirth, W.W.; Lam, B.L. Gene therapy for Leder hereditary optic neuropathy: Low and medium-dose visual results. *Ophthalmology* **2017**, *124*, 1621–1634. [CrossRef]

260. Vignal, C.; Uretsky, S.; Fitoussi, S.; Galy, A.; Blouin, L.; Girmens, J.-F.; Bidot, S.; Thomasson, N.; Bouquet, C.; Valero, S.; et al. Safety of rAAV2/2-ND4 gene therapy. *Ophthalmology* **2018**, *125*, 945–947. [CrossRef] [PubMed]

261. Russell, S.; Bennett, J.; Wellman, J.A.; Chung, D.C.; Yu, Z.F.; Tillman, A.; Wittes, J.; Pappas, J.; Elci, O.; McCague, S.; et al. Efficacy and safety of voretigene neparvovec (AAV2-hRPE65v2) in patients with RPE65-mediated inherited retinal dystrophy: A randomised, controlled, open-label, phase 3 trial. *Lancet* **2017**, *390*, 849–860. [CrossRef]

262. Maguire, A.M.; Bennett, J.; Aleman, E.M.; Leroy, B.P.; Aleman, T.S. Clinical Perspective: Treating RPE65-Associated Retinal Dystrophy. *Mol. Ther.* **2021**, *29*, 442–463. [CrossRef]

263. Flotte, T.R.; Afione, S.A.; Conrad, C.; McGrath, S.A.; Solow, R.; Oka, H.; Zeitlin, P.L.; Guggino, W.B.; Carter, B.J. Stable in vivo expression of the cystic fibrosis transmembrane conductance regulator with an adeno-associated virus vector. *Proc. Natl. Acad. Sci. USA* **1993**, *90*, 10613–10617. [CrossRef] [PubMed]

264. Conrad, C.K.; Allen, S.S.; Afione, S.A.; Reynolds, T.C.; Beck, S.E.; Fee-Maki, M.; Barazza-Ortiz, X.; Adams, R.; Askin, F.B.; Carter, B.J.; et al. Safety of single-dose administration of an adeno-associated virus (AAV)-CFTR vector in the primate lung. *Gene Ther.* **1996**, *3*, 658–668.

265. Cao, H.; Ouyang, H.; Grasemann, H.; Bartlett, C.; Du, K.; Duan, R.; Shi, F.; Estrada, M.; Seigel, K.E.; Coates, A.L.; et al. Transducing Airway Basal Cells with a Helper-Dependent Adenoviral Vector for Lung Gene Therapy. *Hum. Gene Ther.* **2018**, *29*, 643–652. [CrossRef] [PubMed]

266. Limberis, M.; Anson, D.S.; Fuller, M.; Parsons, D.W. Recovery of airway cystic fibrosis transmembrane conductance regulator function in mice with cystic fibrosis after single-dose lentivirus-mediated gene transfer. *Hum. Gene Ther.* **2002**, *13*, 1961–1970. [CrossRef]

267. Cooney, A.L.; Abou Alaiwa, M.H.; Shah, V.S.; Bouzek, D.C.; Stroik, M.R.; Powers, L.S.; Gansemer, N.D.; Meyerholz, D.K.; Welsh, M.J.; Stoltz, D.A.; et al. Lentiviral-mediated phenotypic correction of cystic fibrosis pigs. *JCI Insight* **2016**, *1*, e88730. [CrossRef] [PubMed]

268. Alton, E.W.F.W.; Beekman, J.M.; Boyd, A.C.; Brand, J.; Carlon, M.S.; Connolly, M.M.; Chan, M.; Conlon, S.; Davidson, H.E.; Davies, J.C.; et al. Preparation for a first-in-man lentivirus trial in patients with cystic fibrosis. *Thorax* **2017**, *72*, 137–147. [CrossRef]

269. Ramasamy, M.N.; Minassian, A.M.; Ewer, K.J.; Flaxman, A.L.; Folegatti, P.M.; Owens, D.R.; Voysey, M.; Aley, P.K.; Angus, B.; Babbage, G.; et al. Safety and immunogenicity of ChAdOx1 nCov-19 vaccine adminsitered in a prime-boost regimen in young and old adults (COV002): A singe-blind, randomised, controlled phase 2/3 trial. *Lancet* **2020**, *336*, 1979–1993. [CrossRef]
270. Lundstrom, K. Viral Vectors for COVID-19 Vaccine Development. *Viruses* **2021**, *13*, 317. [CrossRef]
271. Zhu, F.C.; Guan, X.H.; Li, Y.H.; Huang, J.Y.; Jiang, T.; Hou, L.H.; Li, J.X.; Yang, B.F.; Wang, L.; Wang, W.J.; et al. Immunogenicity and safety of a recombinant adenovirus type-5-vectored COVID-19 vaccine in healthy adults aged 18 years or older: A randomised, double-blind, placebo-controlled, phase 2 trial. *Lancet* **2020**, *396*, 479–488. [CrossRef]
272. Logunov, D.Y.; Dolzhikova, I.V.; Shcheblyakov, D.V.; Tukhvatulin, A.I.; Zubkova, O.V.; Dzharullaeva, A.S.; Kovyrshina, A.V.; Lubenets, N.L.; Grousova, D.M.; Erokhova, A.S.; et al. Safety and efficacy of an rAd26 and rAd5 vector-based heterologous prime-boost COVID-19: An interim analysis of a randomised controlled phase 3 in Russia. *Lancet* **2021**, *397*, 671–681. [CrossRef]
273. Callaway, E. Russia's fast-track coronavirus vaccine draws outrage over safety. *Nature* **2020**, *584*, 334–335. [CrossRef]
274. Stephenson, K.E.; Le Gars, M.; Sadoff, J.; de Groot, A.M.; Heerwegh, D.; Truyers, C.; Atyeo, C.; Loos, C.; Chandrashekar, A.; McMahan, K.; et al. Immunogenicity of the Ad26.COV2.S vaccine for COVID-19. *JAMA* **2021**, *325*, 1535–1544. [CrossRef]
275. Henao-Restrepo, A.M.; Longini, I.M.; Egger, M.; Dean, N.E.; Edmunds, W.J.; Camacho, A.; Carroll, M.W.; Doumbia, M.; Draguez, B.; Duraffour, S.; et al. Efficacy and effectiveness of an rVSV-vectored vaccine expressing Ebola surface glycoprotein: Interim results from the Guinea ring vaccination cluster-randomised trial. *Lancet* **2015**, *386*, 857–866. [CrossRef]
276. Henao-Restrepo, A.M.; Camacho, A.; Longini, I.M.; Watson, C.H.; Edmunds, W.J.; Egger, M.; Carroll, M.W.; Dean, N.E.; Diatta, I.; Doumbia, M.; et al. Efficacy and effectiveness of an rVSV-vectored vaccine in preventing Ebola virus disease: Final results from the Guinea ring vaccination, open-label, cluster-randomised trial (Ebola Ca Suffit!). *Lancet* **2017**, *389*, 505–518. [CrossRef]
277. Ollmann Saphire, E. A vaccine against Ebola virus. *Cell* **2020**, *181*, 6. [CrossRef]
278. Wan, X.; Pei, H.; Zhao, M.-J.; Yang, S.; Hu, W.K.; He, H.; Ma, S.Q.; Zhang, G.; Dong, X.Y.; Chen, C.; et al. Efficacy and safety of rAAV2-ND4 treatment for Leber's hereditary optic neuropathy. *Sci. Rep.* **2016**, *6*, 2016. [CrossRef]
279. O'Sullivan, B.P.; Freedman, S.D. Cystic fibrosis. *Lancet* **2009**, *373*, 1891–1904. [CrossRef]
280. Räty, J.K.; Pikkarainen, J.T.; Wirth, T.; Ylä-Herttuala, S. Gene therapy: The first approved gene-based medicines, molecular mechanisms and clinical indications. *Curr. Mol. Pharmacol.* **2008**, *1*, 13–23. [CrossRef]
281. Zhang, W.W.; Li, L.; Li, D.; Liu, J.; Li, X.; Li, W.; Xu, X.; Zhang, M.J.; Chandler, L.A.; Lin, H.; et al. The first approved gene therapy product for cancer ad-p53 (Gendicine): 12 years in the clinic. *Hum. Gene Ther.* **2018**, *29*, 160–179. [CrossRef]
282. Ylä-Herttuala, S. Glybera's second act: The curtain rises on the high cost of therapy. *Mol. Ther.* **2015**, *23*, 217–218. [CrossRef]

MDPI
St. Alban-Anlage 66
4052 Basel
Switzerland
www.mdpi.com

Viruses Editorial Office
E-mail: viruses@mdpi.com
www.mdpi.com/journal/viruses